Energy Metabolism
and the Regulation of
Metabolic Processes
in Mitochondria

Proceedings of a Symposium held at the
University of Nebraska Medical School
Omaha, Nebraska, May 3-4, 1971

Symposia on Metabolic Regulation

Energy Metabolism and the Regulation of Metabolic Processes in Mitochondria

Edited by

Myron A. Mehlman

*Department of Biochemistry
University of Nebraska College of Medicine
Omaha, Nebraska*

Richard W. Hanson

*Fels Research Institute and Department
of Biochemistry
Temple University Medical School
Philadelphia, Pennsylvania*

Academic Press   New York and London   1972

COPYRIGHT © 1972, BY ACADEMIC PRESS, INC.
ALL RIGHTS RESERVED
NO PART OF THIS BOOK MAY BE REPRODUCED IN ANY FORM,
BY PHOTOSTAT, MICROFILM, RETRIEVAL SYSTEM, OR ANY
OTHER MEANS, WITHOUT WRITTEN PERMISSION FROM
THE PUBLISHERS.

ACADEMIC PRESS, INC.
111 Fifth Avenue, New York, New York 10003

United Kingdom Edition published by
ACADEMIC PRESS, INC. (LONDON) LTD.
24/28 Oval Road, London NW1

LIBRARY OF CONGRESS CATALOG CARD NUMBER: 70-187235

PRINTED IN THE UNITED STATES OF AMERICA

CONTENTS

CONTRIBUTORS xi
PREFACE xv

Mechanism and Control of Oxidative Phosphorylation 1
    Efraim Racker and L. L. Horstman

Kinetic Control of Electron Flow 27
    Britton Chance, Maria Erecińska, Edwin M. Chance,
    Alberto Boveris, and Michael Wagner

Thermodynamic Control of Mitochondrial Energy Coupling . . . 39
    David F. Wilson and P. Leslie Dutton

Replenishment and Depletion of Citric Acid Cycle
Intermediates in Muscle 53
    John M. Lowenstein

Anion Transport and the Regulation of the Citric Acid Cycle . . . 63
    G. R. Williams, J. L. Orr, and G. S. Wong

Is There an Organization of Krebs Cycle Enzymes in
the Mitochondrial Matrix? 79
    Paul A. Srere

Critic Acid Cycle Dynamics and Substrate Compartmentation
in Mitochondria from Rabbit Heart and Brain 93
    R. W. Von Korff

The Significance of Mitochondrial Phosphoenolpyruvate
Formation in the Regulation of Gluconeogenesis in
Guinea Pig Liver 109
    Alan J. Garber, F. J. Ballard, and Richard W. Hanson

Production and Utilization of ATP and the Regulation of
Lipogenesis in Adipose Tissue 137
    J. P. Flatt

CONTENTS

Liver Metabolite Content, Redox and Phosphorylation States in Rats Fed Diets Containing 1,3-Butanediol and Ethanol 171
    R. L. Veech and Myron A. Mehlman

Feedback Control of the Citric Acid Cycle 185
    John R. Williamson, Colleen M. Smith, Kathryn F. LaNoue and Jadwiga Bryla

Sources and Dispostion of Aerobically-Generated Intermediates in Heart Muscle 211
    E. Jack Davis, Reneé C. Lin, and David Li-Shan Chao

Hormonal Regulation of Pyruvate Metabolism in Rat Liver Mitochondria 239
    Robert C. Haynes, Jr.

Pyruvate Dehydrogenase Complex: Structure, Function, and Regulation 253
    Lester J. Reed, Tracy C. Linn, Flora H. Pettit, Robert M. Oliver, Ferdinand Hucho, John W. Pelley, Douglas D. Randall, and Thomas E. Roche

Reductive Carboxylation of α-Oxoglutarate by Mitochondria from Livers of Normal, Diabetic, and Fat-Fed Rats 271
    Carl R. Mackerer

Properties of Mitochondrial ATPase 287
    Henry A. Lardy and David Lambeth

SUBJECT INDEX 289

CONTRIBUTORS

F. J. Ballard, Fels Research Institute and Department of Biochemistry, Temple University Medical School, Philadelphia, Pennsylvania 19140

Alberto Boveris, Johnson Research Foundation, Department of Biophysics and Physical Biochemistry, University of Pennsylvania, Philadelphia, Pennsylvania 19104

Jadwiga Bryla, Johnson Research Foundation, Department of Biophysics and Physical Biochemistry, University of Pennsylvania, Philadelphia, Pennsylvania 19104

Britton Chance, Johnson Research Foundation, Department of Biophysics and Physical Biochemistry, University of Pennsylvania, Philadelphia, Pennsylvania 19104

Edwin M. Chance, Department of Biochemistry, University College London, England

David Li-Shan Chao,[*] Department of Biochemistry, Indiana University School of Medicine, Indianapolis, Indiana 46202

E. Jack Davis, Department of Biochemistry, Indiana University School of Medicine, Indianapolis, Indiana 46202

P. Leslie Dutton, Johnson Research Foundation, Department of Biophysics and Physical Biochemistry, University of Pennsylvania, Philadelphia, Pennsylvania 19104

Maria Erecińska, Johnson Research Foundation, Department of Biophysics and Physical Biochemistry, University of Pennsylvania, Philadelphia, Pennsylvania 19104

J. P. Flatt, Department of Nutrition and Food Science, Massachusetts Institute of Technology, Cambridge, Massachusetts 02139

*Present address: Department of Biophysical Sciences, University of Houston, Houston, Texas 77025

CONTRIBUTORS

Alan J. Garber, Fels Research Institute and Department of Biochemistry, Temple University Medical School, Philadelphia, Pennsylvania 19140

Richard W. Hanson, Fels Research Institute and Department of Biochemistry, Temple University Medical School, Philadelphia, Pennsylvania 19140

Robert C. Haynes, Jr., Department of Pharmacology, University of Virginia, Charlottesville, Virginia

L. L. Horstman, Section of Biochemistry and Molecular Biology, Cornell University, Ithaca, New York 14850

Ferdinand Hucho,* Clayton Foundation, Biochemical Institute and Department of Chemistry, University of Texas at Austin, Austin, Texas 78712

David Lambeth, Department of Biochemistry and Institute for Enzyme Research, University of Wisconsin, Madison, Wisconsin 53706

Kathryn F. LaNoue, Johnson Research Foundation, Department of Biophysics and Physical Biochemistry, University of Pennsylvania, Philadelphia, Pennsylvania 19104

Henry A. Lardy, Department of Biochemistry and Institute for Enzyme Research, University of Wisconsin, Madison, Wisconsin 53706

Reneé C. Lin, Department of Biochemistry, Indiana University School of Medicine, Indianapolis, Indiana 46202

Tracy C. Linn,† Clayton Foundation, Biochemical Institute and Department of Chemistry, University of Texas at Austin, Austin, Texas 78712

John M. Lowenstein, Graduate Department of Biochemistry, Brandeis University, Waltham, Massachusetts 02154

Carl R. Mackerer, Department of Toxicology, G. D. Searle & Company, P. O. Box 5110, Chicago, Illinois 60680

Myron A. Mehlman, Department of Biochemistry, University of Nebraska College of Medicine, Omaha, Nebraska 68105

Robert M. Oliver, Clayton Foundation, Biochemical Institute and Department of Chemistry, University of Texas at Austin, Austin, Texas 78712

*Present address: Fachbereich Biologie der Universität, Kostanz, Germany
†Present address: Department of Biochemistry, University of Texas Southwestern Medical School, Dallas, Texas 75216

CONTRIBUTORS

J. L. Orr, Department of Biochemistry, University of Toronto, Toronto, Ontario, Canada

John W. Pelley, Clayton Foundation, Biochemical Institute and Department of Chemistry, University of Texas at Austin, Austin, Texas 78712

Flora H. Pettit, Clayton Foundation, Biochemical Institute and Department of Chemistry, University of Texas at Austin, Austin, Texas 78712

Efraim Racker, Section of Biochemistry and Molecular Biology, Cornell University, Ithaca, New York 14850

Douglas D. Randall, * Clayton Foundation, Biochemical Institute and Department of Chemistry, University of Texas at Austin, Austin, Texas 78712

Lester J. Reed, Clayton Foundation, Biochemical Institute and Department of Chemistry, University of Texas at Austin, Austin, Texas 78712

Thomas E. Roche, Clayton Foundation, Biochemical Institute and Department of Chemistry, University of Texas at Austin, Austin, Texas 78712

Colleen M. Smith, Johnson Research Foundation, Department of Biophysics and Physical Biochemistry, University of Pennsylvania, Philadelphia, Pennsylvania 19104

Paul A. Srere, Basic Biochemistry Unit, Veterans Administration Hospital and University of Texas Southwestern Medical School, Dallas, Texas 75216

R. L. Veech, Section on Neurochemistry, NIMH, IR, AMRDN, and The National Institute of Alcohol Abuse and Alcoholism, St. Elizabeth's Hospital, Washington, D. C. 20032

R. W. Von Korff, Maryland Psychiatric Research Center and Friends Medical Science Research Center, Baltimore, Maryland 21228

Michael Wagner, Johnson Research Foundation, Department of Biophysics and Physical Biochemistry, University of Pennsylvania, Philadelphia, Pennsylvania 19104

G. R. Williams, Department of Biochemistry, University of Toronto, Toronto, Ontario, Canada

*Present address: Department of Agricultural Chemistry, University of Missouri, Columbia, Missouri 65201

CONTRIBUTORS

John R. Williamson, Johnson Research Foundation, Department of Biophysics and Physical Biochemistry, University of Pennsylvania, Philadelphia, Pennsylvania 19104

David F. Wilson, Johnson Research Foundation, Department of Biophysics and Physical Biochemistry, University of Pennsylvania, Philadelphia, Pennsylvania 19104

G. S. Wong, Department of Biochemistry, University of Toronto, Toronto, Ontario, Canada

PREFACE

For those of us interested in metabolic regulation, the current methods for the exchange of information and new ideas seem, at times, inadequate. This area is so diverse, ranging as it does from the control of mRNA synthesis in bacteria to the regulation of specific metabolic pathways in man, that it is easy to understand the difficulty in developing a program for the meaningful discussion of new research ideas within the general framework of a conventional biochemical meeting. This problem is compounded by the wide range of journals that publish articles dealing directly with, or related to, metabolic regulation. Because of these difficulties we have organized a yearly symposium on metabolic regulation, which has been held during the past three years at the University of Nebraska Medical School in Omaha, Nebraska. Each year a specific topic is covered, and the speakers include scientists already well known in the area, as well as younger investigators with new ideas and perhaps a different approach. Wherever possible we present alternative views on a subject and allow time for free discussion of concepts arising from the formal presentation. Most importantly, this symposium is open for all to attend. We set no limit on attendance, and encourage graduate students, postdoctoral fellows, and scientists to join us in this symposium. Our experience over the past three years has shown us that we have a good deal to learn about the intricacies of programming scientific meetings. However, we remain convinced of the wisdom in our original intention of holding an open forum-type of meeting.

In choosing speakers for this, and for previous symposia, we have relied on the suggestions of a committee of scientists including Drs. Sidney Weinhouse, Joseph Katz, Henry Lardy, Howard Katzen, George Cahill, John R. Williamson, and Ronald Estabrook. Their judgment has been particularly important to us in the selection of a timely topic and a varied and representative program. We hope to continue this practice in subsequent symposia but would, of course, welcome suggestions from interested scientists who might wish to contact us directly.

The decision to publish this symposium at a time when the number of journals and books is already overwhelming was not taken lightly. A volume of this type must fill a need to justify its publication, and only time will tell whether other scientists working in this area share our enthusiasm for this type of book. We have drawn together chapters from a number of people actively

PREFACE

engaged in some aspects of metabolic regulation directly concerned with mitochondrial function. Their contributions vary in style, scope, and method of presentation, and we have not attempted to dictate in any way the content of these chapters. Although potential problems, such as a variation in the level of scientific presentation and a certain unavoidable discontinuity between chapters, may arise, our aim has been to allow each contributor the broadest possible discretion. The responsibility for the scientific content of each chapter lies with the individual authors, although we have tried wherever possible to clarify and strengthen them by careful and critical review.

We have been greatly aided in our editorial function by Drs. John Lowenstein, Merton Utter, Henry Lardy, and John Williamson, each of whom has spent considerable time and effort in reading and commenting on the manuscripts presented in the session in which they acted as chairmen. Their help and advice in the preparation of this book have been invaluable, and we are indebted to them.

We wish to express special appreciation to Dr. and Mrs. Ramon Tate for their invaluable editorial assistance in helping to prepare these manuscripts for publication. At the University of Nebraska Medical Center we also had the kind assistance of Dr. W. R. Ruegamer, Chairman, Department of Biochemistry, the Department of Continuing Education, and Dr. R. B. Tobin.

We are also particularly indebted to Bill O'Conner and Darlene Patton of I.B.M. for their generous assistance during the preparation of the manuscripts for composition.

We would like to acknowledge the fine financial support of the following: Celanese Chemical Company; Hoffman-La Roche, Inc.; G. D. Searle and Co.; The Upjohn Company; The Eli Lilly Research Laboratories; Schering Corporation; Smith, Kline and French Laboratories; Sandoz Pharmaceuticals; Geigy Pharmaceuticals; Sterling Drug, Inc.; Ciba Pharmaceutical Company; W. H. Rorer; Merck, Sharp and Dohme; Riker Laboratories, Inc.; Boehringer Mannheim Corporation; Lakeside Laboratories; Mead Johnson; Charles Pfizer and Company; The Mogul Corporation; and the Kroc Foundation. The generosity of these institutions has made the symposium possible.

Myron A. Mehlman
Richard W. Hanson

Energy Metabolism
and the Regulation of
Metabolic Processes
in Mitochondria

MECHANISM AND CONTROL OF OXIDATIVE PHOSPHORYLATION

Efraim Racker and L.L. Horstman

Introduction

Some time ago (1) we attempted to score the evidence in favor of the two current hypotheses of oxidative phosphorylation. At that time, the chemical hypothesis, which includes the conformational hypothesis first proposed by Boyer (2), scored in our books about as high as the chemiosmotic hypothesis of Mitchell (3). Meanwhile, the score board has become larger and it looks as if the balance is shifting in favor of the chemiosmotic hypothesis. I feel confident that Dr. Chance will meet this challenging statement admirably and supply us with suitable counter weight.

Let us look at the score board (Table 1) and particularly note some of the new features. The role of the membrane is still counted in favor of the chemiosmotic hypothesis. A closed compartment capable of separating charges is a key feature of the chemiosmotic hypothesis, but is not essential for the chemical hypothesis. A compartment closed off by a membrane which is relatively impermeable to protons appears in fact to be required for oxidative phosphorylation. I shall discuss later experiments on respiratory control, phosphorylation and proton translocation which show a parallel loss of these functions in the presence of increasing amounts of cholate or of uncoupling agents. However, should someone succeed in demonstrating electron flow associated with phosphorylation in a soluble cytochrome-containing system sensitive to uncouplers such as FCCP,* the case for the chemiosmotic hypothesis would be greatly weakened. Such a claim has been made by Painter and Hunter (4), but confirmation of this work has not been forthcoming in spite of considerable efforts in several laboratories, including our own. Other chemical model systems with key features resembling those of oxidative phosphorylation (*i.e.*, oxidation preceding phosphorylation) have been constructed in membrane-free systems (*cf.* 5), but no evidence for any of these mechanisms operating in mitochondria is available. On the other hand, Hinkle (6) has assembled with liposomes an attractive model conforming to

---

*Abbreviations: FCCP, trifluoromethoxycarbonylcyanide phenylhydrazone; DCCD,N,N'-dicyclohexylcarbodiimide; BSA, bovine serum albumin.

some aspects of the chemiosmotic hypothesis which I shall discuss later. However, attempts to show that this model system is capable of ATP generation have thus far not been successful. In view of these questions of relevance, we only want to draw your attention to these chemical and chemiosmotic models, without giving them a vote on the score board.

The mode of action of uncouplers can be more convincingly explained by the chemiosmotic hypothesis, as pointed out previously (1). Experiments on the effect of valinomycin on submitochondrial particles in the presence of nigericin and ammonia (7, 8) also lend support to Mitchell's formulation.

Recent work on proton translocation at Cornell has yielded data which strengthens the Mitchell hypothesis. It was shown by Hinkle and Horstman (9) that the inward flow of protons in resolved submitochondrial particles is dependent on the same coupling factors which are required for oxidative phosphorylation. Moreover, it was shown that respiratory control and proton translocation were affected by uncouplers in a parallel fashion. Similarly, we have shown in collaboration with Peter Hinkle that submitochondrial particles exposed to cholate lose in a parallel manner proton translocation and respiratory control (Fig. 1). Phosphorylation, measured as $^{32}P_i$-ATP exchange in the presence of coupling factors, was also similarly affected by cholate.

The most recent information in support of Mitchell's formulation has come from studies of the topography of the inner mitochondrial membrane (10, 11). There can be little doubt that the organization is asymmetric, with the coupling factors and succinate dehydrogenase on the M-side (facing the matrix in mitochondria) and with cytochrome c and cytochrome c_1 on the opposite side (C-side). Cytochrome oxidase is found on both sides of the membrane (Fig. 2). The localization of these membrane components has been established by three different experimental approaches: a) resolution and reconstitution; b) interaction with antibodies; c) interaction with a radioactivity labelled compound, diazobenzenesulfonic acid. The latter two approaches depend on the experimentally established facts that the M-side is on the outer surface of submitochondrial particles, that the C-side is on the outer surface of the inner membrane of mitochondria and that neither antibodies nor diazobenzenesulfonic acid can penetrate from one side of the membrane to the other. Since these studies are being presented elsewhere (12), I shall refrain from discussing them in detail. It is clear, however, that the asymmetric assembly of the oxidation enzymes across the membrane is consistent with their role in the separation of charges according to the chemiosmotic hypothesis.

Experiments on K^+ translocation in mitochondria, on Ca^{++} translocation in submitochondrial particles and on the partial reactions of oxidative phosphorylation are more readily explained in terms of the chemical hypothesis as was pointed out before (1). The experiments of Painter and Hunter (4) provide a strong argument in favor of the chemical hypothesis, but in view of lack of reproducibility their significance remains doubtful.

In discussing these score points, one difficult aspect should be re-emphasized (13). Once we agree that there is a proton translocation pump operative in mitochondria and in submitochondrial particles, the question whether the pump operates in series, as postulated by Mitchell, or in parallel, as proposed by the proponents of the chemical hypothesis, becomes very difficult to answer. An unambiguous answer could be obtained if it were possible to separate the pump from the coupling device responsible for phosphorylation, but this has not been achieved.

In any case, the presence of a proton pump and the formation of a membrane potential across the inner mitochondrial membrane is of considerable interest in relation to the problem of respiratory control even if they should not represent the driving force for the dehydration of phosphate and the generation of ATP. It was suggested by Mitchell (3) that the membrane potential is responsible for the phenomenon of respiratory control in mitochondria; we shall come back to this point later. Now I should like to discuss some of the more general aspects of control of the generation of biological energy.

Results

Oxidation control and regulation of glycolysis

The generation of ATP from ADP and phosphate is controlled in nature by the simple and ingenious device of tightly coupling phosphorylation to oxidation; unless phosphate and ADP are present, oxidations do not take place. Since this mechanism is operative not only in mitochondria but also in other systems of ATP generation, *e.g.*, in glycolysis and in photophosphorylation, it seems appropriate to refer to the phenomenon as oxidation control rather than respiratory control. The mechanism of oxidation control in glycolysis has been analyzed and consists of multiple components (14). A key requirement is the stability of the thiol ester intermediate which is formed by glyceraldehyde-3-P dehydrogenase in the first step of oxidation of glyceraldehyde-3-phosphate and which is cleaved by inorganic phosphate to yield 1,3-diphosphoglycerate. The transformation of the latter to 3-phosphoglycerate is dependent on the availability of ADP. As shown in Table 2 the oxidation step is compulsorily linked to phosphorylation. This is possible because of the stability of the thiol ester intermediate. On aging of glyceraldehyde-3-phosphate dehydrogenase, the stability of the thiol ester intermediate diminishes, and the enzyme acquires the ability to hydrolyze acyl phosphates.

Similarly, mitochondria which are kept frozen lose the phenomenon of oxidation control. Some of the response of respiration to ADP can be restored. As can be seen in Fig. 3, mitochondria which were kept frozen for one day lost the response of respiration to ADP, but incubation for 30 min at 20° or 15 min at 30° restored the response. The effect of this incubation was to lower the rate of respiration, with ADP restoring it to the rate of the control mitochondria kept at 0°. The restoration was usually faster at 37°, but with some mitochondrial preparations it was lost again on prolonged incubation. It can be seen that mitochondria which showed no response to ADP after thawing became responsive after several hours of incubation at 0°. The lack of response to ADP is clearly linked to the freezing process, since mitochondria which were never frozen and were kept for one day at 4° were markedly stimulated by ADP. As shown in Fig. 4, the optimum pH of this response is close to neutrality. Several attempts to influence the extent of restoration after freezing by addition of substrates, ATP, CoA or bovine serum albumin failed to yield a clue to the mechanism of the restoration process. It may be related to damage due to freezing of phospholipids which we described previously (15).

Submitochondrial particles show no oxidation control. To demonstrate phosphorylation with a high P:O ratio, a powerful trapping system for ATP (glucose and hexokinase) and in some cases additions of excess coupling factors are required. It was shown by Lee and Ernster (16) that the oxidation of DPNH in submitochondrial particles that were obtained by sonication of mitochondria at an alkaline pH in the presence of EDTA (17) was inhibited by oligomycin and restored by FCCP. This phenomenon, which resembles the release of respiration in mitochondria by ADP, has been intensively studied in our laboratory. Since it differs from the mitochondrial phenomenon inasmuch as ADP cannot substitute for FCCP, we refer to it as oxidation control induced by energy transfer inhibitors (I-I oxidation control).

It can be seen from Fig. 5 that, as in the case of the mitochondria, the most pronounced effects of rutamycin and FCCP are noted in the pH range between 6.4 and 7.2. The presence of sucrose increased the magnitude of these effects; for convenience of mixing, 0.25 M sucrose was added in the assay system, although somewhat higher control ratios were observed at higher concentrations (0.5 M).

Susceptibility of submitochondrial particles to induction of oxidation control by energy transfer inhibitors

It was observed several years ago that the oxidation of DPNH in submitochondrial particles prepared by sonication of mitochondria at pH 7.4 in the presence of 10 mM pyrophosphate was not inhibited by rutamycin or DCCD. We have recently carried out a systematic analysis of the effect of pH and salt during sonication.

It can be seen from Table 3 that at pH 9.2 during sonication of mitochondria the inhibitory effect of rutamycin was considerably greater than in the presence of Tris buffer at pH 7.4. The presence of pyrophosphate during sonication at either pH substantially decreased both the inhibitory effect of rutamycin and the stimulation of the inhibited respiration by FCCP. It can be seen from Table 4 that ammonia is not essential, KOH at the same pH being equally effective. It is apparent from these experiments that there are two important features in the preparation of these particles: the low ionic strength and the alkaline pH of the medium. All these experiments were carried out without EDTA because it is not essential. In fact, it was found clearly detrimental to the phenomenon of I-I oxidation control when used at concentrations above 5 mM. It is therefore obvious that these particles, which in the original description (17) were called A-particles (alkali-particles), should not be renamed EDTA-particles (16).

A more careful evaluation of the effect of pH during sonication is shown in Table 5. It can be seen that with increasing pH during sonication of the mitochondria the ratio of the rates of DPNH oxidation before and after addition of rutamycin became larger. The ratio of the oxidation rates after and before FCCP addition also became larger, but less strikingly so, resulting in a decrease of the ratio of the rates of oxidation elicited by FCCP over the initial rate. This indicates that in the particles prepared at an extreme alkaline pH, FCCP is either not fully effective in reversing the inhibition of respiration by rutamycin, or it is inhibitory to respiration under these conditions.

These findings appear to resolve some of the puzzling discrepancies observed when different particles (*e.g.* SMP or ETP_H) were used instead of A-particles to study the I-I oxidation control. However, it still remains to be explained what basic changes take place in particles which are sonicated at an alkaline pH without salt to render respiration sensitive to oligomycin. Numerous attempts have therefore been made to convert particles which catalyze DPNH oxidation which is insensitive to energy transfer inhibition to particles in which oxidation is inhibited. It has been shown previously that coupling factors are required for respiratory control (7) and for proton translocation (9) in resolved submitochondrial particles. It was therefore considered likely that the lack of response of respiration to rutamycin may be at least partly a reflection of the presence of sufficient coupling factors in these particles. If this were the case, a stimulation of respiration by uncouplers without addition of rutamycin would be expected. Such a stimulation was consistently observed with all particles (SMP, ETP_H, *etc.*) in which respiration was insensitive to rutamycin. However, the observed stimulation of respiration by FCCP was variable and in most instances not higher than 100%, whereas stimulation by FCCP in rutamycin-treated A-particles was 300% or more (*cf.* Table 5). It is possible that the resistant particles may have retained components (salts) during sonication that interfere with the full expression of respiratory control. A more direct approach to this problem was to remove coupling factors from resistant submitochondrial particles. As shown in Table 6, considerable increase in I-I oxidation control was observed after exposure of such particles to either Sephadex-urea or silicotungstate, treatments known to remove coupling factors (18, 19). Although respiration following such treatments, particularly with silicotungstate, varied widely, the particles were invariably sensitized to the inhibitory action of rutamycin. However, the I-I oxidation ratio was never as high as that observed in A-particles. Thus the non-responsiveness of submitochondrial particles to inhibition of respiration to energy-transfer inhibitions appears to be caused partly by the presence of coupling factors, which by themselves contribute to the control of respiration, and partly by the presence of extraneous compounds such as pyrophosphate or other salts which decrease the sensitivity of particles to rutamycin.

Effect of trypsin on I-I oxidation control

We have observed previously (20) that F_1 which was exposed to trypsin lost its ability to serve as a coupling factor. A very short treatment with trypsin rendered F_1 inactive as a coupling factor as measured by the stimulation of the ATP-dependent reduction of DPN^+ by succinate. However, the same exposure of F_1 to trypsin had no effect, or even slightly stimulated ATPase activity. Moreover, on recombination of trypsin-treated F_1 with ASU-particles, a sensitivity of the ATPase activity to rutamycin, though diminished, was still observed. It was therefore of interest to relate these findings to the role of F_1 in oxidation control. As shown in Table 7, oxidation control can be induced in ASU-particles by F_1, but not by trypsin-treated F_1. Thus the effect of trypsin on the capability of F_1 to induce oxidation control correlates with its coupling activity rather than with the sensitivity of the reconstituted APTase to rutamycin. P. Hinkle also observed that trypsin eliminated the ability of

F_1 to induce respiration-driven proton translocation in ASU-particles. The data in Table 7 show that the trypsin preparation was effective even after removal of trace amounts of chymotrypsin. This experiment was carried out because of the observation that the activation of the ATPase activity of CF_1 in chloroplasts by trypsin is different, depending on whether the enzyme preparation is free of chymotrypsin or not (21, 22). We should also like to draw attention to the observation recorded in Table 7 that a preparation of F_1 low in ATPase inhibitor (20) was less effective in inducing oxidation control in ASU-particles. This is of interest in connection with the question of the role of the inhibitor in oxidation control which we shall discuss later.

In contrast to the marked effect of trypsin on the ability of F_1 to induce oxidation control, there was no effect of trypsin on the ability of submitochondrial particles to exhibit I-I oxidation control. ASU-particles were exposed to trypsin under conditions (45 min at 30° with 30 µg trypsin/10 mg particles/ml) which markedly diminished rutamycin-sensitivity of added F_1 (23). The I-I oxidation control ratio in trypsin particles was either undiminished or slightly increased (Table 7). The oxidation of DPNH inhibited by rutamycin was restored by FCCP.

Oxidation control induced by coupling factors

Considerable variations have been encountered in the control of oxidation induced by coupling factors (*cf.* Table 7). It was observed previously (24) that I-I oxidation control was diminished when cytochrome *c* was present on the M-side of submitochondrial particles. With these observations in mind, the effects of addition of polylysine, which inhibits oxidation of external ferrocytochrome *c*, as well as of ATPase inhibitor and of low concentrations of rutamycin were investigated. As shown in Table 8, an oxidation control ratio of 2.0 was increased to 5.0 by addition of polylysine, rutamycin and ATPase inhibitor. Each of them made some contribution to this improvement, as could be seen when they were individually deleted. Of particular interest is the possible role of the ATPase inhibitor in oxidation control. Although a function in control of ATP-linked reactions has been suggested previously (20, 25, 26), a more direct participation at the level of the oxidation chain or of ion transport is suggested by the current study.

We have explored various methods of eliciting a stimulation of respiration in submitochondrial particles (obtained by sonication) by addition of ADP. These attempts were only moderately successful. Control ratios approaching 2 were observed in some particles reconstituted with coupling factors without addition of inhibitors. With other preparations no effect on ADP addition was observed; under some conditions (*e.g.* deletion of Mg^{++}) a distinct inhibition of the rate was noted upon addition of ADP. As shown in Table 9, control ratios of about 1.5 were induced in these particles upon addition of low concentrations of rutamycin, polylysine and ATPase inhibitor. The relative contribution of these compounds was quite variable, the most consistent effect being observed with polylysine.

Discussion

The role of XOH and YH in the oxidation process

The mechanism of control of both electron transport-linked oxidative phosphorylation and substrate level oxidative phosphorylation is intimately linked to the mechanism of the coupling process. It is now generally accepted that in both cases a) oxidation precedes phosphorylation, and b) that the energy of oxidation is conserved in a relatively stable form. In substrate level phosphorylation a thiol ester intermediate is formed; in oxidative phosphorylation an unknown $X \sim Y$ is formed. In mitochondria the latter is stable enough to be used for ion transport and energy-dependent reductions; in submitochondrial particles it is less stable, but can be stabilized by energy-transfer inhibitors such as oligomycin. This fact has been instrumental in establishing not only the existence of $X \sim Y$ but its function in reactions other than ATP generation.

$X \sim Y$ is an intermediate in the chemical as well as the chemiosmotic hypothesis, but neither has formulated a clear picture of its chemical nature explaining its stability on the one hand and its high reactivity in different energy-dependent reactions on the other. There are some interesting differences in the formulation of the mechanism of $X \sim Y$ formation and in its utilization for ATP generation. The utilization of $X \sim Y$ in the chemical hypothesis consists, in analogy to glyceraldehyde-3-phosphate dehydrogenase action, of a phosphorolytic cleavage to $X \sim Y$ which transfers its $\sim P$ to ADP. Mitchell, however, proposes a concerted reaction catalyzed by the ATPase at the M-side of the membrane, in which $X \sim Y$,* ADP and phosphate interact to yield ATP, XO^- and Y^-. The two negatively charged products are repulsed from the negatively charged M-side and move towards the C-side, thereby driving the reaction toward ATP formation. A membrane potential thus becomes a driving force not only for the formation of $X \sim Y$ but also for the generation of ATP. The two hypotheses differ even more fundamentally in the mechanism of $X \sim Y$ formation. In the chemiosmotic hypothesis, XOH and YH are components

---

*The reason for the slight deviation in the designations $X \sim Y$, XOH and YH from those used by Mitchell (3) is explained in a previous publication (1).

of the oligomycin-sensitive ATPase and are not directly associated with the respiratory chain. In contrast, in the chemcial hypothesis XOH and YH are cofactors of respiration which do not undergo oxidoreduction. To permit the process to continue, XOH and YH must be regenerated either by hydrolysis or by an energy-consuming process.

$$A_{red} + B_{ox} + XOH + YH \rightleftharpoons A_{ox} + B_{red} + X{\sim}Y + H_2O$$

$$X{\sim}Y + H_2O \rightarrow XOH + YH$$

It is rather surprising therefore how little effort has been directed to identify XOH and YH as possible components of the respiratory chain. Instead, it has been generally assumed that the so-called uncoupled state of respiration is a short-circuited oxidoreduction reaction between the respiratory carriers. Although this is very likely true for the artificial systems studied with respiratory enzymes removed from the membrane, it is quite conceivable that membrane-linked respiration is dependent on the availability of XOH and YH as well as on the hydrolysis of $X{\sim}Y$ to XOH and YH. Recent experiments in our laboratory by Dr. Nishibayashi-Yamashita (*cf.* 27) have shown the requirement for an unknown component in the reconstituted system of succinoxidase. Thus far no pigment has been detected in the purified protein, and it is conceivable that it may represent a compulsory component of the coupling device. If this is true, it would represent the first member of the coupling device which operates on the C-side of the membrane, since in contrast to succinate dehydrogenase, the new component is not active when added to the formed complex of succinoxidase.

According to the chemiosmotic hypothesis, the transformation of a low-energy form X–Y to the high-energy form $X{\sim}Y$ takes place by moving from the positively charged C-side to the negatively charged M-side. Thus the force for this transformation is the proton gradient and the membrane potential. Neither XOH nor YH play a cofactor function in the respiratory chain. This interesting difference between the two formulations is rather fundamental. Thus, if a member of the coupling device

$$X{\sim}Y \xrightarrow{P_i,\ ADP} ATP$$

could be shown to be required for respiration, this would demand a serious revision of the chemiosmotic hypothesis.

A model for the Mitchell hypothesis

The key features of the Mitchell hypothesis are the proton-motive force and the membrane potential which develop during respiration via the loops of the oxidation chain. There is little doubt that proton translocation of the type formulated by Mitchell takes place. Although the identity of the hydrogen and proton carriers have not been established, his concept of the membrane potential as a control mechanism of respiration is very attractive. It seems appropriate therefore to devote a few minutes to an ingenious model for respiratory control designed by Peter Hinkle (6).

Liposomes were prepared from soybean phospholipids by sonication in the presence of ferricyanide. This internally trapped electron acceptor could not be reduced by externally added ascorbate, which could not penetrate through the lipid membrane. However, on addition of ferrocene and tetraphenylboron a slow rate of ferricyanide reduction (100 nmoles per min) was observed. When FCCP or other uncouplers were added, the rate of reduction was increased to 530 nmoles per min and the electron movements were accompanied by proton movements into the vesicles similar to those observed during electron flow in submitochondrial particles or chloroplasts. Experiments with black membranes indicated that electron flux with ferrocene is associated with the formation of a membrane potential. These experimental observations are explained as shown in Fig. 6. Ascorbate interacts with the ferricinium cation to form ferrocene, which in turn is oxidized by ferricyanide. The tetraphenylboron anion is visualized to facilitate movement of the sluggish ferricinium cation. The negative membrane potential building up inside slows the flux of electrons. FCCP, which facilitates proton flux into the vesicles, can therefore abolish the membrane potential and accelerate electron flux. Dr. Hinkle and Dr. Kagawa have attempted to incorporate the coupling device (oligomycin-sensitive ATPase and coupling factors) into this model system with ascorbate inside and promote ATP formation, but thus far these attempts have yielded ambiguous results.

I would like to discuss briefly now the role of oligomycin, which was discovered by Lardy and his collaborators to be representative of a group of energy transfer inhibitors. Oligomycin has proved immensely useful in studies of oxidative phosphorylation and of its partial reactions.

According to the chemical hypothesis, we can visualize oligomycin to interact with X in a rather selective manner. As indicated below, oligomycin alters X so that it ($X_\bullet$) is still capable of forming $X \sim Y$ and of participating in energy-dependent reductions and ion translocations (a). At the same time oligomycin must render $X \sim Y$ incapable of interacting with either water (b) or phosphate (c). Moreover, at low concentrations it preferentially inhibits

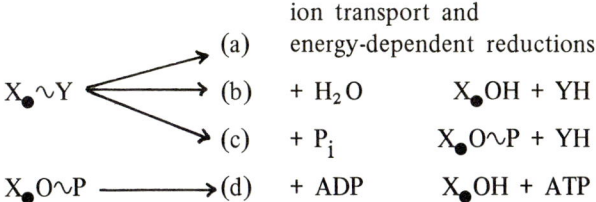

the reaction with water while allowing phosphate to interact. In the reverse reaction (d), catalyzed by ATPase, XOH modified by oligomycin cannot accept the phosphate group from ATP or, alternatively, cannot catalyze the transfer of $X \sim Y + YH \rightarrow X \sim Y + P_i$. The latter formulation, without inhibition of $X \sim P$ formation, assumes a stable $X \sim Y$ which has not as yet been detected. It is apparent that this formulation of a selective modification of X by oligomycin represents a useful working hypothesis which should stimulate attempts to isolate X or its derivative after its interaction with an energy transfer inhibitor. Oligomycin or rutamycin are probably not suitable for such attempts since they were shown to interact in a reversible manner (28). On the other hand, the interaction with DCCD was not found to be reversible (29), and ^{14}C-labeled DCCD has not been used in attempts to isolate the membrane component with which it interacts (30).

According to Mitchell (31) one of the effects of oligomycin in depleted submitochondrial particles is to decrease the permeability of the membrane to protons, a function normally fulfilled by coupling factors (9). Since the impermeability to protons is very likely associated with the phospholipid components of the membrane, the question arises whether energy transfer inhibitors interact with a protein or a phospholipid. Earlier experiments by Beechey and his associates pointed to a lipid or proteolipid (30), while the recent studies localized the radioactivity in a protein (32). These findings are most interesting in view of unpublished observations by Montal and Kagawa in our laboratory which indicated that, at the concentrations of energy transfer inhibitors used to block ATPase, phospholipids as well as insoluble protein components of the membrane interacted, while under the same conditions none of the soluble coupling factors were affected (33). It may also be remembered that inhibition of ATPase activity by rutamycin was reversed by washing of the treated particles with phospholipids (28). It is therefore possible that the action of energy transfer inhibitors is complex and may concern more than one constituent of the membrane. This would help to explain the curious biphasic effect of energy transfer inhibitors on A-particles, stimulating phosphorylation at low concentrations and inhibiting at higher concentrations. On the other hand, interaction of transfer inhibitors with more than one component complicates the interpretation of their mode of action in the light of either the chemical or the chemiosmotic hypothesis. Nevertheless, we should like to propose that energy transfer inhibitors do indeed affect proton permeability by interaction with phospholipid components, but also interact with X, a protein component of the membrane. Experiments by M. Montal in our laboratory on the effect of energy transfer inhibitors on model membranes consisting of phospholipid bilayers indicated that energy transfer inhibitors induced both a stabilization and a decreased conductivity of the membranes. Unfortunately the experiments were extremely variable and not reproducible enough to warrant firm conclusions.

Finally, I cannot leave the subject of respiratory control without mentioning long chain unsaturated fatty acids. It has been known since the work of Pressman and Lardy (34) that these naturally occurring compounds serve as uncouplers of oxidative phosphorylation. Recent findings in our laboratory (35) suggest that fatty acids are associated with a specific protein of the mitochondrial membrane. Although released as free fatty acids by solvents, the association is so firm that a large excess of bovine serum albumin cannot displace them from the native protein. These discoveries have led us into exciting speculations which I shall spare you. However, we look upon fatty acids now not only as natural regulators of respiration, but as potential proton carriers participating in the mechanism of coupling process proper.

Summary

The control of ATP generating systems is closely linked to the mechanism of the coupling process and dependent on the stability of the high-energy intermediate $X \sim Y$. Experimental data on oxidation control induced by energy transfer inhibitors and by coupling factors are presented which point to a relationship between respiratory control and proton translocation. Current hypotheses of oxidative phosphorylation and experimental model systems have proved most useful in designing experiments which may establish the role of the unknown intermediate XOH and YH in oxidoreduction and hydration-dehydration processes. The mode of action of energy transfer inhibitors and the possible role of unsaturated fatty acids in oxidative phosphorylation are discussed.

Presented by E. Racker. The experimental work reported in this paper was supported by United States Public Health Grant CA-08964 from the National Cancer Institute and an Albert Einstein Award.

References

1. Racker, E. In: E. Racker (Editor), Membranes of mitochondria and chloroplasts, Van Nostrand Reinhold Company (1970), pp. 127-171.
2. Boyer, P.C. In: T.E. King, H.S. Mason and M. Morrison (Editors), Oxidases and related redox systems, Vol. 2 (1965), p. 994.
3. Mitchell, P. Chemiosmotic coupling in oxidative and photosynthetic phosphorylation. Biol. Rev. Cambridge Phil. Soc. 41:445-502(1966).
4. Painter, A.A. and F.E. Hunter, Jr. Phosphorylation coupled to oxidation of thiol groups (GSH) by cytochrome c with disulfide (GSSG) as an essential catalyst II. Demonstration of ATP formation from ADP and HPO_4^{-2}. Biochem. Biophys. Res. Commun. 40:369-377(1970).
5. Lardy, H.A. and S.M. Ferguson. Oxidative phosphorylation in mitochondria. Ann. Rev. of Biochem. 38:991-1034(1969).
6. Hinkle, P. A model system for mitochondrial ion transport and respiratory control. Biochem. Biophys. Res. Commun. 41:1375-1381(1970).
7. Cockrell, R.S. and E. Racker. Respiratory control and potassium transport in submitochondrial particles. Biochem. Biophys. Res. Commun. 35:414-419(1969).
8. Montal, M., B. Chance and C.P. Lee. Uncoupling and charge transfer in submitochondrial particles. Biochem. Biophys. Res. Commun. 36:428-434(1969).
9. Hinkle, P. and L.L. Horstman. J. Biol. Chem., in press.
10. Racker, E. Topography of coupling factors in oxidative phosphorylation. In: B. Chance (Editor), The energy-linked function of mitochondria, Academic Press, Inc., New York (1963), pp. 75-85.
11. Racker, E. The two faces of the inner mitochondrial membrane. In: P.N. Campbell and F. Dickens (Editors), Essays in biochemistry, Vol. 6 (1970), pp. 1-22.
12. Schneider, D.L. and E. Racker. In: T.E. King, H.S. Mason and M. Morrison (Editors), Oxidases and related redox systems, in press.
13. Racker, E. Resolution and reconstitution of the inner mitochondrial membrane. Fed. Proc. 26:1335-1340(1067).
14. Racker, E. Mechanisms in bioenergetics, Academic Press, Inc., New York (1965), 259 p.
15. Livine, A. and E. Racker. Partial resolution of the enzymes catalyzing phosphorylation V. Interaction of coupling factor 1 from chloroplasts with ribonucleic acid and lipids. J. Biol. Chem. 244:1332-1338(1969).
16. Lee, C.P. and L. Ernster. Energy transfer system of submitochondrial particles II. Effects of oligomycin and aurovertin. European J. Biochem. 3:391-400(1968).
17. Fessenden, J.M. and E. Racker. Partial resolution of the enzymes catalyzing oxidative phosphorylation XI. Stimulation of oxidative phosphorylation by coupling factors and oligomycin, inhibition by an antibody against coupling factor 1. J. Biol. Chem. 241:2483-2489(1966).
18. Racker, E. and L.L. Horstman. Partial resolution of the enzymes catalyzing oxidative phorphorylation XIII. Structure and function of submitochondrial particles completely resolved with respect to coupling factor 1. J. Biol. Chem. 242:2547-2551(1967).
19. Racker, E., L.L. Horstman, D. Kling and J.M. Fessenden-Raden. Partial resolution of the enzymes catalyzing oxidative phosphorylation XXI. Resolution of submitochondrial particles from bovine heart mitochondria with silicotungstate. J. Biol. Chem. 244:6668-6674(1969).

20. Horstman, L.L. and E. Racker. Partial resolution of the enzymes catalyzing oxidative phosphorylation XXII. Interaction between mitochondrial adenosine triphosphatase inhibitor and mitochondrial adenosine triphosphatase. J. Biol. Chem. 245:1336-1344(1970).
21. Lynn, W.S. and K.D. Straub. Isolation and properties of a protein from chloroplasts required for phosphorylation and H^+ uptake. Biochemistry 8:4789-4793(1969).
22. Lien, S. and E. Racker. J. Biol. Chem., in press.
23. Racker, E. A mitochondrial factor conferring oligomycin sensitivity on soluble mitochondrial ATPase. Biochem. Biophys. Res. Commun. 10:435-439(1963).
24. Racker, E. Bloomington Symposium, in press.
25. Pullman, M.E. and G.C. Monroy. A naturally occurring inhibitor of mitochondrial adenosine triphosphatase. J. Biol. Chem. 238:3762-3769(1963).
26. Asami, K., K. Juntti and L. Ernster. Possible regulatory function of a mitochondrial ATPase inhibitor in respiratory chain-linked energy transfer. Biochim. Biophys. Acta 205:307-311(1970).
27. Cunningham, C., H. Nishibayashi-Yamashita and E. Racker. J. Biol. Chem., submitted for publication.
28. Kagawa, Y. and E. Racker. Partial resolution of the enzymes catalyzing oxidative phosphorylation IX. Reconstruction of oligomycin-sensitive adenosin triphosphatase. J. Biol. Chem. 241:2467-2474(1966).
29. Bulos, B. and E. Racker. Partial resolution of the enzymes catalyzing oxidative phosphorylation XVIII. The masking of adenosinetriphosphatase in submitochondrial particles and its reactivation by phospholipids. J. Biol. Chem. 243:3901-3905(1968).
30. Knight, I.G., C.T. Holloway, A.M. Robertson and R.B. Beechey. The chemical nature of the site of action of dicyclohexylcarbodiimide in mitochondria. Biochem. J. 109:27P(1968).
31. Mitchell, P. Proton-translocation phosphorylation in mitochondria, chloroplasts, and bacteria: natural fuel cells and solar cells. Fed. Proc. 26:1370-1379(1967).
32. Cattell, K.J., I.G. Knight, C.R. Lindop and R.B. Beechey. The isolation of dicyclohexylcarbodiimide-binding proteins from mitochondrial membranes. Biochem. J. 117:1011-1013(1970).
33. Knowles, A.F., R.J. Guillory and E. Racker. Partial resolution of the enzymes catalyzing oxidative phosphorylation XXIV. A factor required for the binding of mitochondrial adenosine triphosphatase to the inner mitochondrial membrane. J. Biol. Chem. 246:2672-2679(1971).
34. Pressman, B.C. and H.A. Lardy. Effect of surface active agents on the latent ATPase of mitochondria. Biochim. Biophys. Acta 21¾458-466(1956).
35. Chan, S. A fatty acid-binding protein in bovine heart mitochondria and its interaction with coupling factors. Fed. Proc. 30:1246(1971).
36. Racker, E. Studies of factors involved in oxidative phosphorylation. Proc. Nat. Acad. Sci. U.S. 48:1659-1663(1962).
37. Steinman, C.R. and W.B. Jakoby. Yeast aldehyde dehydrogenase I. Purification and crystallization. J. Biol. Chem. 242:5019-5023(1967).

TABLE 1

REVISED SCOREBOARD

	Hypothesis	
	Chemiosmotic	Chemical
Role of membrane	+	−
Model systems	−	−
Uncouplers and ionophores	+	−
Proton translocation	+	−
Topography of oxidation chain	+	−
K^+ and Ca^{++} transport	−	+
The Painter and Hunter experiments	−	+
Exchange reactions	−	+

TABLE 2

OXIDATION CONTROL IN GLYCOLYSIS

Glyceraldehyde-3-P + NAD-E $\rightleftharpoons$ Acyl-E + $NADH_2$
Acyl-E + P_i $\rightleftharpoons$ 1,3-diphosphoglycerate
1,3-diphosphoglycerate + ATP $\rightleftharpoons$ phosphoglycerate + ATP

Sum:
  Glyceraldehyde-3-phosphate + P_i + ATP + NAD $\rightleftharpoons$ phosphoglycerate + ATP + $NADH_2$
P-Cycle: ATP $\xrightarrow{\text{ATPase}}$ ADP + P_i
H-Cycle: $NADH_2$ + pyruvate $\rightleftharpoons$ NAD + lactate

TABLE 3

I-I OXIDATION CONTROL IN A-PARTICLES (pH 9.2) AND SMP (pH 7.4)

A suspension of 30 ml of bovine heart mitochondria (20 mg protein per ml) in 0.25 M sucrose was adjusted with freshly diluted 1 N ammonia to pH 9.2 and exposed to sonication in a Raytheon oscillator for 2 min at maximum output with ice water cooling. Samples of mitochondrial suspension containing sodium pyrophosphate or Tris sulfate at 10 mM final concentration adjusted to the indicated pH were sonicated in the same manner. After centrifugation for 20 min at 26,000 x g, the supernatants were centrifuged for 90 min at 105,000 x g and the particles collected, washed with 0.25 M sucrose and resuspended in 0.25 M sucrose. Respiration was measured at 22° with 150 to 500 µg particles in a final volume of 1.2 ml in the presence of 0.25 M sucrose, 25 mM KP_i, pH 7.0, 1.2 mg bovine serum albumin. The reaction was initiated by addition of 1 µmole of DPNH; later 2 µg of rutamycin and then 50 nmoles of FCCP was added and rates of oxidation and control ratios calculated.

Preparation of particles	Initial rate of oxidation	Ratios of rates	
		−rutamycin / +rutamycin	+ FCCP / +rutamycin
	µAtoms oxygen/min/mg		
Ammonia pH 9.2	0.56	2.76	2.76
Ammonia + pyro-P	0.34	1.48	1.42
Tris sulfate pH 7.4	0.33	1.33	1.67
Tris sulfate + pyro-P	0.39	1.15	1.48

TABLE 4

EFFECT OF PRESENCE OF AMMONIA AND SALT PRESENT DURING SONICATION ON I-I OXIDATION CONTROL

Preparation of particles and measurement of I-I induced respiratory control was carried out as described in Table 3 except that in one sample KOH was used instead of ammonia to adjust to pH 9.2, in the third sample 10 mM Tris sulfate pH 7.4 was used and in the fourth sample 20 mM KCl was present during sonication.

Preparation of particles	Initial rate of oxidation	Ratios of rates	
		−rutamycin +rutamycin	+ FCCP +rutamycin
	$\mu Atoms\ oxygen/min/mg$		
Ammonia pH 9.2	0.60	3.10	2.85
KOH pH 9.2	0.60	2.74	2.86
Tris sulfate pH 7.4	0.39	1.17	1.50
Ammonia pH 9.2 + 20 mM KCl	0.56	1.26	1.69

TABLE 5

EFFECT OF pH DURING SONICATION (AT LOW IONIC STRENGTH) ON I-I OXIDATION CONTROL

The submitochondrial particles were prepared as described in Table 3 except that only sucrose and no buffer was present during sonication. The pH was adjusted with ammonia or HCl respectively. Respiration and oxidation control were measured as described in Table 3.

Preparation of particles	Initial rate of respiration	Ratios of rates		
		−rutamycin +rutamycin	+ FCCP +rutamycin	+ FCCP −rutamycin
pH	$\mu Atoms/min/mg$			
6.80	0.34	2.00	2.42	1.23
7.40	0.37	2.00	2.28	1.12
7.95	0.37	2.40	2.68	1.12
8.80	0.41	3.44	3.08	0.90
9.70	0.46	4.50	3.54	0.80
10.20	0.50	5.35	3.60	0.67

TABLE 6

CONFERRAL OF I-I OXIDATION CONTROL TO INSENSITIVE SUBMITOCHONDRIAL PARTICLES

SMP (pyrophosphate) were prepared as described previously (36). SMP (Tris-sulfate) were prepared in the same manner except that 25 mM tris sulfate - 25 mM KCl (pH 7.4) were used instead of pyrophosphate. Treatment with Sephadex followed by urea was as described previously (18); some particles were treated with 1% silicotungstate (19) after passage through Sephadex. Measurements of respiration and of oxidation control ratios were carried out as described in Table 3 except that in the case of particles treated with STA three times as much bovine serum albumin was added.

Particles	Initial rate of oxidation	Ratio of rates	
		−rutamycin +rutamycin	+ FCCP +rutamycin
	μAtoms oxygen/min/mg		
Experiment 1			
SMP (pyrophosphate)	0.58	1.10	1.68
SMP treated with Seph-urea	1.39	2.20	1.95
SMP treated with Seph-STA	0.145	2.10	2.80
Experiment 2			
SMP (Tris-sulfate)	0.82	1.53	1.65
SMP treated with Seph-urea	0.50	3.30	2.46
SMP treated with Seph-STA	0.072	1.90	2.40

TABLE 7

EFFECT OF TREATMENT OF F_1 WITH TRYPSIN ON OXIDATION CONTROL

Reconstitution of ASU-particles with factors was carried out essentially as described previously (7). Trypsin-treated F_1 was prepared by incubating 2 mg of F_1 in 0.25 M sucrose, 10 mM Tris sulfate, 2 mM EDTA, 4 mM ATP at pH 8.2 with 40 µg of trypsin. After 10 min at 23° the reaction was stopped with 200 µg of trypsin-inhibitor (soybean). The F_1 was reprecipitated twice by addition of an equal volume of saturated ammonium sulfate and stored at 0°. The inhibitor-low F_1 was prepared as described by Horstman and Racker (20).

Additions to ASU-part. + OSCP	Initial rate of oxidation	Oxidation control ratio: +FCCP/−FCCP
	nmoles/min/mg	
Experiment 1		
Native F_1	435	2.3
F_1 treated with TPCK-trypsin (Free of chymotrypsin)	714	1.2
Experiment 2		
Native F_1	470	1.96
Inhibitor low-F_1	560	1.41
F_1 treated with trypsin	840	1.20
Experiment 3		
Native F_1	280	3.8
F_1 treated with trypsin	870	1.3

TABLE 8

EFFECT OF POLYLYSINE, RUTAMYCIN AND ATPase INHIBITOR ON OXIDATION CONTROL IN SUBMITOCHONDRIAL PARTICLES RELEASED BY FCCP

A-particles were reconstituted with F_1 and OSCP as described (7) and then passed through a Sephadex G-50 (coarse) column (0.8 cm x 12 cm) with 50 mM Tris sulfate, pH 7.4 - 0.25 M sucrose as eluant. To 162 µg of reconstituted A-particles 0.028 µg of rutamycin, 100 µg of polylysine (M.W. 100,000) 10 µg of ATPase inhibitor (20) and 2 mg of defatted bovine serum albumin were added. Oxidation control was measured as described in Table 3 except that a NADH regenerating mixture consisting of 10 µg alcohol dehydrogenase, 0.16 unit yeast aldehyde dehydrogenase (37), 10 µmoles ethanol and 1 µmole dithiothreitol was used. The reaction was initiated by addition of 1 µmole NAD^+.

System	NADH oxidation	Control ratio +FCCP/−FCCP
	nmoles/min/mg	
Complete	220	5.0
Omit inhibitor	300	3.3
Omit rutamycin	450	2.3
Omit polylysine	430	2.3
Omit all three	490	2.0

TABLE 9

OXIDATION CONTROL IN SUBMITOCHONDRIAL PARTICLES RELEASED BY ADP

The experimental condition were as described in Table 8 except that control was released by 3 μmoles ADP instead of FCCP.

Order of additions to reconstituted particles	NADH oxidation	Control ratio $\frac{\text{before ADP}}{\text{after ADP}}$
	nmoles/min/mg	
Experiment 1		
NADH regenerating system	1.111	
Rutamycin	0.966	
Inhibitor	0.933	
Polylysine	0.649	
BSA	0.488	
ADP	0.781	1.6
Experiment 2		
NADH regenerating system	0.933	
BSA	0.933	
Rutamycin	0.915	
Polylysine	0.675	
Inhibitor	0.538	
ADP	0.800	1.5

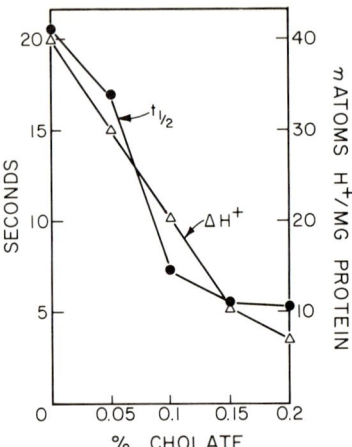

Fig. 1 A. *Effect of cholate on H^+ translocation following O_2 pulse.* In a final volume of 1 ml containing 50 mM KCl, 180 mM sucrose, 1 mM K phosphate buffer pH 6.5, 4 μg rutamycin, 0.5 μg valinomycin, 0.2 μmole DPNH, 100 μg alcohol dehydrogenase, 0.1 M ethanol, 500 μg of A-particles and the indicated concentrations of cholate, uptake of protons was measured (3) after addition of 2 μl of oxygen-saturated ethanol. The extent (ΔH^+) and the rate of decay ($t_{1/2}$) of proton translocation were expressed as described by Mitchell (3).

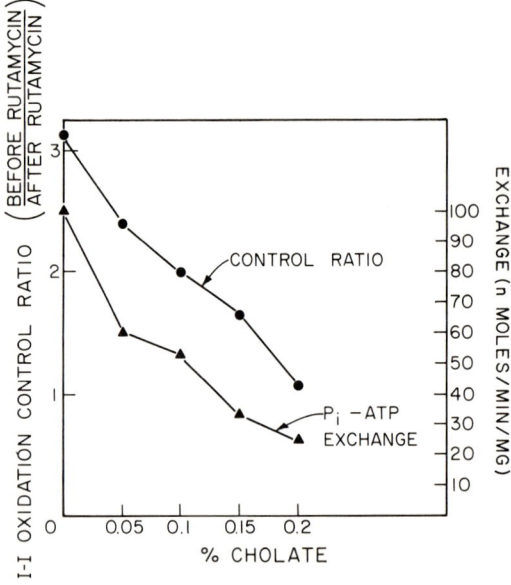

Fig. 1 B. *Effect of cholate on I-I oxidation control and on $^{32}P_i$-ATP exchange.* I-I oxidation control was measured esssentially as described in Table 3 except that the conditions of pH and ionic strength were maintained similar to those used for measuring H^+ translocation (Fig. 1A). For comparative purposes the $^{32}P_i$-ATP exchange was also measured under these conditions except that A-particles (500 μg) were used after reconstitution with F_1 and OSCP (7).

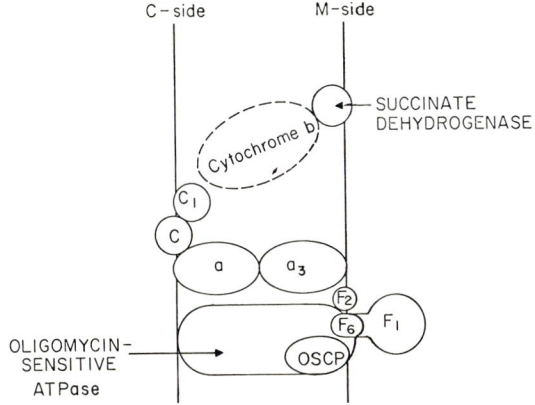

Fig. 2. *Topography of the inner mitochondrial membrane.*

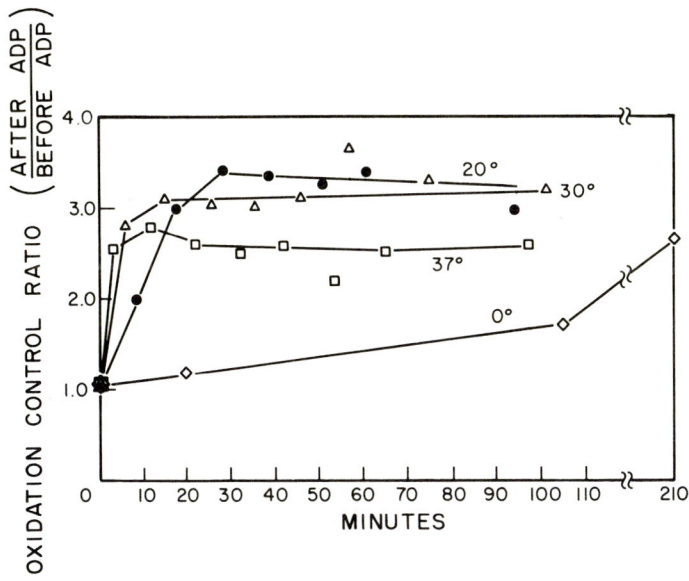

Fig. 3. *Reactivation of oxidation control after freezing of bovine heart mitochondria.* Heavy layer mitochondria prepared the previous day and kept frozen at -70° were thawed slowly without allowing the suspension to warm up. Samples were then exposed to elevated temperatures for the times indicated and analyzed for oxidation control in a Gilson Oxygraph. In a final volume of 1.1 ml the assay mixture contained 250 mM sucrose, 1 mM $MgSO_4$, 100 mM K-phosphate buffer (pH 6.9), mitochondria (1.4 mg) and pyruvate-malate (10 mM each). Control ratios are expressed in terms of the respiration rate after and before addition of ADP (4 μmoles).

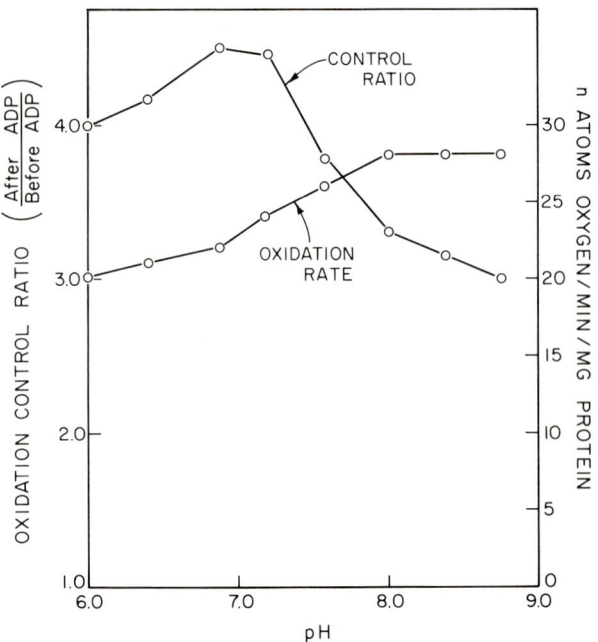

Fig. 4. *Effect of pH on oxidation control in bovine heart mitochondria.* Heavy layer mitochondria prepared the previous day and kept at 0° overnight were used in this experiment. The pH of the assay mixture was varied by changing the proportion of KH_2PO_4 and K_2HPO_4. The conditions of assay were as described in the legend of Fig. 3.

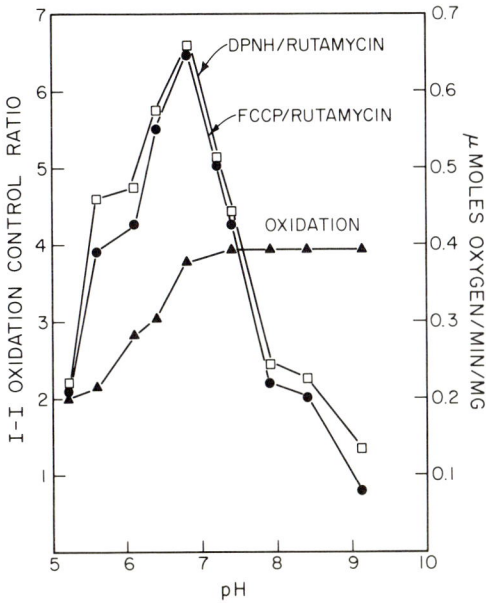

Fig. 5. *Effect of pH on I-I oxidation control of submitochondrial particles.* A set of buffers were prepared at room temperature by adding 1 M acetic acid to a solution containing 25 mM unneutralized Tris, 0.25 M sucrose and 0.1% bovine serum albumin. I-I oxidation control was measured with DPNH as substrate as described in Table 3.

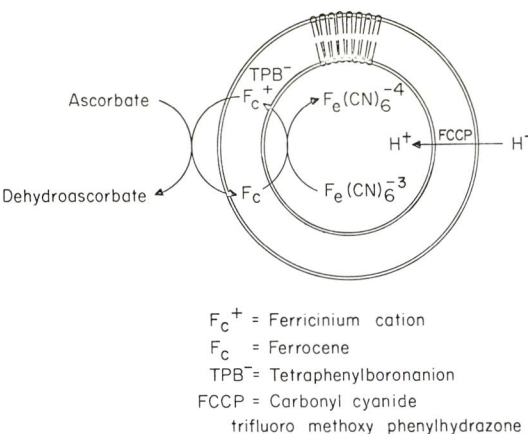

F_c^+ = Ferricinium cation
F_c = Ferrocene
TPB^- = Tetraphenylboronanion
FCCP = Carbonyl cyanide trifluoro methoxy phenylhydrazone

Fig. 6. *Oxidation control in liposomes.*

KINETIC CONTROL OF ELECTRON FLOW

*Britton Chance, Maria Erecińska, Edwin M. Chance,
Alberto Boveris and Michael Wagner*

Introduction

The phenomenological expression of the operation of the respiratory chain is the utilization of substrates and the consumption of oxygen which leads to formation of H_2O and CO_2 with concomitant synthesis of ATP. In other words, the transfer of electrons from a negative substrate to a highly positive oxygen is coupled to the synthesis of high energy bond. The direct conclusion is that the basic phenomena underlying the concept of the respiratory chain are of thermodynamic nature. Since thermodynamics itself does not tell us anything about the actual reaction pathways or rates, the only way to fully understand the operation of the respiratory chain is by careful analysis of both the thermodynamic and kinetic parameters of the system. The former have been elegantly presented in the preceding paper (1). This contribution attempts to analyze the kinetic data and thus by successful application of both approaches afford a basis for a new concept of coupled electron flow.

The two phenomena relevant to the concept of the respiratory chain, but best expressed kinetically, are respiratory control and endothermic reversed electron flow. The idea of respiratory control can be traced back to the original studies of Lardy and Wellman (2), Loomis and Lipmann (3), Ochoa (4) and ourselves (5) and is most simply described as the enhanced oxygen uptake in the presence of phosphate acceptor. The low rate of oxygen uptake in the absence of phosphate acceptor can be correlated with a high degree of reduction of the respiratory chain carriers [mostly pyridine nucleotides and cytochrome b (6)]. Upon transition from the controlled state 4 (absence of phosphate acceptor) to the active state 3 (in the presence of P_i + ADP) the increased oxidation rate is accompanied by an increase in the oxidized steady state level of the carriers.

The reversed electron flow effect demonstrated initially as ATP-linked reduction of NAD^+ by succinate (7, 8) was later confirmed to occur at all levels of the chain from cytochrome a_3 to pyridine nucleotide (9-13) but not oxygen. This suggests ready reversibility of all intercarrier reactions along the respiratory chain and affords the experimental support for the postulated equilibrium condition required by thermodynamic studies.

Kinetic studies of electron transfer embrace two basic types of experiments which complement each other. In one type of study where anaerobic mitochondria are pulsed with oxygen, the carriers make an abrupt transition to oxidized steady state and are reduced when oxygen is exhausted. In the second type of study, the aerobic mitochondria are pulsed with reductant and, after the reduced steady state has been achieved and the substrate utilized, the carriers return to the oxidized state. In both methods, repeated additions of either oxidant or reductant can be carried out. Both approaches require that the oxidant or reductant react at a rate comparable to that of the intercarrier reactions within the chain. Oxygen diffusion is adequately fast to permit its use in kinetic studies, but it has also been found recently that one of the quinols, durohydroquinone, is able to react with the respiratory chain rapidly enough to be able to serve as a pulsed reductant (14, 15).

The experimental systems can be simplified in such a way that the kinetic measurements are limited to one or two of the three operative phosphorylation sites of the respiratory chain. The most convenient system is afforded by the rapid initiation of electron transfer of the reduced carriers by an oxygen pulse applied to the antimycin A blocked system in which only site III is operative. A second alternative is to investigate control mechanisms at sites III and II with the exclusion of those at site I as in the rotenone inhibited system. Particularly interesting phenomena occur under this case since energy conserved at site III in the initial phases of the oxygen pulse response can be transferred to site II and thereby afford a picture of the development of control and coupling at site II in response to site III. Finally, by using pulses of durohydroquinone, a specific reductant of cytochromes b (16) in cytochrome c depleted mitochondria, it is possible to limit kinetic changes to cytochrome b–ubiquinone–cytochrome c region of the respiratory chain (Fig. 1).

Results

Kinetic studies on phosphorylation site III.

The very rapid oxidation reactions of the cytochrome c_1, c, a, and a_3 can be accurately studied using a recently described flow flash method (17) demonstrated by the experimental tracing of Fig. 2. A continuous profile of concentration against time from 50 μsec onwards can be obtained by laser-induced photolysis of the cytochrome a_3-CO compound in the presence of oxygen. The experiment is controlled first by photolysis in the absence of oxygen, which establishes a baseline for completely reduced a_3. This then is followed by photolysis in the presence of oxygen in which rapid oxidation of reduced a_3 occurs. Oxidation of cytochrome a_3 initiates the oxidation of the other carriers along the chain, which can be recorded at appropriate wavelengths as illustrated by trace B of Fig. 2 in which cytochrome c oxidation has been followed. It is noteworthy that the 60 Hz dual wavelength spectrophotometer can now be adapted to record these very rapid reactions (down to 20 μsec) provided the laser flash is appropriately synchronized with the flash of the measuring wavelength of the spectroscopic system.

The recorded curves for cytochromes a_3, a, c_1 and c can be expressed in numerical terms either as a half-time of reaction or initial slope. Alternatively, the data can be used for further mathematical evaluation using computing devices which not only lead to accurate determination of the actual rate constants (it is essential to employ computer devices to this purpose because the concentrations of cytochromes are so nearly equal that pseudo-first order approximations are not valid), but also, when fitted to an appropriate model, can give us information on reaction mechanisms. The experimental curves for cytochromes a_3, a, c_1 and c at two oxygen concentrations (64 μM and 1.8 μM) under conditions similar to that of state 4 were fitted into a model consisting of a sequence of first order reversible reactions. Previous computer solutions have emphasized kinetic recordings at oxygen tensions stoichiometrically less than the amounts of the respiratory carriers. In that case (18) unsatisfactory fits were obtained with a linear sequence of irreversible reactions, but acceptable fits were obtained using what was termed the "oxygen control model" (18, 19). The model employed here (Fig. 3) differs from the "oxygen control model" in that reversible bimolecular reactions are represented (20). A number of constraints are imposed upon the solution (2): (a) molar concentrations of the reactions are employed; (b) large numbers of data points are included (between 200 and 500); (c) equilibria for the oxygen reaction for the $a_3 + a$, $a + c$ and for the $c + c_1$ reactions are available; (d) the flux through the system is identified; (e) the configuration of the system, *i.e.*, the sequence of electron transfer reactions is specified.

Under these conditions, 500 data points (10 unknown constants) over time ranges from 100 μsec to roughly 200 sec may be fitted with a sum of squares of all errors of 2%, the sum of squares per point being extremely small. This obviously does not identify the only fit to the experimental data, but it identifies one of the known characteristics.

The computer derived constants (Table 1) illustrate the reaction velocity constants that are relatively rapid in both directions and nearly equal. The approximate equality of the forward and reverse velocity constants for the interchain reactions suggests that the total system from cytochrome a_3 to c_1 is close to equilibrium. In the coupled mitochondria this would easily be expected on the basis of Wilson and Dutton's (21) data in which the measured midpoint potential of cytochrome a_3 is very close to that of the other cytochromes.

The computer, however, chooses the similar equilibrium conditions for the mitochondria in the uncoupled state in which the measured midpoint potential of cytochrome a_3 is +385 mV; thus the two cytochromes would not be expected to equilibrate as rapidly. In order to satisfy the computer postulated equilibrium constant and reconcile it with the experimentally measured midpoint potential of cytochrome a_3, we put forward (2) the hypothesis that cytochrome a_3 reacts with cytochrome a in its low potential form and with oxygen in its high potential form. A transition from the high to the low potential form can represent a step in energy conservation due to the alteration of some structural feature of the system, most probably in the vicinity of the heme. Similarly, the low potential reduced cytochrome a_3 undergoes an energy yielding reaction in which the intermediate capable of phosphorylating ADP can be formed. Oxygen remains bound to cytochrome oxidase during this cycle of reactions and the process is repeated three more times in order to reduce oxygen to H_2O. This hypothetical scheme satisfies both the thermodynamic requirement for an isopotential electron transfer between cytochrome a and a_3 and the computer requirement for reversibility of the cytochrome a_3-a reaction. Furthermore, it permits the most economical conversion of energy in the cytochrome a_3-oxygen span, *i.e.*, when ΔG is minimal (1).

The near equilibrium of oxygen to cytochrome a_3 obtained in computer solutions does not agree with the experimental observations of Schindler (22) in which a high irreversibility of oxygen and a_3 as well as dependence of cytochrome oxidase on flux through the chain have been found. Furthermore, no evidence has been obtained for oxygen evolution from cytochrome oxidase under the influence of ATP (22). For this reason a constraint on the reverse reaction for cytochrome a_3-oxygen has been put into the computer solution. The fit (Table 1, column 2) was found to be satisfactory, and the slight readjustment of individual rates did not change the main characteristics of the system. Thus, the irreversible cytochrome a_3-oxygen reaction can be accepted by the computer without affecting the overall fit. The data obtained by combined experimental and computer study leads us to interesting conclusions. In the coupled state, the reaction velocity constants for the forward and reverse reactions at site III are so fast compared with the flux values of state 4 ($\sim$1-10/sec), that the assumption of equilibrium at this site is justified in state 4. In this condition, thermodynamic considerations can be fully applied.

Studies on phosphorylation site II

Studies on phosphorylation site II have been of interest for a long time due to initial observations of kinetic peculiarities of cytochrome b (23). For this reason it is much more than coincidence that cytochrome b itself has been found to be an energy conserving cytochrome (24, 25).

It has been previously demonstrated that in the coupled mitochondria, cytochrome b kinetics are biphasic (24, 25). Plot of the fast (0-200 msec) and slow (200-2000 msec) phases of the oxidation against wavelength demonstrated that the fast phase had an absorption maximum at 560 nm, while the slow phase at 564 nm. On the basis of these data, as well as the potentiometric titrations of Wilson and Dutton (26), the existence of two cytochromes $b-b_K$ (short wavelength) and b_T (long wavelength)—has been concluded. Addition of an uncoupler caused both types of cytochrome b to respond monotonically to an oxygen pulse and the spectrum obtained followed closely the sum of the slow and fast phases of the coupled mitochondria.

Elegant studies of Sato (28) in our laboratory led to a detailed spectral characterization of both cytochromes b. Cytochrome b_K exhibits a single peak at room temperature with maximum at 560 nm, whereas cytochrome b_T shows a broad peak with a maximum at 564 nm and a shoulder at 557 nm. The low temperature spectra show the characteristics of both cytochromes more clearly: a sharp peak at 558 of b_T and a split peak of b_K at 562.5 and 555 nm.

In order to correlate kinetic changes of cytochrome b with those of their immediate neighbors, ubiquinone and cytochrome c_1, oxidation traces of all three components were obtained for the same mitochondrial sample in the coupled and uncoupled state. The experimental results shown in Fig. 4 demonstrate that the rate of oxidation of cytochrome b and ubiquinone are markedly enhanced in the uncoupled state. In order to obtain more information on the possible reaction mechanism, the data were subjected to a detailed mathematical evaluation, using a model similar to that presented above for the fast cytochromes. Fitting of experimental data (29) demonstrated that upon transition from the coupled to uncoupled state, the only constants really changed were those between cytochrome b and c_1. Calculation of the apparent equilibrium constants defined as the ratio of the rate constants enabled us to show that the reaction between oxidized cytochrome b and reduced cytochrome c_1 is undetectable by kinetic computer study in the uncoupled mitochondria but is shifted in the opposite direction (towards cytochrome c_1 oxidation) in the coupled state.

This leads us to several interesting conclusions. a) The apparent equilibrium constants can be calculated purely on the basis of kinetic data. b) The most sensitive rate constant in the coupled to uncoupled transition is the reaction between cytochrome b and c_1, a step predicted by thermodynamic studies as the energy coupling reaction (27). c) The apparent equilibrium constant between cytochrome b and c_1 in the coupled mitochondria is near unity, indicating reversibility of the reaction in state 4. d) The calculated ratio of the apparent equilibrium constants in the coupled to uncoupled transition 5 X 10^4 (28) corresponds to the midpoint potential change of one of the two components by about 200 mV which is in good agreement with the value obtained by potentiometric titration (27) and theoretical prediction (1) and a sufficient free energy change for ATP synthesis.

Another approach to the studies of cytochrome b is afforded by using pulses of durohydroquinone, a specific reductant of b cytochromes (16). In the intact mitochondria, durohydroquinone reduces cytochrome b, which equilibrates rapidly with ubiquinone and is then oxidized by the cytochrome chain. Since the rates of oxidation in the cytochrome system are very rapid, the cycles of transient reduction are short. A convenient way to study this region of the chain is by using a kinetically less complex system which is afforded by depletion of the mitochondria of their endogenous cytochrome c content. In the c-depleted system, even in the uncoupled state, the rate limiting reaction of the overall electron transfer is at the cytochrome $c_1 \rightarrow c$ site, and the c extraction does not impair energy coupling *per se* as demonstrated by the fact that reincorporation of cytochrome c restores both electron transfer and energy coupling (28). Due to the existence of the rate limiting step at the cytochrome c level, the reducing equivalents from durohydroquinone are distributed to varying degrees between cytochromes b, ubiquinone and cytochrome c_1. The degree of reduction of individual components is governed by the reaction rates within the system, which in turn depend upon the state of the mitochondria. In c-depleted mitochondria in the presence of endogenous substrate, cytochrome c_1 is about 40% reduced prior to durohydroquinone addition. Addition of ATP results in oxidation of cytochrome c_1 and rapid reduction of one of the b cytochromes, cytochrome b_T. Since under these conditions the midpoint potential of cytochrome b_T is 25 mV more positive (+245 mV) than that of cytochrome c_1 (220 mV) (31), it follows that under equilibrium conditions of state 4 the degree of reduction of cytochrome b_T is higher than that of c_1, which is indeed observed experimentally. Upon addition of a pulse of durohydroquinone, kinetic changes followed at cytochrome b wavelengths correspond primarily to reduction of cytochrome b_K, since b_T is over 80% reduced prior to durohydroquinone addition. The reaction is biphasic, the initial slow phase of reduction being speeded up after about 300 msec by a factor of 2. The speeding of the reduction of cytochrome b_K coincides with the time in which the ubiquinone pool becomes reduced and indicates the extremely rapid rate of electron transfer between cytochrome b_K and ubiquinone towards ubiquinone in the presence of ATP. Addition of uncoupler causes disappearance of the biphasicity and a speeding up of the initial phase of b_K reduction which might indicate that electron transfer from reduced cytochrome b_K to oxidized ubiquinone is considerably slowed down in the uncoupled state.

Reduction of cytochrome b_T concomitant with oxidation of cytochrome c_1 which occurs upon addition of ATP suggests ready reversibility of the cytochrome b_T-c_1 reaction in the coupled mitochondria. Determination of the actual reduction rates of cytochrome c_1 by the durohydroquinone shows that the rate is 4-fold faster in the uncoupled than in the coupled case, thus supporting the data obtained by the oxygen pulse technique (25, 28) in which enhanced rate of oxidation of cytochrome b_T in the uncoupled state could be demonstrated. Thus studies with pulsed reductant indicate that in state 4, in the presence of ATP, the equilibrium of the b_T-c_1 reaction is shifted towards b_T reduction. Furthermore, the high degree of reduction of ubiquinone supports the suggestion that state 4 is a true equilibrium state.

The data presented enable us to draw, on the basis of purely kinetic studies, some general conclusions about the nature and function of the respiratory chain. The results clearly indicate the functionality of the reverse reactions on transient responses of the electron transport systems. From thermodynamics alone it would be impossible to tell whether the approximate equality of the forward and reverse reactions involves reaction velocity constants of sufficient magnitude to be of significance in these rapid reactions. The experimental data coupled with computer solutions show that they are significant insofar as the transient response of the electron transport system is concerned.

In steady state operation of the respiratory chain in energy coupling reactions, the electron flux is small as compared to the rates of forward and reverse reactions. This suggests that most of the system remains at equilibrium in state 4 and these reactions themselves are not rate limiting. Thus, state 4 can be called a thermodynamically controlled state, while state 3 remains a controlled activation.

Summary

Two general conclusions relevant to electron transport and energy coupling emerge from our studies. The first is that the rigid concept of a respiratory chain need no longer be held. Instead one finds three redox potentials set by three groups of carriers. The several carriers of a group are capable of rapid equilibrium compared to the flux demand at state 4. The second is that at site III (a, Cu, c and c_1) the actual sequence of reactions now assumes much less importance than it did under the previous concepts of the sequential chain. Insofar as the function of the chain in energy coupling and electron flow are concerned, it would seem to make little difference whether c_1, c, a or copper reacted on the one hand to reduce cytochrome a_3 or on the other hand to oxidize cytochrome b_T. The rationale of the respiratory chain as a chemical sequence of carriers must then be based on questions of specificity of electron flow and structure of the membrane bound components.

The key components of the respiratory chain are then the "T" or transducing components currently identified as cytochrome a_3, cytochrome b_T and those which have yet to be identified for site I. These are the "work horses" of the respiratory chain, the ones that have unique properties and the ones capable of the unique process of alternating oxidation-reduction states. It is their chemistry, their thermodynamics and their ability to transduce from electron flow to a high energy covalent compound that epitomize the unique features of the respiratory chain.

Presented by Britton Chance.

References

1. Wilson, D.F. and P.L. Dutton. Thermodynamic control of mitochondrial energy coupling. This volume.
2. Lardy, H.A. and H. Wellman. Oxidative phosphorylations: Role of inorganic phosphate and acceptor systems in control of metabolic rates. J. Biol. Chem. 195:215-224(1952).
3. Loomis, W.F. and F. Lipmann. Reversible inhibition of the coupling between phosphorylation and oxidation. J. Biol. Chem. 173:807-808(1948).
4. Ochoa, S. a-Ketoglutaric dehydrogenase of animal tissues. J. Biol. Chem. 155:87-100(1944).
5. Chance, B. and G.R. Williams. A simple and rapid assay of oxidative phosphorylation. Nature 175:1120-1121(1955).
6. Chance, B. and G.R. Williams. The respiratory chain and oxidative phosphorylation. Advan. Enzymol. 17:65-134(1957).
7. Chance, B. and G. Hollunger. Succinate-linked pyridine nucleotide reduction in mitochondria. Fed. Proc. 16:163(1957).
8. Chance, B. and G. Hollunger. Energy-linked reduction of mitochondrial pyridine nucleotide. Nature 185:666-672(1960).
9. Klingenberg, M. and W. Slenczka. Pyridinnucleotide in Leber-Mitochondrien. Biochem. Z. 331:486-517(1959).

10. Klingenberg, M., W. Slenczka and H. Ritt. Vergleichende Biochemie der Pyridinnucleotid-Systeme in Mitochondrien verschiedener Organe. Biochem. Z. 332:47-66(1959).
11. Packer, L. and M.D. Denton. Reduction of pyridine nucleotide from a terminal phosphorylating segment of the respiratory chain. Fed. Proc. 21:53(1962).
12. Tager, J.M., J.L. Howland, E.C. Slater and A.M. Snoswell. Synthesis of glutamate from a-oxoglutarate and ammonia in rat-liver mitochondria. IV. Reduction of nicotinamide nucleotide coupled with the aerobic oxidation of tetramethyl-p-phenylenediamine. Biochim. Biophys. Acta 77:266-275(1963).
13. Penefsky, H.S. The reduction of ubiquinone associated with the oxidation of tetramethyl-p-phenylenediamine (1). Biochim. Biophys. Acta 58:619-621(1962).
14. Boveris, A., R. Oshino, M. Erecińska and B. Chance. Reduction of Mitochondrial Components by Durohydroquinone. Biochem. Biophys. Acta 245:1-16(1971).
15. Ruzicka, F.J. and F.L. Crane. Four quinone reduction sites in the NADH dehydrogenase complex. Biochem. Biophys. Res. Commun. 38:249-254(1970).
16. Boveris, A., R. Oshino, M. Erecińska and M. Wagner. Biochem. Biophys. Acta, in press.
17. Chance, B. and M. Erecińska. Flow flash kinetics of the cytochrome a_3-oxygen reaction in coupled and uncoupled mitochondria (2). Arch. Biochem. Biophys. 143:675-687(1971).
18. Chance, B. and M. Pring. Die Zwanzigste Coll. Gesell. Biolog. Chemie, Mosbach/Baden, 1969, Springer Verlag, Berlin, p. 102.
19. Pring, M. In: B. Chance, C.P. Lee, J.K. Blasie, T. Yonetani and A.S. Mildvan (Editors). Fifth Johnson Foundation Colloq., Probes of structure and function of macromolecules and membranes, Acad. Press, N.Y., Vol. II, 453-467(1971).
20. Chance, B., M. Erecińska and E.M. Chance. In: T.E. King, H.S. Mason and M. Morrison (Editors). Second symposium on oxidases and related redox systems, J. Wiley and Sons, N.Y.(1971).
21. Wilson, D.F. and P.L. Dutton. The oxidation-reduction potentials of cytochromes a and a_3 in intact rat liver mitochondria. Archiv. Biochem. Biophys. 136:583-584 (1970).
22. Schindler, F. Oxygen kinetics in the cytochrome oxidase-oxygen reaction. Ph.D. dissertation, Univ. of Pennsylvania (1964).
23. Chance, B. Spectra and reaction kinetics of respiratory pigments of homogenized and intact cells. Nature 169:215-221(1952).
24. Chance, B., D.F. Wilson, P.L. Dutton and M. Erecińska. Energy-coupling mechanisms in mitochondria: kinetic, spectroscopic, and thermodynamic properties of an energy-transducing form of cytochrome b. Proc. Natl. Acad. Sci. U.S. 66:1175-1182(1970).
25. Slater, E.C., C.P. Lee, J.A. Berden and H.J. Wegdam. High-energy forms of cytochrome b. Nature 226:1248-1249(1970).
26. Erecińska, M. and B. Chance. In: E.C. Slater, E. Quagliariello and S. Papa (Editors). Coll. on Bioenergetics, Pugnochiuso, 1970, Adriatica Ed., Bari, in press.
27. Wilson, D.F. and P.L. Dutton. Energy dependent changes in the oxidation-reduction potential of cytochrome b. Biochim. Biophys. Res. Commun. 39:59-64(1970).
28. Sato, N., D.F. Wilson and B. Chance. Two b cytochromes of pigeon heart mitochondria. FEBS Letters 15:209-212(1971).
29. Wagner, M., M. Erecińska and B. Chance. In: E.C. Slater, E. Quagliariello and S. Papa (Editors). Coll. on Bioenergetics, Pugnochiuso, 1970, Adriatica Ed., in press.

30. Jacobs, E.E. and D.R. Sanadi. The reversible removal of cytochrome c from mitochondria. J. Biol. Chem. 235:531-534(1960).
31. Dutton, P.L., D.F. Wilson and C.P. Lee. Oxidation-reduction potentials of cytochromes in mitochondria. Biochem. 9:5077(1970).

TABLE I

STRAIGHT AND BRANCHED CHAINS WITH REVERSIBLE REACTIONS FOR PIGEON HEART MITOCHONDRIA

Experiment	131 c_3	131 e
$[O_2]$ M	64 & 1.8	64 & 1.8
Configuration	Straight	Straight
Constraint	None	$k_2 = 0.01$
Coupled	Yes	Yes
k_1/k_2	70/150	60/0.01
k_3/k_4	2100/1900	2300/2000
k_5/k_6	410/800	730/1700
k_7/k_8	850/1500	260/440
k_9	0.85	0.90
SSQ	2.3	3.1
Data Points	540	540
SSQ/Data Points	0.0042	0.0057

Straight Chain $\quad O_2 \underset{k_2}{\overset{k_1}{\rightleftarrows}} a_3 \underset{k_4}{\overset{k_3}{\rightleftarrows}} a \underset{k_6}{\overset{k_5}{\rightleftarrows}} c \underset{k_8}{\overset{k_7}{\rightleftarrows}} c_1 \underset{k_{10}}{\overset{k_9}{\rightleftarrows}}$

k_1 is $\mu M^{-1} \times sec^{-1}$; others are sec^{-1}

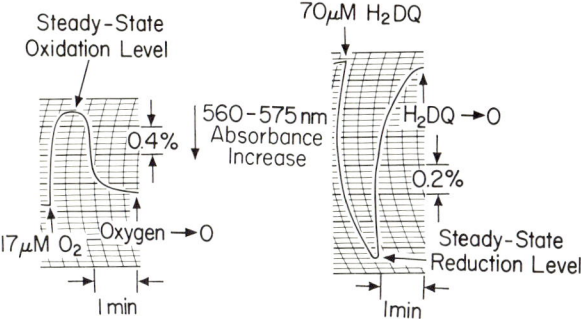

Fig. 1. *Schematic diagram of the oxygen and reductant pulse technique.*

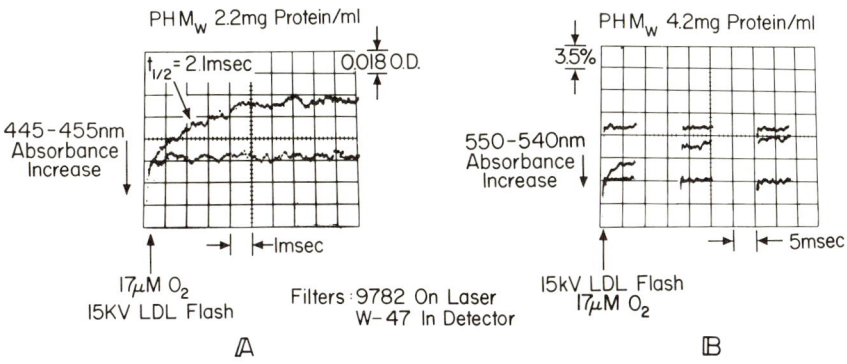

Fig. 2. *Oxidation kinetics of cytochrome* a + a_3 *(A) and cytochrome* c *(B) following the addition of* 17 μM O_2. Medium: 0.225 M mannitol, 0.075 M sucrose, 0.050 M morpholinopropane sulphonate (MOPS) buffer, pH 6.8, 6.0 mM succinate, 1.5 mM glutamate, 0.1 µg antimycin A/mg protein; (A) 2.2 mg protein/ml, and (B) 4.2 mg protein/ml. Baseline in both cases is anaerobic photolysis of cytochrome a_3-CO compound.

THERMODYNAMIC CONTROL OF MITOCHONDRIAL ENERGY COUPLING

David F. Wilson and P. Leslie Dutton

Introduction

The concept of a thermodynamic control of mitochondrial energy coupling is not entirely correct. Thermodynamics can tell us many useful things about the possible behavior of a system, but it cannot directly tell us very much about the reaction pathways or the rates at which the reactions occur. The latter questions of reaction pathways and rates are considered in detail in this volume (1).

The mitochondrial respiratory chain is a series of oxidation-reduction components which has dual functions: (a) that of transferring electrons from a reduced substrate to molecular oxygen and (b) that of utilizing the energy available in the oxidation-reduction reactions for ATP synthesis. A thermodynamic analysis requires a knowledge of the properties of both the oxidation-reduction components and of ATP. The behavior of oxidation-reduction components is best analyzed according to their oxidation-reduction potentials. Each component may be regarded as one-half of an electrical cell (equation 1) with its electron donating tendency described by equation 2. E_h expresses the tendency of the couple to donate electrons

1) $Ox + ne^- \rightleftharpoons Red$

2) $E_h = E'_o + \dfrac{RT}{nF} \ln \dfrac{[Ox]}{[Red]}$

to a standard hydrogen electrode; E'_o is the characteristic half reduction potential for the couple; R, T and F are the gas constant, absolute temperature and the Faraday constant, respectively. The n is the number of electrons transferred when the component is oxidized or reduced. The thermodynamic properties of electron transfer in the general reaction:

3) $A_{red} + B_{ox} \rightleftharpoons A_{ox} + B_{red}$

is described in equation 4: $\Delta E_{(A-B)}$ is the potential of the electrical cell formed

4) $\Delta E_{(A-B)} = \Delta E'_{o(A-B)} + \frac{RT}{nF} \ln \frac{[A_{ox}][B_{red}]}{[A_{red}][B_{ox}]}$

by the oxidation-reduction couples of A and B, $\Delta E'_{o(A-B)}$ is the difference between the characteristic half cell potentials of the two couples; the remaining symbols have been previously defined. It should be noted that the activity coefficients for the oxidized and reduced forms of a given compound are usually equal (2) and thus may be assumed to be one in calculations (they divide out when the ratios are calculated).

A knowledge of the characteristic midpoint potential for the oxidation-reduction components of the respiratory chain is essential to any thermodynamic discussion of the reactions involved because of the relationship that $\Delta G = (-nF)\Delta E$. Experimentally, a number of techniques have been used to measure the half-reduction potentials of oxidation-reduction components (2). These include polarographic techniques, the method of mixtures and a potentiometric technique. The last has a much broader applicability and generally greater precision than the other techniques. For this reason we have developed appropriate instrumentation to permit potentiometric techniques to be used to measure the oxidation-reduction midpoint potentials of components in suspensions of biological materials (3, 4, 5). The methodology has been published in detail. It is sufficient here to note that the method is designed to measure electrometrically the oxidation-reduction potential of an anaerobic sample and either to measure simultaneously by optical techniques the reduction of the component or to transfer aliquots of the suspension anaerobically to electron paramagnetic resonance (EPR) sample tubes. The EPR sample tubes are then immersed in liquid nitrogen in order to trap the oxidation-reducation state and to permit EPR measurements at 77°K.

Results

The oxidation-reduction potentials of the optically measurable components in mitochondria.

We have set out to determine systematically the midpoint potentials and n values of the oxidation-reduction components of the respiratory chain of intact mitochondria. In order to permit analysis, we will present some typical experimental results and then a complete tabulation of the measured values for the components of pigeon heart mitochondria.

In Fig. 1A the oxidation-reduction potential dependence of the reduction of the heme a components in intact pigeon heart mitochondria at pH 7.2 (measured at 445 nm minus 455 nm) is presented graphically with the logarithm of the ratio of the oxidized to reduced form on the abscissa and the measured oxidation-reduction potential on the ordinate. As may be seen from equation 2, on such a plot a single component would be expected to appear as a straight line with a slope of 59.3 mV per log unit (**n**=1.0) or 29.7 mV per log unit (**n**=2.0). The plotted experimental data for the absorbance wavelengths of 445-455 nm actually appears as a sigmoid curve in which the extremities approach the slopes expected for an **n** value of 1.0, thus indicating the existence of more than one component. Mathematical resolution of the curve into two components with **n** values of 1.0 (Fig. 1B) indicates that the two have midpoint potentials of 385 mV (cytochrome a_3) and 210 mV (cytochrome a). The identification of the two components as cytochromes a_3 and a, respectively, was confirmed by the observation that in the presence of carbon monoxide the +385 mV component is not present (a_3-CO has an absorption maximum at 430 nm and is not measured at these wavelengths). As may also be seen in Fig. 1A, when ATP was added to the mitochondrial suspensions, the titration curve was shifted such that the oxidation-reduction potential required to reduce the cytochrome a_3 became much more negative (4) and the titration curves no longer permitted the two to be separated. At pH 7.9 (not shown) the cytochrome a_3 half reduction potential was shifted to values sufficiently more negative than that of cytochrome a to permit resolution. The ATP-induced shift at pH 7.9 (6) was 230 mV for cytochrome a_3, but there was no effect of ATP on the midpoint potential of cytochrome a.

Similar experiments have been made measuring the c-cytochromes (7), the b cytochromes (4, 7, 8) and the flavoproteins (9). In Fig. 2 the measured values of the midpoint potentials of these components at pH 7.2 as determined in our laboratory are presented schematically on a vertical scale of the oxidation-reduction potential as rectangles centered on the midpoint potential and extending from the potential at which the component is 91% oxidized to the potential at which it is 91% reduced. The cytochromes all have **n** values of 1.0 and the respective midpoint potentials for cytochromes a, a_3, c, c_1, b_K and b_T are 210 mV, 385 mV, 235 mV, 225 mV, 30 mV and -30 mV in the uncoupled mitochondria. As noted in Fig. 2, in the presence of ATP the respective measured midpoint potentials of cytochromes a_3 and b_T in coupled mitochondria become approximately 165 mV and 245 mV (5, 8), respectively. The midpoint potential (225 mV) for cytochrome c_1 determined *in situ* is in good agreement with the value reported by Green and associates (10) and confirmed by Dutton and associates (7) for the isolated cytochrome c_1; this is the only cytochrome for which the isolated cytochrome has the same value that it has in mitochondria. The midpoint potential of cytochrome a has been reported to be near 260 mV in isolated cytochrome oxidase (11-14) and more positive than 330 mV in intact mitochondria (15). Cytochrome c in the soluble form is between 250 mV and 280 mV, depending on the assay medium (7). When soluble horse heart cytochrome c is bound inside phospholipid vesicles (16) or is in its place in the mitochondrial membrane (7), the midpoint potential is shifted to 230 mV and 235 mV, respectively.

The existence of more than one b cytochrome in intact mitochondria has been recognized only recently (5, 7, 8); the values in the previous literature are reported for beef heart submitochondrial particles assuming a single component (17-19) instead of three components (7).

The other components which can be measured by optical techniques include the flavoproteins and the copper responsible for the 830 nm absorbance. Three flavoproteins have been measured in pigeon heart mitochondria (9) with midpoint potentials of -45 mV, -160 mV and -220 mV and n values of 2.0. Of these three only the -45 component may resonably be considered to be a part of the respiratory chain (succinate dehydrogenase) and is included in Fig. 2. The -220 mV "component" was heterogeneous and was considered to be a mixture of lipoic dehydrogenase and NADH dehydrogenase. The optical change due to the NADH dehydrogenase which is expected to have a midpoint potential of near -300 mV [*cf.* (20) for the value in yeast mitochondria] was too small to be reliably measured. The copper responsible for the 830 nm absorption band of cytochrome oxidase has an n value of 1.0 and a midpoint potential of 240 mV (21). Urban and Klingenberg (19) have measured the midpoint potential of coenzyme Q in beef heart mitochondria and this value is presented.

The oxidation-reduction potentials of the EPR measurable components in mitochondria.

The reduction of the iron-sulfur proteins and the copper in cytochrome oxidase can be measured by electron paramagnetic resonance techniques; indeed EPR seems to be the only reliable approach for measurement of the iron-sulfur proteins in complex biological materials. In order to measure these components an apparatus has been used which allows aliquots of a suspension at a defined oxidation-reduction potential to be transferred under strictly anaerobic conditions into quartz EPR sample tubes. These aliquots are then frozen by immersion in liquid nitrogen and their EPR signals measured. In Fig. 3 three EPR spectra, each for the different oxidation-reduction potential, are presented as the derivative of the absorbance as is conventional for these instruments. The signal at g = 2.0 is the organic free radical which is present on partial reduction of the phenazines dyes and on complete reduction of the viologen dyes, two compounds which were added as oxidation-reduction mediators. The signal of the iron-sulfur proteins is centered at approximately g = 1.94. At an oxidation-reduction potential of 105 mV there is no detectable signal due to reduced iron-sulfur protein. As the potential was lowered to -218 mV the signal typical of the "high potential" iron sulfur protein became apparent, and at the still slower potential of -420 mV the signal changed in shape and increased in size as the "low potential" iron-sulfur protein was reduced. When the height of the respective signals is assumed to be proportional to the concentration of the reduced form and plotted as previously described (Fig. 4), the resulting plots are typical for oxidation-reduction components with **n** values of 1.0 and midpoint potentials of 25 mV and -305 mV for the high and low potential iron-sulfur proteins, respectively. Similar techniques have been used to determine the midpoint potential of the EPR detectable copper (6, 22). The measured values for pigeon heart mitochondria of 250 mV and **n** = 1.0 are very close to the values for the copper responsible for the 830 nm absorption band (Em = 240 mV, **n** = 1.0), in agreement with reports (23, 24) that the same component is responsible for both of the EPR signals and the 830 nm absorbance.

A thermodynamic profile of the respiratory chain components and the energy conservation sites.

An examination of Fig. 2 shows that the midpoint potentials of the components of the respiratory chain form four groups: the first at about -300 mV, the second at about 0 mV, the third near 220 mV and the fourth near 400 mV. These divisions roughly correlate with the potential spans utilized for ATP synthesis. The actual oxidation-reduction potentials of each component can be estimated from its steady state percent reduction, midpoint potential and **n** value.

In mitochondria in state 4 the estimated oxidation-reduction potentials of each of the components in a group is very close to that for all other members of the group; thus the components may be regarded as being essentially in equilibrium. For example, cytochromes a, c and c_1 all have oxidation-reduction potentials near 260 mV. The respiratory chain may be thought of as a series of isopotential groups of components with energy conservation associated with electron transfer between isopotential groups. We have estimated the oxidation-reduction potential for each of the four isopotential groups for mitochondria in state 4, and the values are designated in the boxes on the left side of Fig. 2. The approximate potential span available at the first, second and third phosphorylation sites (I, II and III) are 330 mV, 310 mV and 320 mV, respectively. The values for the first and second sites are readily estimated from direct measurements, but the value for site III makes use of the measured second order rate constant for the reaction of O_2 with reduced cytochrome a_3 of 8×10^7 M^{-1} sec^{-1} (25), an assumed oxygen concentration of 200 μM, and a measured half-time for the oxidation of cytochrome a by cytochrome a_3 of 6 msec. These values permit us to estimate that in state 4 the cytochrome a_3 is only 0.1% reduced.

When electrons are transferred from the pyridine nucleotide-linked substrate to cytochrome a_3, the potential change (ΔE) is approximately 1 volt (960 mV). The Gibbs free energy available for ATP synthesis on transfer of 2 electrons through this potential span may be calculated from the relationship

5) $\Delta G = -nF\Delta E$

and is approximately -45 Kcal. When this available energy is distributed to three phosphorylation sites, each individual site can supply approximately -15 Kcal. Cockrell and associates (26) have measured the maximum phosphate potential which could be formed by mitochondria under state 4 conditions and found it to be 15.6 Kcal. It is apparent then that adequate energy is available for ATP synthesis without using the potential span from cytochrome a_3 to oxygen.

The nature of the energy transducing reaction.

The energy transduction reaction itself may be evaluated profitably in thermodynamic terms since such an evaluation has led to experimentally useful conclusions (4, 5, 8). It is important to note that the oxidation-reduction reactions involve electron transfer, while ATP synthesis from ADP and inorganic phosphate does not (27, 28). The general equations containing the required information for an energy conservation site are described in equations 6-8:

6) $A_{red} + B_{ox} \rightleftharpoons A_{ox}^{\sim} + B_{red}$

7) $A_{ox}^{\sim} + ADP + P_i \rightleftharpoons A_{ox} + ATP$

8) $A_{ox} + C_{red} \rightleftharpoons A_{red} + C_{ox}$

We have designated the energy transducer as a 2 electron carrier A and the other electron carriers as B and C in order to avoid mechanistic arguments which are special cases and must conform to the general evaluation. The first equation is the energy transduction reaction in which a "high energy" form of the oxidized carrier is generated as A is oxidized by a component of the high potential pool (an analogous equation may be written for a "high energy" reduced form). The second equation is the synthesis of ATP utilizing the available negative free energy of the "high energy" intermediates, and the third equation provides net ATP synthesis under normal operating conditions by regenerating reduced A using electrons from the low potential pool. This set of equations has been used by several investigators, including Chance and Williams (28) and Slater (27), to describe energy conservation.

The ΔG for each reaction must approach zero for an efficient energy transduction. Thus the oxidation-reduction couple

9) $A_{ox}^{\sim} + 2\ e^- \rightleftharpoons A_{red}$

must be the same oxidation-reduction potential as the high potential acceptor in equation 6, while the oxidation-reduction potential of the couple

10) $A_{ox} + 2\ e^- \rightleftharpoons A_{red}$

must be at the same potential as that of the low potential donor in equation 8. The oxidation-reduction potentials of the couples are shown in equations 11 and 12, respectively:

11) $E_A = E'_{oA} + \dfrac{RT}{nF} \ln \dfrac{[A_{ox}^{\sim}]}{[A_{red}]}$

12) $E_A = E'_{oA} + \dfrac{RT}{nF} \ln \dfrac{[A_{ox}]}{[A_{red}]}$

The difference between equations 11 and 12 is:

13) $340\ mV = \Delta E = \Delta E'_o + \dfrac{RT}{nF} \ln \dfrac{[A_{ox}^{\sim}]}{[A_{ox}]}$

The required potential span may be obtained with no difference in the midpoint potential of the two couples only if the ratio of $[A_{ox}^{\sim}]$ to $[A_{ox}]$ is greater than 10^{11}. This ratio can be excluded on a kinetic basis. The turnover number of the cytochromes in state 4 is greater than $1\ sec^{-1}$. If we assume an equivalent homogeneous reaction mechanism, this corresponds to a second order velocity constant for electron transfer of approximately

$10^6 M^{-1} sec^{-1}$ for the reduction of A_{ox} by C_{red} if the total concentrations of A and C are equal to the cytochrome concentration. If the ratio of $[A_{ox}^{\sim}]$ to $[A_{ox}^{\sim}]$ is 10^{11} then the required rate constant is $10^{17} M^{-1} sec^{-1}$, an unrealistic number. The assumption that this rate constant cannot be greater than $10^9 M^{-1} sec^{-1}$ (diffusion limited) requires that in equation 13 the $\Delta E_o'$ term must be at least 240 mV. Since the E_o term is the midpoint potential of the indicated oxidation-reduction couple at the selected pH (7.0), the energy transduction reaction must be associated with an electron carrier which can be reversibly oxidized and reduced through two different specific reaction pathways which give rise to two chemically different species as the oxidized products. The oxidation-reduction couples which give rise to these different species must have midpoint potentials at physiological pH which differ by more than 240 mV. Optimum kinetic conditions are achieved when electron transfer occurs between oxidation-reduction couples with approximately equal midpoint potentials.

Clark (2) has described in considerable detail the behavior of a two component oxidation-reduction system in which a chemical equilibrium between the reduced or the oxidized forms is dependent on a non-electron transferring reagent (in this case ATP, ADP and inorganic phosphate).

The properties of the energy transducing reaction may therefore be summarized as follows:

1) The primary energy conservation event is the formation of a new chemical derivative of the energy transducing electron transport component. This new species is part of an oxidation-reduction couple having a different midpoint potential from that of the original compound.

2) The experimentally measured midpoint potential of the energy transducing element is dependent on the phosphate potential.

3) This measured midpoint potential changes by more than 240 mV and is limited at one extreme by the value for infinite phosphate potential.

4) A measured midpoint potential which has a more negative value with high phosphate potential implies a "high energy" reduced form, while a more positive value implies a "high energy" oxidized form.

The only components for which we have been able to demonstrate an energy dependence of the measured oxidation-reduction midpoint potentials are cytochromes b_T and a_3 (4, 5, 8); this appears to establish these two cytochromes as the oxidation-reduction components most likely to be directly involved in energy transduction at phosphorylation sites II and III, respectively. An alternate interpretation of the experimentally measured ATP shift in the midpoint potentials has been reported (29).

The relation between the thermodynamic and kinetic parameters

Thermodynamics can tell us the behavior of a chemical system in the limiting case of equilibrium and the direction of the net flux through the system. It can be readily demonstrated that

$$14)\ \Delta G = RT\ \ln\frac{V_1}{V_{-1}}$$

where V_1 and V_{-1} are the forward and reverse rates for the system. It is important to note that the free energy change is related to the ratio of the flux rates and not to the absolute values of the flux. Experimentally, however, the net flux is usually measured, and the net flux is simply the difference between the forward and reverse flux:

$$15)\ V_{net} = V_1 - V_{-1}$$

These two equations are the mathematical expressions stating that if the ΔG is equal to zero there is no net flux, if the ΔG is negative a net forward flux will occur and if the ΔG is positive a net reverse flux will occur. To the experimentalist they indicate that if two of these four parameters can be measured (for example, the ΔG and V_{net}), the other two (V_1 and V_{-1}) can be calculated. A ΔG of approximately -1400 calories corresponds to a flux ratio of 10 and a net flux which is 9 times greater than the reverse flux.

The mitochondrial respiratory chain is an interesting example which can be analyzed by this technique. The ΔG for the transfer of two electrons from reduced pyridine nucleotide to cytochrome a_3 is approximately -45 Kcal. If we assume a stoichiometry of 3 ATP synthesized for each two electrons and a ΔG for ATP synthesis of +15.6 Kcal (total +47 Kcal), it is apparent that the two are equal within experimental error and the efficiency is very high for the process (net ΔG near zero). Unfortunately the available data were not obtained under strictly analogous conditions, and an apparent discrepancy of 2 out of 46 Kcal is not significant. This conclusion is in good agreement with the observed ability of ATP to drive reversed electron transport. The reversibility of each of the individual reactions of a sequence of consecutive reactions must be even greater than that for the overall system, because the overall ΔG is the summation of the ΔG's for each of the individual reactions. The individual reactions of a mandatory sequence of reactions (single pathway) must in the steady state have ΔG values of the same sign as the overall system because they must have a net flux equal in magnitude and direction to that of the entire system in the steady state.

In concluding we would like to stress the need for precise thermodynamic data relevant to the mitochondrial system, such as *in vitro* measurements of the ADP/O ratio as a function of the energy required for the ATP synthesis (to values approaching state 4), with simultaneous measurement of the steady state reduction of the respiratory carriers. If the mitochondrial phosphate potential, the oxidation-reduction potential of the mitochondrial NAD^+ and the cellular oxygen tension were also known for physiological conditions, then it would be possible to describe more clearly the function of mitochondria in the cell by comparison with the properties of the isolated mitochondria.

Presented by David F. Wilson, recipient of U.S.P.H.S. Career Development Award 1-K04-GM-18151. The experimental work reported in this paper was supported by United States Public Health Service Grant 12202 and National Science Foundation Grant GB-28125.

References

1. Chance, B., M. Erecińska, E. Chance, A. Boveris and M. Wagner. Kinetic control of electron flow. This volume.
2. Clark, W.M. Oxidation-reduction potentials of organic systems. Baltimore, Md., Waverly Press (1960).
3. Dutton, P.L. Oxidation-reduction potential dependence of the interaction of cytochromes, bacteriochlorophyll and carotenoids at 77°K in chromatophores of *Chromatium* D and *Rhodopseudomonas gelatinosa*. Biochim. Biophys. Acta 226:63-80(1971).
4. Wilson, D.F. and P.L. Dutton. Oxidation-reduction potentials of cytochromes *a* and a_3 in intact rat liver mitochondria. Arch. Biochem. Biophys. 136:583-585(1970).
5. Wilson, D.F. and P.L. Dutton. Energy dependent changes in the oxidation-reduction potential of cytochrome *b*. Biochem. Biophys. Res. Commun. 39:59-64(1970).
6. Wilson, D.F., J.F. Leigh, Jr., J.G. Lindsay and P.L. Dutton. Some thermodynamic properties and interactions of the hemes and copper of native cytochrome oxidase. In: T.E. King, H.S. Mason and M. Morrison (Editors), Oxidases and related redox systems II, John Wiley and Sons, New York, in press.
7. Dutton, P.L., D.F. Wilson and C.P. Lee. Oxidation-reduction potentials of cytochromes in mitochondria. Biochemistry 9:5077-5082(1970).
8. Chance, B., D.F. Wilson, P.L. Dutton and M. Erecińska. Energy-coupling mechanisms in mitochondria: kinetic, spectroscopic, and thermodynamic properties of an energy-transducing form of cytochrome *b*. Proc. Nat. Acad. Sci. U.S. 66:1175-1182(1970).
9. Erecińska, M., D.F. Wilson, Y. Mukai and B. Chance. Oxidation-reduction midpoint potentials of the mitochondrial flavoproteins. Biochem. Biophys. Res. Commun. 41:386-392(1970).
10. Green, D.E., J. Järnefelt and H.D. Tisdale. Studies on the electron transport system XIV. The isolation and properties of soluble cytochrome c_1. Biochim. Biophys. Acta 31:34-46(1960).

11. Ball, E.G. Oxidation and reduction of the three cytochrome components. Biochem. Z. 295:262-264(1938).
12. Wainio, W.W. Reduction of cytochrome oxidase with ferrocytochrome c. J. Biol. Chem. 216:593-599(1955).
13. Minnaert, K. Measurement of the equilibrium constant of the reaction between cytochrome c and cytochrome a. Biochim. Biophys. Acta 110:42-56(1965).
14. Tzagoloff, A. and D.C. Wharton. The reaction of cytochrome oxidase with carbon monoxide. J. Biol. Chem. 240:2628-2633(1965).
15. Caswell, A.H. Potentiometric determination of interrelationships of energy conservation and ion gradients in mitochondria. J. Biol. Chem. 243:5827-5836(1968).
16. Kimelberg, H.K. and C.P. Lee. Interactions of cytochrome c with phospholipid membranes II. Reactivity of cytochrome c bound to phospholipid liquid crystals. J. Membrane Biol. 2:252-262(1970).
17. Holton, F.A. and J.P. Colpa-Boonstra. Spectrophotometric observations relating to the oxidation-reduction potential of cytochrome b in nonphosphorylating heart-muscle particles. Biochem. J. 76:179-189(1960).
18. Feldman, D. and W.W. Wainio. Isolation, purification, and some properties of mammalian cytochrome b. J. Biol. Chem. 235:3635-3639(1960).
19. Urban, P.F. and M. Klingenberg. Redox potentials of ubiquinone and cytochrome b in the respiratory chain. Eur. J. Biochem. 9:519-525(1969).
20. Erecińska, M., D.F. Wilson, Y. Mukai and T. Ohnishi. (Unpublished experiments).
21. Erecińska, M., B. Chance and D.F. Wilson. FEBS Letters 16:284-286(1971).
22. Tsudzuki, T. and D.F. Wilson. Arch. Biochem. Biophys. 145:149-154(1971).
23. Tzagoloff, A. and D.H. MacLennon. The copper protein component of cytochrome c oxidase. In: J. Peisach, P. Aisen and W.E. Blumberg (Editors), The biochemistry of copper, Academic Press, New York (1966), pp. 253-265.
24. Wharton, D.C. and A. Tzagoloff. Studies on the electron transfer system LVII. The near infrared absorption band of cytochrome oxidase. J. Biol. Chem. 239:2036-2041(1964).
25. Chance, B., M. Erecińska and E.M. Chance. Second internatl. symposium on oxidases and related redox reactions, Memphis, Tenn., in press.
26. Cockrell, R.S., E.J. Harris and B.C. Pressman. Energetics of potassium transport in mitochondria induced by valinomycin. Biochemistry 5:2326-2335(1969).
27. Slater, E.C. Mechanism of phosphorylation in the respiratory chain. Nature 172:975-978(1953).
28. Chance, B. and G.R. Williams. The respiratory chain and oxidative phosphorylation. Adv. Enzymol. 17:65-134(1956).
29. DeVault, D. Energy transduction in electron transport. Biochim. Biophys. Acta 226:193-199(1971).

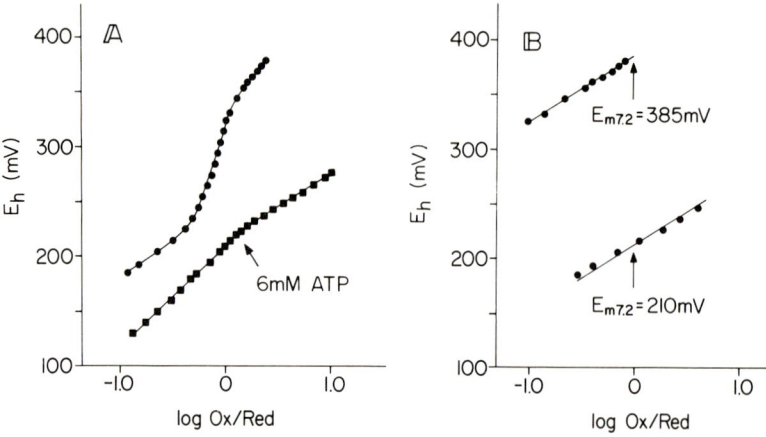

Fig. 1 A. *The oxidation-reduction midpoint potentials of cytochromes* a *and* a_3 *in pigeon heart mitochondria at pH 7.2.* Pigeon heart mitochondria were suspended at 1.8 mg/ml in a medium containing 0.22 M-mannitol, 0.05 M-sucrose and 50 mM-morpholinopropane sulphonate buffer (MS-MOPS medium), pH 7.2. The reduction of cytochromes a and a_3 were measured at 444 nm minus 455 nm. TMPD (20 µM) and DAD (20 µM) were added as redox mediators and aliquots of ascorbate were added until anaerobiosis was achieved as measured by cytochrome reduction and an E_h of less than 300 mV. ATP (6 mM) was then added and re-oxidation was achieved by adding ferricyanide (40 µM). A reductive titration was carried out by the combined slow donation of electrons from endogenous substrate and the addition of small aliquots of dihydroascorbate. It was customary to make samples aerobic after the titration to check the total excursion. (●) 6 mM ATP and 1 µM 5-chloro-1,3-(p-chlorophenyl)-2,4,5-trichlorosalicylanilide added at anaerobiosis (■) 6 mM ATP.

Fig. 1 B. *The resolution of the sigmoid titration curve of 1 A into its two component parts (5).*

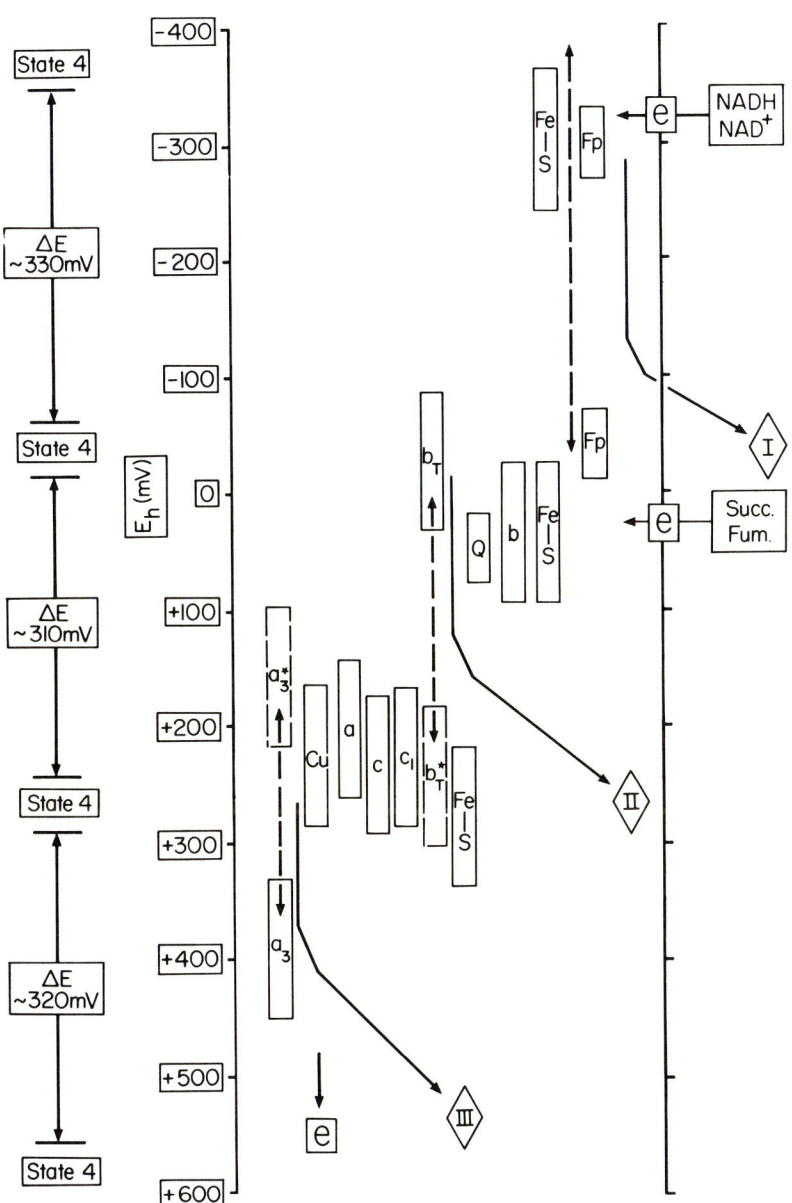

Fig. 2. *Oxidation-reduction potentials of carriers in the mitochondrial respiratory chain.* The blocks represent the potentials over which the carriers become 10-90% oxidized or reduced at pH 7.2 in the uncoupled state. The dashed lines and blocks indicate the ATP-induced changes in the midpoint potential.

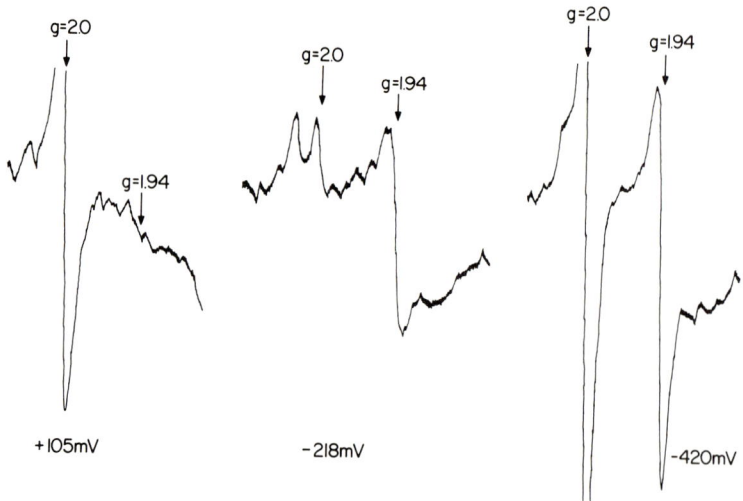

Fig. 3. *The iron-sulfur protein EPR signal in pigeon heart mitochondria as a function of the oxidation-reduction potential of the sample before freezing.* The pigeon heart mitochondria were suspended at approximately 35 mg protein/ml in a 0.25 M sucrose medium buffered with 40 mM morpholinopropane sulfonate, pH 7.0. Oxidation-reduction mediators added for the high potential region were 25 μM diaminodurene, 40 μM duroquinone, 6 μM pyocyanine and 10 μM 2-hydroxynaphthoquinone. For the low potential region the mediators were 11 μM 2-hydroxynaphthoquinone, 60 μM phenosafranine, 72 μM benzylviologen and 130 μM methylviologen. The oxidation-reduction potential of the sample used for each spectrum is given in the figure. The spectra were measured at 77°K using a Varian Model E3 EPR spectrometer with a power setting of 20 mW, modulation amplitude of 10 gauss.

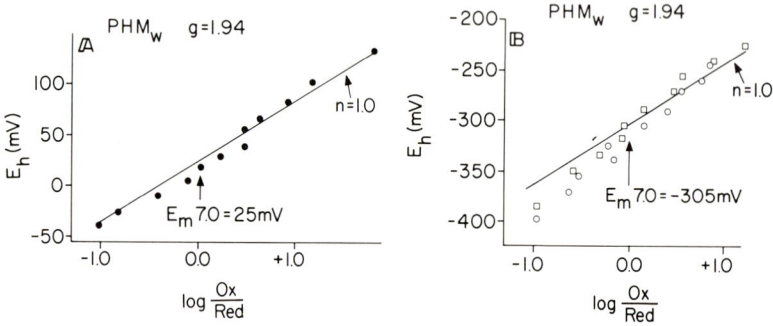

Fig. 4. *The oxidation-reduction potential dependence of the reduction of the iron-sulfur protein of pigeon heart mitochondria.* The experimental conditions were the same as for Fig. 3. The height of the EPR signal was assumed to be proportional to the concentration of reduced form. The data from two experiments (*1* and *3*) are included to indicate the scatter in the low potential region.

REPLENISHMENT AND DEPLETION OF CITRIC ACID CYCLE INTERMEDIATES IN MUSCLE

John M. Lowenstein

Introduction

The total amount of citric acid cycle intermediates present in a given tissue depends on the metabolic conditions. For example, the levels of citrate, isocitrate, α-ketoglutarate and malate increase two to five-fold when perfused rat heart is deprived of insulin. Large changes in the levels of citric acid cycle intermediates are observed when the fuel burned by the heart is changed from carbohydrate to fatty acids or ketones. Large changes in the levels of citric acid cycle intermediates also occur when an aerobic tissue becomes anaerobic and vice versa (1-4). Most tissues show very large increases in the level of citrate during fluoroacetate or fluorocitrate poisoning (5-7). Corresponding decreases in citrate levels occur when the poisoning is overcome. These changes imply that there must be mechanisms which make possible a net increase or decrease in the tissue content of the citric acid cycle intermediates.

Replenishment of citric acid cycle intermediates

Purine nucleotide cycle

It has been proposed that the purine nucleotide cycle (Fig. 1) is the main mechanism whereby citric acid cycle intermediates can be replenished in tissues which possess little or no pyruvate carboxylase. The reactions of the cycle are as follows:

1) $AMP + H_2O \rightarrow IMP + NH_3$

2) $IMP + aspartate + GTP \rightarrow adenylosuccinate + GDP + P_i$

3) $adenylosuccinate \rightarrow AMP + fumarate$

4) Sum: aspartate + GTP + H_2O → fumarate + GDP + P_i + NH_3

Extracts of rat skeletal muscle catalyze a conversion of aspartate to fumarate and ammonia which is dependent on the initial presence of either AMP, or IMP, or adenylosuccinate (8, 9). The experiment shown in Table 1 demonstrates that the requirement is catalytic. The conversion of aspartate to fumarate and ammonia also shows an absolute requirement for GTP.

The enzyme system requires GTP for the adenylosuccinate synthetase reaction (equation 2). One of the products of this reaction, namely GDP, acts as a powerful feedback inhibitor. The successful demonstration of the operation of the purine nucleotide cycle therefore depends on the removal of GDP, preferably as rapidly as it is formed. This can be achieved by employing a regenerating system which converts GDP back to GTP. Either phosphoenolpyruvate and pyruvate kinase, or creatine phosphate and creatine phosphokinase are suitable for this purpose.

Two alternative reactions which might serve as sources of dicarboxylic acids for the citric acid cycle require comment. The first of these is the glutamate dehydrogenase reaction and the second the pyruvate carboxylase reaction.

Glutamate dehydrogenase and adenylate deaminase

The currently accepted scheme for the deamination of aspartate to oxaloacetate involves transamination to yield glutamate (equation 5), and deamination of the resulting glutamate by the glutamate dehydrogenase reaction (equation 6):

5) aspartate + α-ketoglutarate ↔ oxaloacetate + glutamate

6) glutamate + NAD^+ ↔ α-ketoglutarate + NADH + NH_4^+

7) Sum: aspartate + NAD^+ ↔ oxaloacetate + NADH + NH_4^+

This is the "transdeamination" scheme of Braustein (10, 11).

A number of considerations indicate that the glutamate dehydrogenase reaction may not be the only, or even the major reaction responsible for the liberation of oxaloacetate and α-ketoglutarate from aspartate and glutamate.

The first consideration is the free energy of the glutamate dehydrogenase reaction. For the reaction written in equation 6 the free energy is +6.5 kcals per mole (12). In other words, the equilibrium of the reaction greatly favors glutamate formation. The deamination sequence via the purine nucleotide cycle which is summarized in equation 4 can be made to correspond closely to the glutamate dehydrogenase reaction by adding the aspartate-glutamate transaminase reaction (equation 8) and the fumarase and malate dehydrogenase reactions (equations 9 and 10):

8) glutamate + oxaloacetate ↔ aspartate + α-ketoglutarate

9) fumarate + H_2O ↔ malate

10) malate + NAD^+ ↔ oxaloacetate + NADH + H^+

The sum of reactions 1, 2, 3, 8, 9, and 10 is

11) glutamate + NAD^+ + GTP + $2H_2O$ → α-ketoglutarate + DPNH + H^+ + GDP + NH_3 + P_i

The free energy change of reaction 11 is −1.0 kcals per mole (calculated for a free energy of hydrolysis of GTP to GDP plus P_i at pH 7.0 of −8.0 kcals per mole). The net reaction shown in equation 11 is similar to the glutamate dehydrogenase reaction shown in equation 6, except that the concomitant hydrolysis of GTP makes the deamination of glutamate (or aspartate) thermodynamically much more favorable.

The second consideration is a comparison of the organ distribution of glutamate dehydrogenase and adenylate deaminase. Glutamate dehydrogenase activity is high in organs such as liver, intermediate in kidney and brain, and low or absent in skeletal muscle (13, 14). Adenylate deaminase activity is highest in skeletal muscle, and is low in liver (15). Muscular work is accompanied by the production of ammonia (16). The organ distribution of glutamate dehydrogenase and adenylate deaminase indicates strongly that muscle converts aspartate to ammonia via the reactions of the purine nucleotide cycle [Scheme 1; (9)].

The third consideration is the intracellular distribution of glutamate dehydrogenase and of the enzymes of the purine nucleotide cycle. Glutamate dehydrogenase is an intramitochondrial enzyme (17). Adenylate deaminase is an extramitochondrial enzyme (18). The complete purine nucleotide cycle can be demonstrated to occur in the high-speed supernatant fraction of a homogenate of skeletal muscle prepared in isotonic sucrose (8). Moreover, the formation of ammonia which is catalyzed by this fraction is accompanied by a stoichiometric production of fumarate plus malate (19).

JOHN M. LOWENSTEIN

Pyruvate carboxylase

In liver, kidney and adipose tissue dicarboxylic acids can be replenished from three carbon compounds by the pyruvate carboxylase reaction [equation 12; (20)]:

12) pyruvate + CO_2 + ATP → oxaloacetate + ADP + P_i

A paper by Ballard, Hanson and Resheff (21) reported that pyruvate carboxylase activity is also present in muscle. However, the activity of pyruvate carboxylase per mg protein in muscle was only about 1% of that found in liver, kidney, and adipose tissue. It seems likely that this small amount of pyruvate carboxylase activity is due to fat cells in the skeletal muscle. It is unlikely that pyruvate carboxylase plays amajor role in replenishing the dicarboxylic acids in tissues such as muscle.

It is concluded that the purine nucleotide cycle is the major mechanism for the replenishment of dicarboxylic acids required for the operation of the citric acid cycle.

Depletion of citric acid cycle intermediates

The total levels of intermediates of the citric acid cycle can decrease as well as increase. Since glutamate dehydrogenase activity in skeletal muscle is low or absent, removal of α-ketoglutarate by reductive amination to glutamate can probably be ruled out. A reaction which is capable of removing oxaloacetate from the citric acid cycle is catalyzed by phosphoenolpyruvate carboxykinase (equation 10). The enzyme occurs in skeletal muscle (22). Phosphoenolpyruvate formed from oxaloacetate can be converted to pyruvate by the pyruvate kinase reaction. Alternatively it can be converted to hexose by the reversal of the reactions of glycolysis. Muscle lacks pyruvate carboxylase; hence it lacks the capacity to convert lactate and pyruvate to hexose at a significant rate. Nevertheless, it contains fructose diphosphatase (23-25), and there is a correlation between the levels of phosphoenolpyruvate carboxykinase and fructose diphosphatase in different types of muscle (22). It follows that muscle possesses the capacity to make hexose from oxaloacetate. However, the main point that I wish to make here is that phosphoenolpyruvate carboxykinase catalyzes a reaction which causes a lowering of the total level of citric acid cycle intermediates.

In conclusion, let me point out that the reactions of the purine nucleotide cycle, combined with a portion of the citric acid cycle, phosphoenolpyruvate carboxykinase, pyruvate kinase, and pyruvate dehydrogenase provide a pathway for converting glutamate and aspartate to acetyl-CoA (Fig. 2). It follows that glutamate and aspartate can be converted to acetyl-CoA in tissues such as muscle (9). Whether these substances actually act as a source of energy in tissues such as skeletal muscle remains to be demonstrated.

The experimental work reported in this paper was supported by National Institute of Health Grant GM-07261. Publication number 900 from the Graduate Department of Biochemistry.

References

1. Bowman, R.H. Effects of diabetes, fatty acids, and ketone bodies on tricarboxylic acid cycle metabolism in the perfused rat heart. J. Biol. Chem. 241:3041-3048(1966).
2. Randle, P.J., P.J. England, and R.M. Denton. Control of the tricarboxylate cycle and its interactions with glycolysis during acetate utilization in rat heart. Biochem. J. 117:677-695(1970).
3. Safer, B., C.M. Smith, and J.R. Williamson. Control of the transport of reducing equivalents across the mitochondrial membrane in perfused rat heart. J. Molecular and Cellular Cardiology 2:111-124(1971).
4. Williamson, J.R. Glycolytic control mechanisms I. Inhibition of glycolysis by acetate and pyruvate in the isolated, perfused rat heart. J. Biol. Chem. 240:2308-2321(1965).
5. Bowman, R.H. Inhibition of citrate metabolism by sodium fluoroacetate in the perfused rat heart and the effect of phosphofructokinase activity and glucose utilization. Biochem. J. 93:13c-15c(1964).
6. Spencer, A.F. and J.M. Lowenstein. Citrate content of liver and kidney of rat in various metabolic states and in fluoroacetate poisoning. Biochem. J. 103:342-348(1967).
7. Williamson, J.R., E.A. Jones and G.F. Azzone. Metabolic control in perfused rat heart during fluoroacetate poisoning. Biochem. Biophys. Res. Commun. 17:696-702(1964).
8. Lowenstein, J.M. and K. Tornheim. Ammonia production in muscle: the purine nucleotide cycle. Science 171:397-400(1971).
9. Lowenstein, J.M. Ammonia production in muscle and other tissues. The purine nucleotide cycle. Physiol. Revs., in press.
10. Braunstein, A.E. and S.M. Bychkov. A cell-free enzymatic model of $\mathcal{L}$-amino acid dehydrogenase ('$\mathcal{L}$-deaminase'). Nature 144:751-752(1939).
11. Braunstein, A.E. Les voies principales de l'assimilation et dissimilation de l'azote chez les animaux. Adv. Enzymol. 19:335-389(1957).
12. Burton, K. and H.A. Krebs. The free energy changes associated with the individual steps of the cycle. In: D.M. Greenberg (Editor), Metabolic pathways, Vol. 1, Academic Press, New York (1960), pp. 187-192.
13. Copenhaver, J.H., W.H. McShan and R.K. Meyer. The determination of glutamic acid dehydrogenase in tissue homogenates. J. Biol. Chem. 183:73-79(1950).

14. Williamson, D.H., P. Lund and H.A. Krebs. The redox state of free nicotinamide-adenine dinucleotide in the cytoplasm and mitochondria of rat liver. Biochem. J. 103:514-527(1967).
15. Conway, E.J. and R. Cooke. LIX. The deaminase of adenosine and adenylic acid in blood and tissues. Biochem. J. 33:479-492(1939).
16. Needham, D.M. Biochemistry of muscle. Methuen and Co., Ltd., London (1932), p. 96.
17. Delbrück, A., H. Schimassek, K. Bartsch and Th. Bücher. Enzym-verteilungsmuster in einigen Organen und in experimentellen Tumoren der Ratte und der Maus. Biochem. Z. 331:297-311(1959).
18. Purzycka, J. AMP and adenosine aminohydrolases in rat tissues. Acta Biochim. Polon. 9:83-93(1962).
19. Tornheim, K. and J.M. Lowenstein. The purine nucleotide cycle II. Demonstration of the cycle in cell-free extracts of skeletal muscle. J. Biol. Chem., in press.
20. Scrutton, M.C. and M.F. Utter. The regulation of glycolysis and gluconeogenesis in animal tissues. Ann. Rev. Biochem. 37:249-302(1968).
21. Ballard, F.J., R.W. Hanson and L. Reshef. Immunochemical studies with soluble and mitochondrial pyruvate carboxylase activities from rat tissues. Biochem. J. 119:735-742(1970).
22. Opie, L.H. and E.A. Newsholme. The activities of fructose 1,6-diphosphatase, phosphofructokinase and phosphoenolpyruvate carboxykinase in white muscle and red muscle. Biochem. J. 103:391-399(1967).
23. Enser, M., S. Shapiro and B.L. Horecker. Immunological studies of liver, kidney, and muscle fructose 1,6-diphosphatases. Arch. Biochem. Biophys. 129:377-386(1969).
24. Fernando, J., S. Pontremoli and B.L. Horecker. Fructose diphosphatase from rabbit muscle II. Amino acid composition and activation by sulfhydryl reagents. Arch. Biochem. Biophys. 129:370-376(1969).
25. Krebs, H.A. and M. Woodford. Fructose 1,6-diphosphatase in striated muscle. Biochem. J. 94:436-445(1965).

TABLE 1

CATALYTIC ROLE OF IMP IN AMMONIA PRODUCTION FROM ASPARTATE

The complete reaction mixture contained 0.52 mM IMP, 0.29 mM GTP, 4mM aspartate, 27 mM imidazole hydrochloride buffer (pH 6.7), 8.3 mM $MgCl_2$, 1.67 mM creatine phosphate, 1.2 units per milliliter of yeast hexokinase (specific activity 70 unit/mg), and rat muscle extract equivalent to 1.0 mg of protein per milliliter. The protein extract contributed 2.5 mM orthophosphate, 47 mM KCl, 0.83 mM ethylenediaminetetraacetate (EDTA), and 17 μM dithiothreitol to the reaction mixture. The reaction was started by adding the protein extract and was run at 31°C. It was followed spectrophotometrically in a Cary 15 recording spectrophotometer. The reference cuvette contained the same reaction mixture but lacked aspartate. The light path was 1 mm. When the formation of adenine nucleotide had ceased, 8.3 μl of 0.2M 2-deoxyglucose per milliliter of reaction mixture was added to give a final concentration of 1.67 mM 2-deoxyglucose. This led to the conversion of adenine nucleotide to IMP.

When the formation of IMP had ceased, another cycle was initiated by addition of a second portion of creatine phosphate. After the IMP had been converted to adenine nucleotide, a second portion of 2-deoxyglucose was added. The additions were repeated twice more.

Portions of the complete reaction mixture and of controls which lacked either IMP or aspartate were analyzed for ammonia at the end of each half cycle of the complete reaction mixture. The analyses showed that ammonia was released only during the AMP → IMP half of the cycle.

The muscle extract was prepared as follows. Rat leg muscle was minced and suspended in three volumes of solution consisting of 90 mM potassium phosphate (pH 6.5), 180 mM KCl. The suspension was blended for 20 seconds in a Waring blendor and stirred in the cold for 1 hour. It was centrifuged first at 31,000g for 10 minutes, and then at 85,000g for 30 minutes. The supernatant from the high speed centrifugation was placed on a column of Sephadex G-50 and eluted with a mixture containing 15 mM potassium phosphate (pH 6.5), 280 mM KCl, 5 mM EDTA, and 0.1 mM dithiothreitol. The pooled peak fractions of the excluded proteins were used in the experiment.

Omissions from reaction mixture	Ammonia concentration at				
	Start of 1st cycle	End of 1st cycle	End of 2nd cycle	End of 3rd cycle	End of 4th cycle
			$\mu moles/ml$		
None	0.04	0.50	0.95	1.34	1.88
IMP	0.04	0.03	0.05	0.06	0.08
Aspartate	0.02	0.01	0.01	0.02	0.03

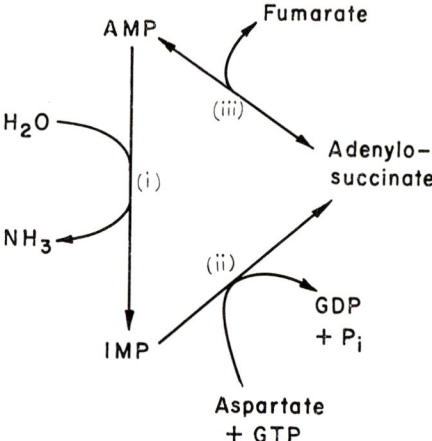

Fig. 1. *The purine nucleotide cycle.* The enzymes involved in the reactions of the cycle are as follows: (i) adenylate deaminase, (ii) adenylosuccinate synthetase, and (iii) adenylosuccinase. One turn of the cycle results in the net reaction:

$$\text{aspartate} + \text{GTP} + \text{H}_2\text{O} \rightarrow \text{fumarate} + \text{GDP} + \text{P}_i + \text{NH}_3$$

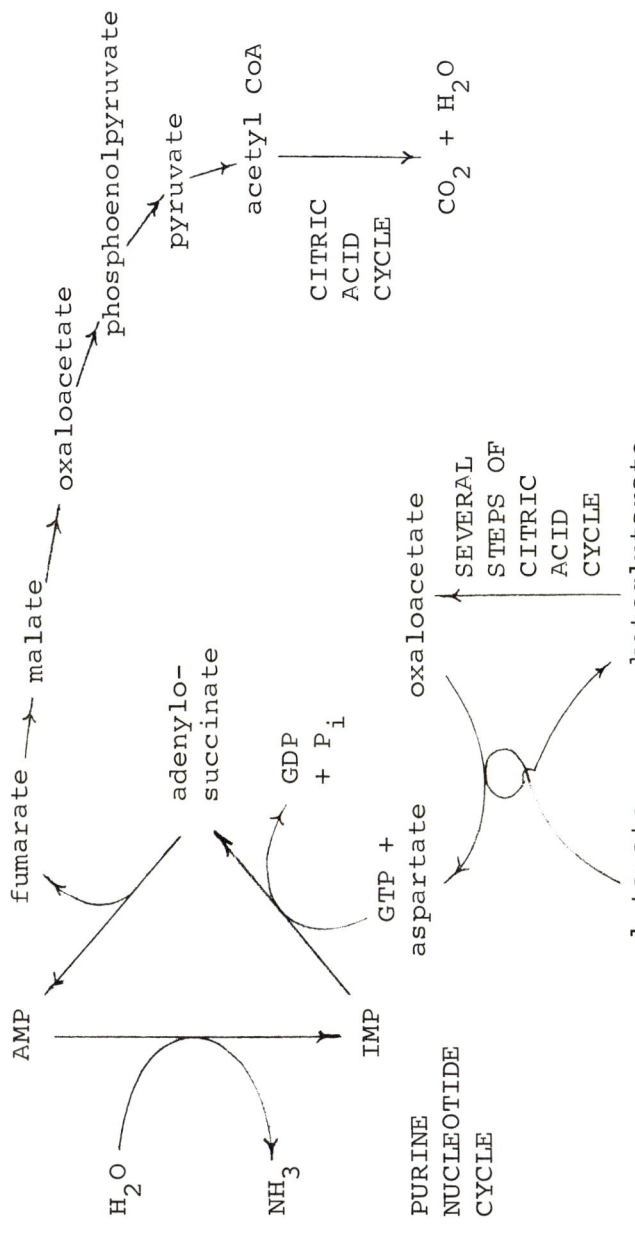

Fig. 2. *Pathway for utilizing glutamate and aspartate as energy source in tissues lacking or low in glutamate dehydrogenase.*

ANION TRANSPORT AND THE REGULATION OF THE CITRIC ACID CYCLE

G.R. Williams, J.L. Orr and G.S. Wong

Introduction

Previous publications from this laboratory have dealt with the relationship between respiratory control and the overall flow of carbon through the tricarboxylic acid cycle as it occurs in mitochondria isolated from rat liver (1) and rat heart (2). As had been expected, it was found that the rate of flow through the cycle increased upon addition of ADP. In fact the ratio of the rates in state 3 to those in state 4 is quite similar to the respiratory control ratio [(2); Table 1].

Although such a relationship may appear so obvious as scarcely to require experimental verification, it nevertheless poses the far from simple problem of how the flow of substrate through the cycle is coordinated to the flow of reducing equivalents through the respiratory chain. Does ADP bring about its effect on the cycle chiefly by acting directly as an allosteric ligand on enzymes involved in and associated with the cycle or is its effect largely mediated indirectly via a changed redox status of the respiratory carriers, particularly the nicotinamide nucleotides? In analyzing this situation, Klingenberg (3) added to these alternatives a third possibility, *i.e.*, "energy control." Since that time a great deal of activity has taken place directed to this additional facet of the problem. The general postulate is that "under certain conditions, the entry of substrate into the mitochondria is the rate-limiting factor of the mitochondrial respiration" and that this movement across the mitochondrial inner membrane is energy-dependent (4, 5). In view of the complexity of the problem it is not surprising that there is not a perfect correlation between accumulation and metabolism (6).

Results and discussion

Little is known about "energy-control" of overall flow through the tricarboxylic acid cycle. Experiments such as those published from this laboratory on the effect of ADP upon the cycle do not cast much light on the subject because the state 4 - state 3 transition involves large changes of rate with a rather small difference in terms of energy demand. It has been shown by Klingenberg, Heldt and Pfaff (7) that the atractyloside sensitive carrier serves to buffer the internal ATP/ADP ratio against changes in the ATP/ADP ratio of the medium [Table 1; (7)]. From a different point of view Mitchell and Moyle (8) have suggested that the chemiosmotic potential in state 3 is 200 mM, only a relatively small change from that calculated for the resting state (230 mV). These considerations suggest that "energy-control" of the tricarboxylic acid cycle should be studied by comparison of the phosphate-acceptor stimulated state 3 with the uncoupled state, and it is the purpose of this paper to report such studies for rat liver mitochondria. The mitochondrial preparations used have been described in previous publications, as has the general methodology. Specific details are reported in the legends to the figures.

The presence of an uncoupling agent such as FCCP does not alter significantly the rate of disappearance of pyruvate from a suspension of rat liver mitochondria under the conditions used (Fig. 1). However, if the pyruvate is labelled in position 2, the yield of $^{14}CO_2$ is much greater in the presence of uncoupling agent than in the presence of a phosphate acceptor system. This is shown in Fig. 2; the remainder of this paper is directed towards understanding this difference. In order for ^{14}C from position 2 of pyruvate to appear as carbon dioxide it must enter the cycle to form citrate labelled in the *pro*-S-carboxyl position which will become C-5 of α-oxoglutarate and thus form carboxyl-labelled dicarboxylic acids. On the second turn of the cycle $^{14}CO_2$ is liberated at the isocitric dehydrogenase and α-oxoglutarate dehydrogenase steps.

Fig. 3 demonstrates that the two conditions differ much less with respect to the metabolism of citrate. Even more strikingly, Fig. 4 shows that if this difference in rate of citrate utilization is allowed for, there is no difference in the number of counts reaching malate. That is to say, at times when mitochondria in state 3 have metabolized amounts of $[1,5-^{14}C]$citrate equal to that metabolized by a similar sample of the same mitochondria in the uncoupled state, one finds approximately equal amounts of radioactive malate in the suspension. Thus the difference between the ADP stimulated condition and the uncoupled condition cannot be accounted for by effects on those enzymic steps lying between citrate and malate. The loss of the effect must therefore lie at the malate dehydrogenase or citrate synthetase level.

This suggestion is supported by the experiment shown in Fig. 5. When rat liver mitochondria which are oxidizing pyruvate are supplied with succinate labelled in the carboxyl group and the incubation mixture is analyzed after

2 min, it is seen (Fig. 5) that roughly equal amounts of succinate remain, but that the distribution of radioactivity between malate and citrate differs greatly between the uncoupled state and state 3. The accumulation of counts in malate in the latter condition taken together with the greater appearance of counts in citrate in the uncoupled mitochondria constitute evidence for an inhibition of the cycle between malate and citrate in state 3.

Such a proposition does, however, raise considerable difficulty. Rat liver mitochondria in state 3 are capable of forming citrate very readily when supplied with pyruvate and exogenous malate (Fig. 6). We therefore performed an experiment analogous to that of Figs. 1 and 2 except that 5 mM malate was added; the result is shown in Fig. 7. It will be observed that the mitochondria can form $[^{14}C]$-malate from $[2-^{14}C]$-pyruvate as readily in state 3 as in the presence of FCCP. It is essential in interpreting this experiment to note that under these conditions the anaplerotic CO_2 fixing pathway via pyruvate carboxylase is inoperative. This was demonstrated by Stuart and Williams (1), who showed that there was no incorporation of radioactivity from [carboxyl-^{14}C]-pyruvate into malate in the presence of either ADP or uncoupling agents and has been confirmed in recent experiments (Fig. 8). We are thus left with no enzymic step which can be shown to be impaired in state 3 relative to the uncoupled state. We therefore attribute the limitation of carbon flow through the cycle to the competition between malate dehydrogenase and malate export. Such a theory has been proposed by De Jong, Hulsmann and Meijer (9) to account for phosphate inhibition of the tricarboxylic acid cycle. It is therefore important to point out that the phenomenon reported here is not related directly to the phosphate content of the medium. In Fig. 9 it is shown that the stimulating effect of FCCP is not counteracted by either ADP or inorganic phosphate or both together. Table 1 makes the same point for a number of concentrations of phosphate.

It therefore appears that there is a greater tendency for rat liver mitochondria to lose their endogenous malate in the relatively energy-rich state 3 than in the uncoupled condition. The effect is reminiscent of the energy-linked expulsion of malate proposed by us earlier for rat heart mitochondria (10) and the energy-linked outward movement of aspartate from the same organelles proposed at last year's Omaha symposium (11). Drs. LaNoue and Williamson have in personal communications queried the significance of our original report on the increased levels of malate in heart mitochondria in state 3 as compared with those in state 4 on the grounds that the malate accumulation might be a consequence of a brief anaerobiosis on the Millipore filter. We have therefore re-analyzed our original data along with those from a number of subsequently performed experiments, estimating the intramitochondrial malate by difference between the whole suspension and the filtrate. The results are ambiguous. Fig. 10 shows the intramitochondrial [^{14}C]-malate as time plot for rat heart mitochondria metabolizing pyruvate and a low level of [1,4-^{14}C]-succinate. Table 2 gives the turnover times calculated from this plot, the intramitochondrial malate measured by enzymic analysis and, in column 4, the derived value for the rate of exit from the intramitochondrial pool of malate. This exit should be the sum of the malate dehydrogenase and malate export rates if there is only a very low rate of regain of radioactive malate from the suspending medium. The absolute values of these figures are approximately correct for state 4 and 4_{ol} (*i.e.*, in the presence of ADP and oligomycin) but too low for state 3.

The partition between export and dehydrogenation in the two states cannot be determined, as it involves four unknowns. Assuming a respiratory control ratio of 5 for the malate dehydrogenase step reduces the number of unknowns to three and one may then establish the limiting solution for an export rate equal to zero in state 3. This solution is presented in column 5 of Table 2 where it is shown that the data are consistent with a very low export rate in state 3 and an export rate in state 4 equal to about 1.5 times the malate dehydrogenase rate. It must be admitted that these calculations do not constitute proof of the hypothesis advanced by McElroy *et al.* (10), but they do indicate a solution to the dilemma posed by the discrepancy between the measured intramitochondrial malate and that calculated from the isotope kinetics.

It is interesting to note that our data lead us to postulate energy-linked expulsion of malate (and probably other anions) as the mechanism which differentiates states 3 and 4 in heart mitochondria but which in the case of liver mitochondria causes the metabolic differences between state 3 and the uncoupled condition. LaNoue and Williamson (11) have reported for heart mitochondria that the flux through citrate synthase is very similar in the uncoupled state and state 3. This is in agreement with our own earlier findings (12) and more recent unpublished observations of Drs. McElroy and Wong.

In any case, we find ourselves in agreement with the statement of LaNoue and Williamson (11) that "Control (of flux in the citric acid cycle) is presumably exerted chiefly by the oxalacetate concentration, which is determined by both the malate concentration and the NADH oxidation-reduction state."

For liver mitochondria the results of Stuart and Williams (1) had already pointed to the NADH/NAD$^+$ ratio available to malate dehydrogenase as the key factor in determining the difference in flow through the cycle between states 3 and 4. Our present postulate is that the limitation in cycle flux in state 3 compared to the uncoupled state (for liver mitochondria) or in state 4 compared to state 3 (for heart mitochondria) reflects a low level of intramitochondrial malate and that this level is kept low by the activity of the malate transporting systems. Any influence therefore which reduces the tendency of these systems to export malate will have as a consequence an increased flow of carbon through the tricarboxylic acid cycle.

If indeed there is a difference in the ability of malate export processes to compete with malate dehydrogenase in different "energy-states," there is no clear cut mechanism which may be invoked to explain such effects. Differential inhibition studies suggest the existence of at least three transport systems which can carry the malate anion (13), but the experiments from which their existence is deduced are carried out under non-metabolizing conditions and it is difficult to estimate their relative activities under more physiological conditions. The inhibition of the carriers by uncoupling agents (14) takes place at much higher concentrations than those used to stimulate respiration, but the effect would be in the right direction and perhaps is a contributory factor. The chemiosmotic potential proposed by Mitchell (15) would also have the correct polarity to cause an anion expulsion. In the developed form of the chemiosmotic theory the antiport carriers are supposed to permit the movement of anionic substrates and products whilst preventing a net anion flux which would dissipate the potential. However, the "revolving door" model of antiport is not the only method for achieving charge neutralization. Harris (16) has shown that valinomycin in the presence of K^+ can induce citrate uptake, presumably with net gain of potassium citrate. We also have observed, in the presence of K^+, ATP, and valinomycin, the reduction of mitochondrial nicotinamide nucleotides upon addition of citrate without a requirement for exogenous malate. Movement of anions such as malate down the chemiosmotic gradient might cause a depletion of intramitochondrial intermediates with the consequence pointed out by De Jong et al. (9).

The physiological significance of the observations is obscure and may even be non-existent. A suspension of mitochondria has at least one feature which is grossly different from the *in vivo* situation, namely the relationship between the intra- and extra-mitochondrial volumes. A molecule of malate expelled *in vitro* into the considerable volume of the suspending medium will have a far lower probability of being regained by the mitochondria than if expelled into the relatively limited volume of the cytosol. However, it is not difficult to produce teleological rationalizations for a control system which ensures that when the cell is in a state of energy demand the mitochondria retain their intermediates in order to preserve a functioning tricarboxylic acid cycle serving the single role of supplying reducing equivalents to the energy yielding respiratory chain.

Presented by G.R. Williams. One of us (G.S.W.) was the recipient of a post-doctoral

Fellowship of the Medical Research Council of Canada and the work reported in this paper was supported by Grant MT-3182 from that body to G.R.W.

References

1. Stuart, S.C. and G.R. Williams. Pyruvate metabolism by isolated rat liver mitochondria as a function of adenosine diphosphate controlled respiratory state. Biochemistry (Wash.) 5:3912-3919(1966).
2. McElroy, F.A. and G.R. Williams. Rate parameters of the tricarboxylic acid cycle. Arch. Biochem. 126:492-502(1968).
3. Klingenberg, M. Control characteristics of the adenine nucleotide system. In: B. Chance, R.W. Estabrook and J.R. Williams (Editors), Control of energy metabolism. Academic Press, New York (1965), pp. 149-155.
4. Quagliariello, E. and F. Palmieri. Influence of substrate uptake on mitochondrial metabolism. Atti del Seminario di Studi Biologici 3:253(1967).
5. Palmieri, F., M. Cisternino and E. Quagliariello. Inhibition of uptake and oxidation of succinate in rat-liver mitochondria. Biochim. Biophys. Acta 143:625-627(1967).
6. Van Dam, K. and C.S. Tsou. Accumulation of substrates by mitochondria. Biochim. Biophys. Acta 162:301-309(1968).
7. Klingenberg, M., H.W. Heldt and E. Pfaff. The role of adenine nucleotide translocation in the generation of phosphorylation energy. In: S. Papa, J.M. Tager, E. Quagliariello and E.C. Slater (Editors), The energy level and metabolic control in mitochondria. Adriatica Editrice, Bari (1969), pp. 237-253.
8. Mitchell, P. and J. Moyle. Estimation of membrane potential and pH difference across the cristae membrane of rat liver mitochondria. Europ. J. Biochem. 7:471-484(1969).
9. De Jong, J.W., W.C. Hülsmann and A.J. Meijer. Phosphate-induced loss of citric acid cycle intermediates from rat-liver mitochondria. Biochim. Biophys. Acta 184:664-666(1969).
10. McElroy, F.A., G.S. Wong and G.R. Williams. Distribution of malate across the mitochondrial membrane as a significant factor in respiratory control. Arch. Biochem. Biophys. 128:563-565(1968).
11. La Noue, K.F. and J.R. Williamson. Interrelationships between malate-aspartate shuttle and citric acid cycle in rat heart mitochondria. Metabolism 20:119-140(1971).
12. Williams, G.R. Dynamic aspects of the tricarboxylic acid cycle in isolated mitochondria. Canad. J. Biochem. 43:603-615(1965).
13. Robinson, B.H., G.R. Williams, M.L. Halperin and C.C. Leznoff. The effects of 2-ethylcitrate and tricarballylate on citrate transport in rat liver mitochondria and fatty acid synthesis in rat white adipose tissue. Europ. J. Biochem. 15:263-272(1970).
14. Papa, S., N.E. Lofrumento, E. Quagliariello, A.J. Meijer and J.M. Tager. Coupling mechanisms in anionic substrate transport across the inner membrane of rat liver mitochondria. J. Bioenergetics 1:287-307(1970).
15. Mitchell, P. Chemiosmotic coupling in oxidative and photosynthetic phosphorylation. Biol. Rev. 41:445-502(1966).
16. Harris, E.J. The dependence on dicarboxylic acids and energy of citrate accumulation in depleted rat liver mitochondria. Biochem. J. 109:247-251(1968).
17. Ambus, T., E.E. Dryden and J.F. Manery. Insulin stimulation of $^{14}CO_2$ production by frog muscle from D(-)- and L(+)-lactate-1-^{14}C. Int. J. Biochem. 1:553-556(1970).

18. McElroy, F.A. and G.R. Williams. Determination of specific radioactivities of citric acid cycle intermediates by enzymic decarboxylation. In: J.M. Lowenstein (Editor), Methods in enzymology, Vol. 13, Academic Press, New York (1969), pp. 528-535.

TABLE 1

PYRUVATE OXIDATION IN UNCOUPLED RAT LIVER MITOCHONDRIA

Rat liver mitochondria (4.5 mg protein) metabolizing 2.3 mM [2-^{14}C] pyruvate in presence of 1 μM FCCP (time = 7 min).

Additions		$^{14}CO_2$ Recovered
P_i	ADP	(as percentage of pyruvate disappearance)
millimolar		
0	0	27
0	0.8	25
4	0	32
10	0	30
10	0.8	26
20	0	28

TABLE 2

KINETICS OF [^{14}C] MALATE POOL IN RAT HEART MITOCHONDRIA METABOLIZING 1.0 mM PYRUVATE AND 10 μM [1,4-$^{14}C_2$] SUCCINATE

See text for a complete description of the experimental conditions.

State	Turnover time*	Malate	Total rate of exit from intramitochondrial malate pool	Minimum rate of export of intramitochondrial malate
	secs	*nmoles/mg protein*	*nmoles/sec/mg protein*	
3	4	0.6	0.15	0
4	8	0.6	0.075	0.045
4	18	1.2	0.075	0.045

*Turnover time = $\dfrac{\left\{[^{14}C]\ \text{malate}\right\}}{\dfrac{d\left\{[^{14}C]\ \text{succinate}\right\}}{dt} - \dfrac{d\left\{[^{14}C]\ \text{malate}\right\}}{dt}}$

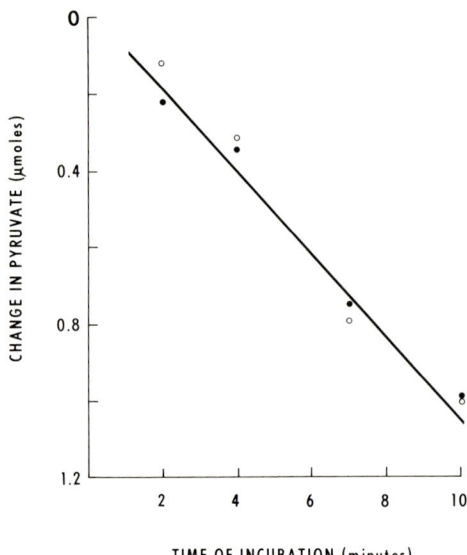

Fig. 1. *Pyruvate utilization as a function of time.* Rat liver mitochondria (5.9 mg protein) prepared in 0.21 M mannitol, 0.07 M sucrose, 1.0 mM Tris, 1.0 mM ethylene glycol bis (amino-ethyl) tetra-acetate, pH 7.4, were suspended in 1.0 ml of 0.125 M KCl, 0.05 M Tris, pH 7.4, and shaken at room temperature (21°). [2-^{14}C]-pyruvate was included, concentration = 1.4 mM, radioactivity = 2.01 x 10^5 dpm. The reaction was terminated at times indicated by addition of 1.0 ml of 0.6 M HClO$_4$. Pyruvate was estimated enzymically using lactic dehydrogenase and NADH in the deproteinized extract after neutralization with 3 M K$_2$CO$_3$, 0.5 M triethanolamine. Additions to the basic system were as follows: •, carbonylcyanide-*p*-trifluoromethoxy-phenylhydrazone (FCCP), 1.0 μM; ○, K$_2$HPO$_4$/KH$_2$PO$_4$ buffer, pH 7.4, 10 mM; ADP, 0.8 mM; glucose, 60 mM; hexokinase, 0.1 mg (Koch-Light).

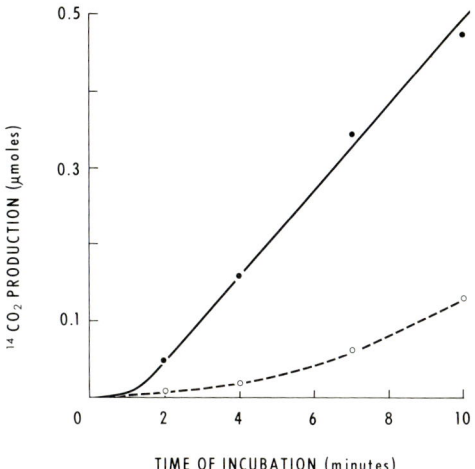

Fig. 2. $^{14}CO_2$ production from $[2-^{14}C]$-pyruvate. Data are drawn from the same experiment as that of Fig. 1 and the symbols represent the same additions, $^{14}CO_2$ was trapped in NCS solubilizer (Amersham-Searle) according to the technique of Ambus, Dryden and Manery (17) and counted as described by these authors.

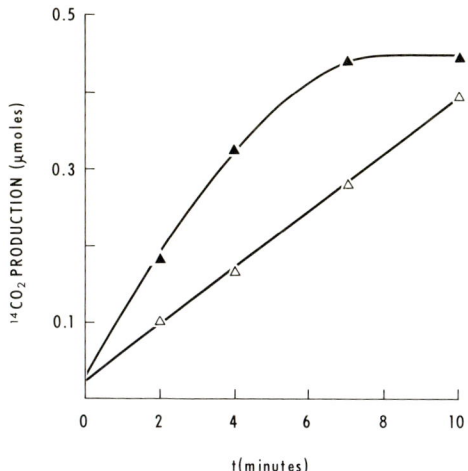

Fig. 3. $^{14}CO_2$ production from $[6-^{14}C]$-citrate. Basic conditions and methods were as in Figs. 1 and 2, except that the radioactive pyruvate was replaced by $[6-^{14}C]$-citrate, concentration = 0.5 mM, radioactivity = 1.0×10^6 dpm; mitochondrial protein = 9.5 mg protein. ▲, uncoupled; △, state 3.

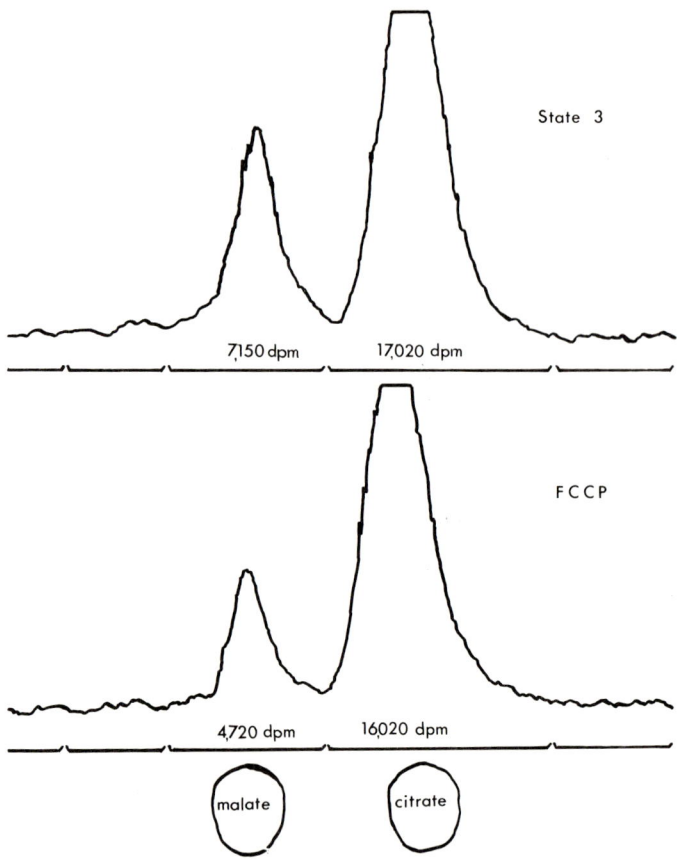

Fig. 4. *Scan of radioactivity after high voltage electrophoresis (HVE) of neutralized, deproteinized extract of rat liver mitochondria metabolizing [1,5-^{14}C]-citrate.* The experiment was similar to that of Fig. 3 except that the radioactivity (1.4 × 10^6 dpm) was in carboxyls 1 and 5 of citrate rather than carboxyl 6. The incubation in state 3 was terminated at 4 min and that in the uncoupled state at 2 min. HVE was performed for 3.5 hours at 3000 V in formic acid/acetic acid/water (1.7:1.7:100, by vol) adjusted to pH 3.1 with pyridine. Citrate moved 46 cm from the origin. Strips were cut from the paper after scanning and counted in the scintillation fluid described by McElroy and Williams (2). Zero-time extracts subjected to similar procedures were recovered with an efficiency of 65%. The values printed on the figure are not corrected for this loss. These values pertain to the 100 µl sample applied to the paper. This sample was taken from 2.01 ml of extract.

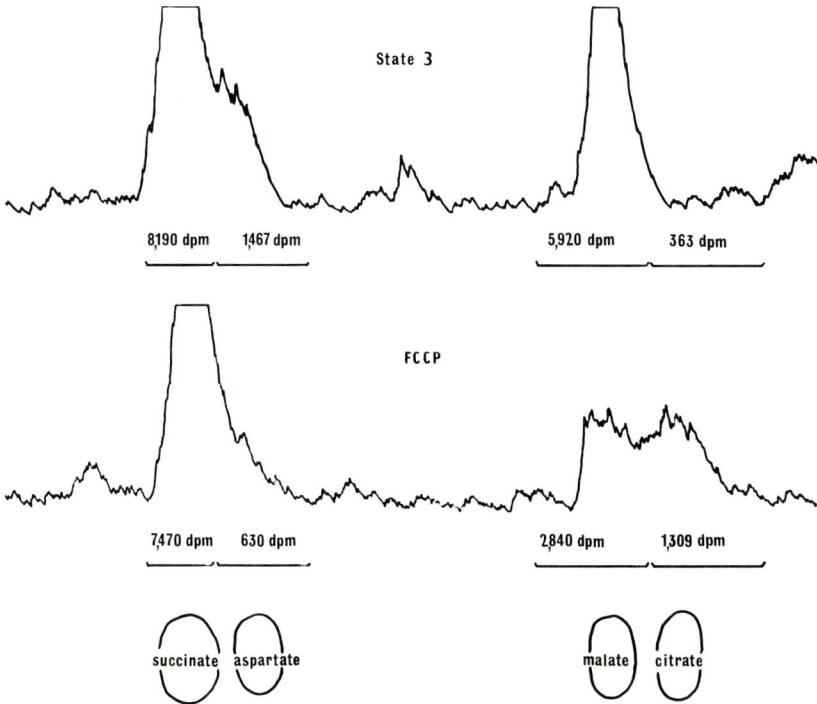

Fig. 5. *High voltage electrophoresis of radioactive intermediates derived from [1,4-^{14}C]-succinate.* The reaction system was that of Fig. 1 except that non-radioactive pyruvate was used (1.8 mM) and [1,4-^{14}C]-succinate was added, concentration = 7.5 µM, radioactivity = 3.3 × 10^5 dpm. Electrophoresis and estimation of radioactivity was performed as in Fig. 4. Sample size, 200 µl; mitochondrial protein = 8.3 mg. Time of incubation, 2 min.

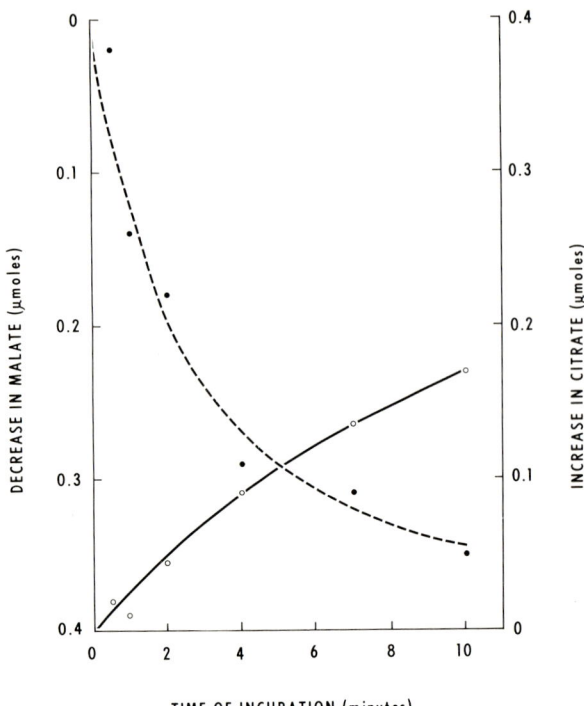

Fig. 6. *Citrate production and malate utilization.* Rat liver mitochondria prepared as for the experiment of Fig. 1 (5.3 mg protein) were suspended in 1.0 ml of 0.050 M KCl, 0.050 M Tris, 0.020 M Na_2HPO_4/NaH_2PO_4, 5.0 mM $MgSO_4$, 0.150 M sucrose, 1.0 mM ADP, 0.050 M glucose, 2.2 units of hexokinase (Boehringer-Mannheim), 2.0 mM pyruvate, 1.2 mM malate. The reaction was terminated as described in the legend to Fig. 1 and the extracts analyzed enzymically for malate using malate dehydrogenase and acetyl-pyridine-NAD^+ in glycine/hydrazine, pH 9.1, and for citrate using citrate lyase and malate dehydrogenase.

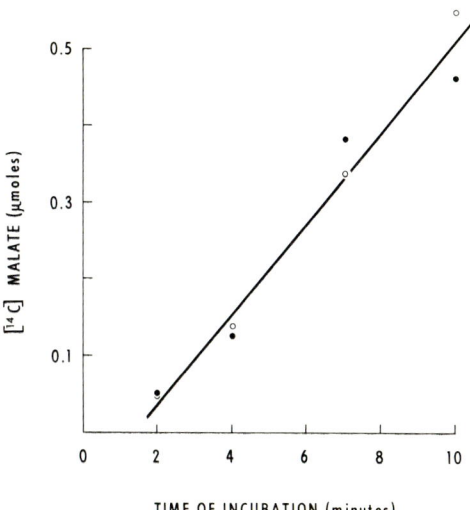

Fig. 7. *Incorporation of radioactivity from [2-^{14}C]-pyruvate into malate.* Experimental conditions and symbols used are described in the legend to Fig. 1 (radioactivity of pyruvate added = 1.74 x 10^5 dpm) except that malate was added to a concentration of 5 mM. [^{14}C]-malate was estimated after high voltage electrophoresis as described in the legend to Fig. 4. Mitochondrial protein, 9.5 mg.

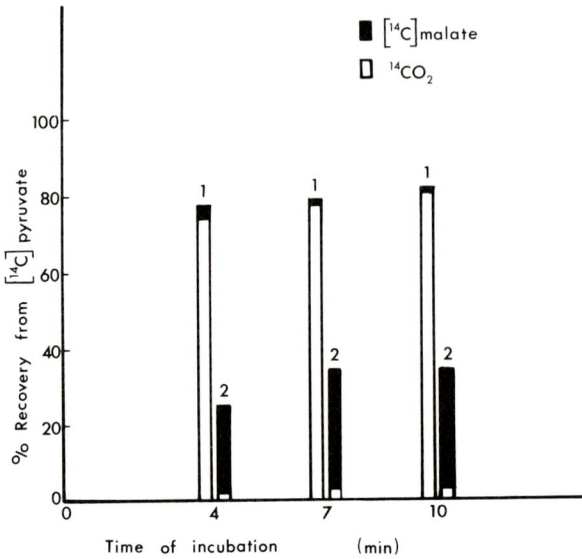

Fig. 8. *Incorporation into malate of radioactivity from pyruvate labelled in positions 1 or 2.* Experimental conditions were as for Fig. 1, ADP/hexokinase/glucose being present in every case. Pyruvate concentration was 2.0 mM, radioactivity of [1-^{14}C]-pyruvate = 1.27 × 10^6 dpm, of [2-^{14}C]-pyruvate = 1.41 × 10^6 dpm. ^{14}CO$_2$ and [^{14}C]-malate determinations were as described for earlier experiments in this paper. The solid bars indicate radioactivity recovered in malate, the open bars that in ^{14}CO$_2$. The number above each bar indicates the position of the label in the radioactive pyruvate added. Recovery is reported as a percentage of pyruvate utilized at the time indicated. Mitochondrial protein, 8.2 mg.

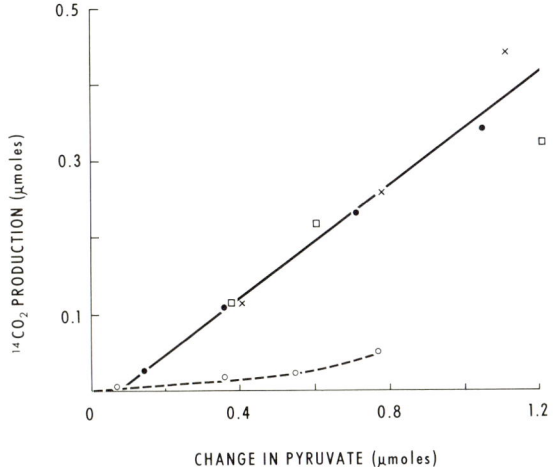

Fig. 9. *Failure of ADP and inorganic phosphate to decrease the high yields of $^{14}CO_2$ from [2-^{14}C]-pyruvate obtained in the uncoupled state.* Basic conditions were as described in the legend to Fig. 1. Pyruvate concentration, 2.3 mM; radioactivity = 2.06 × 10^5 dpm. Mitochondrial protein, 5.3 mg; ○, state 3; ●, FCCP alone; ×, FCCP plus 0.8 mM ADP; □, FCCP plus 0.8 mM ADP and 10 mM P_i.

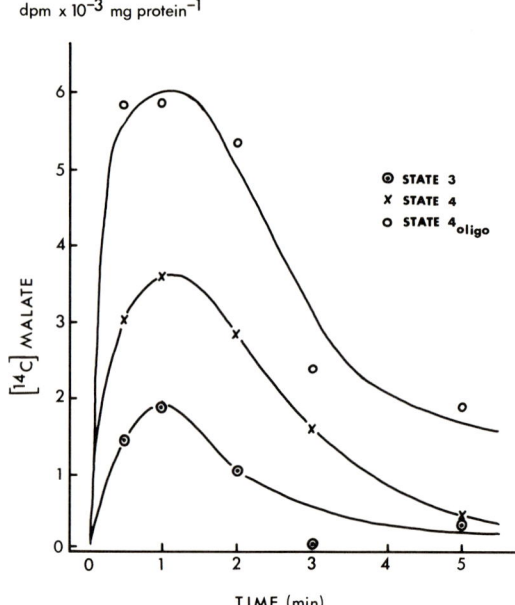

Fig. 10. *Turnover of intramitochondrial $[^{14}C]$-malate derived from $[1,4$-$^{14}C]$-succinate added to rat heart mitochondria.* The experiments on which these plots are based are similar to that shown in Fig. 5 and have been described in detail elsewhere (2). Intramitochondrial malate is calculated by subtracting malate found after rapid Millipore filtration from malate determined in the whole suspension. $[^{14}C]$-Malate was estimated by the enzymic decarboxylation method of McElroy and Williams (18).

IS THERE AN ORGANIZATION OF KREBS CYCLE ENZYMES IN THE MITOCHONDRIAL MATRIX?

Paul A. Srere

Control mechanisms

It is probable that every enzyme in a metabolic pathway has a maximal capacity greater than the demands of that pathway. In such a situation, if there is not a large difference in the V_{max} of each enzyme, control is possible at each step in that sequence. This occurs since the slowest step will eventually result in an adjustment of substrate levels for all other steps so that their rates are equal to or oscillate around that of the rate limiting step. The rate of each step can be modified by changing the concentration of the enzyme, its substrates or its effectors. In recent years discoveries of rate modifying reactions for enzymes of the Krebs cycle have increased enormously. An ATP effect on citrate synthase that can be modified by substrates or Mg^{++} has been found. Aconitase can be inhibited by an iron-oxaloacetate chelate, while isocitrate dehydrogenase (NAD^+) is affected by ADP, NADH and ATP. α-Ketoglutarate dehydrogenase can be modified by a kinase-phosphatase system which, in turn, is affected by ATP and Mg^{++} concentrations. Succinate dehydrogenase is inhibited by oxaloacetate and fumarase is inhibited by MgATP complex. Which, if any, of these interactions are involved in control of the cycle is not at all clear. It is possible, however, that the activity of a series of enzymes might be affected by a fourth parameter: their spatial relationships. This presentation will consider the likelihood of a fixed spatial relation between the Krebs cycle enzymes.

Location of Krebs cycle enzymes

Our present knowledge of mitochondrial structure has been reviewed and summarized by many workers in that area (1, 2, 3, 4, 5). Fig. 1 is a diagram of the compartments of a mitochondrion. Each of the membranes is believed to have a constant structure. Oxidative phosphorylation is carried out by the inner membrane and many detailed proposals concerning its structure-function relation have been presented. One reason for this interest has been the observation of "knobs" (ATPase) and the tripartite structure of the inner membrane seen by the use of electron microscopy. The matrix presents a uniform and uninteresting aspect to the microscopist; however, it contains most of the enzymes of the Krebs cycle. I would like to present the view that these enzymes are not in random disarray in the matrix but are organized in assemblies with a fixed relation to each other and the inner membrane.

This idea is not new and such an arrangement was visualized by A.G. MacLeod (6) in a publication of the Upjohn Co. Certain aspects of such an arrangement can be tested with data already in the literature. In addition to its relation to the control of the Krebs cycle, such a proposal also has implications concerning mitochondrial biogenesis.

Tte enzyme of the Krebs cycle are shown in Table 1. There are, in addition, a number of enzymes closely related to the cycle which are found in the mitochondria: pyruvate dehydrogenase complex, transaminases and fatty acid oxidizing enzymes. I did not indicate whether the NAD^+ or $NADP^+$-specific ICDH is involved in the cycle. In either case, the transfer of electrons from $NADP^+$ to NAD^+ can take place either by way of a transhydrogenase or by way of the NAD^+, $NADP^+$-glutamate dehydrogenase.

It has been clear for some time that each of the enzymes of the Krebs cycle can be found in the mitochondria of cells, although some may occur in other parts of cells as indicated in Table 2. One of them, succinate dehydrogenase, has been known to be an integral part of the inner membrane for a number of years. The mitochondrial location of the other enzymes had been a subject of controversy, but the experiments of Schnaitman and Greenawalt (7) definitely located them in the matrix compartment. In their experiments, Schnaitman and Greenawalt treated rat liver mitochondria with digitonin to remove only the outer membranes and the components of the intermembrane space. They were then able to isolate inner-membrane matrix particles. These spherical particles retained all of the Krebs cycle functions of the intact mitochondria. The contention that PDC, KGDC and STK are part of the inner membrane has been abandoned. Thus, the best available data indicate that the enzymes of the Krebs cycle, with the exception of SDH, are in the matrix. That one exception bothers me. Why should a series of enzymes which has a continuous and integrated function have an apparent discontinuity in its morphology?

Size of mitochondria and their compartments

The location of the enzymes in the matrix raises a question concerning comparative mitochondrial structure. Fig. 2 shows electron micrographs of rat liver, heart and kidney. The magnification is the same in each, and, roughly speaking, the same volume of each tissue is occupied by mitochondria. On the other hand, in Fig. 3 it can be seen that the morphology of each mitochondrion is quite different. Looking at the extremes, we see that liver mitochondria have relatively little inner membrane and a great deal of matrix; the situation is reversed in heart mitochondria. The oxidative capacity of these mitochondria, or the cytochrome content, is correlated to the amount of inner membrane of the mitochondria, in agreement with the location of these proteins in the inner membrane. I will show later that the amounts of Krebs cycle enzymes are also correlated with the amount of inner membrane and not with the amount of matrix, the compartment in which they are located.

In order to test the relation of enzyme content to compartment size of mitochondria, accurate morphometric measurements are needed. The data of Loud (8) and those of Weibel *et al.* (9) indicated that a rat liver mitochondrion has a volume of about 0.75 μ^3 and about 30 μ^2/μ^3 inner membrane. On the other hand, a heart mitochondrion is somewhat smaller, 0.55 μ^3, but has about 3 times the inner membrane surface, $90\mu^2/\mu^3$ (10, 11). A comparison of the total volume occupied by mitochondria in heart versus that in liver shows that rat heart mitochondria occupy about 33% of the cell volume while rat liver mitochondria occupy about 21% of the cell volume (9). Since rat heart contains 70 units/g citrate synthase and liver has 14 units/g, and since per g of tissue, heart contains 1.6 times more mitochondria, the difference in CS per gram of mitochondria is (70/14)(1/16) or 3.1. This ratio is the same as the ratio between inner membrane areas of the two mitochondria. It will be of interest to see if the same correlation exists in mitochondria from other tissues. One notes with interest that kidney mitochondria appear to have an amount of inner membrane surface somewhere between heart and liver and a citrate synthase content also intermediate between heart and liver.

Relation of biogenesis of mitochondria to their organization

Next let us consider the question of biosynthesis of mitochondria as it might relate to the possible organization of the Krebs cycle enzymes in the matrix. It is well established that mitochondria carry both the information and apparatus to synthesize proteins (12, 13). A number of experiments have indicated, however, that matrix enzymes like MDH (14) and even the membrane-bound succinate dehydrogenase (15) are synthesized outside the mitochondria. One must postulate some mechanism to gather these extramitochondrially synthesized proteins into one package. Whether this is a self association of these proteins and/or an aggregation on some mitochondrially coded organizer protein or a transfer from cytosol into an empty mitochondrion is not at all clear.

The work of Pette, Klingenberg and Bücher (16) has indicated that the enzymes of the Krebs cycle constitute a constant proportion group of enzymes. That is, the ratio of activities of Krebs cycle enzymes is remarkably constant among a wide variety of cell types which would indicate that the relative amounts of each are constant. We have demonstrated recently that the citrate synthases of rat liver and rat heart are in all likelihood identical proteins (17). Further immunological studies with citrate synthases from other rat tissues indicate that they also may be identical to the enzyme from rat heart and liver (17). If we assume that similar results will be found for other Krebs cycle enzymes, and since the same proportions of these exist in each cell, then it would appear that each cell type of the rat reads the same part of its nuclear chromosome to make its mitochondrial enzymes. Then the differences in mitochondria from cell type to cell type will depend upon each cell's mitochondrial DNA and the ability of some mitochondrially coded organizer to assemble the proteins of the Krebs cycle. To indulge in an even wilder flight of image-ination (to use Buckminster Fuller's word), an inner membrane protein may have a designated number of sites for the tightly bound succinate dehydrogenase which acts as an organizer for the other Krebs cycle enzymes. Since the proportion of enzyme to inner membrane may be constant, the unique information from tissue to tissue is the amount of this membrane that is enclosed in an outer membrane.

From the biosynthetic point of view it appears that some organization step is mandatory for mitochondrial formation.

Size and stoichiometry of Krebs cycle enzymes

What I have tried to do so far is to present the oddments of information that have led me to postulate an organization of the Krebs cycle enzymes in the mitochondrial matrix. What other experimental evidence exists either for or against such a postulate?

The matrices of many mitochondria have been examined by electron microscopy. A uniform field of particles which have been sized from 40 to 60 Å (2) has been reported. No evidence of structures or arrays in or near the membrane or in any part of the matrix has been reported. What would we expect to see? Most of the enzymes of the Krebs cycle have been purified from pig or beef heart. Their molecular weights, subunit structure and size (either calculated or seen by electron microscopy) are shown in Table 3 (18).

The assumption that they are spherical has been made and in several cases guesses are included. With the exception of the KGDC then, the size of 40 to 60 Å particles seen in the matrix is not unusual.

Since the volume of the matrix fraction of mitochondria is not correlated to the amount of Krebs cycle enzyme, it seems reasonable to assume that they do not fill the matrix. If a stoichiometry of one active site per assembly is made, and one enzyme per assembly, then we find that the area occupied by 1 Krebs cycle unit is about $(8)[\pi(3 \times 10^{-3})^2]$ or $2.3 \times 10^{-4} \mu^2$. A rat mitochondrion has 27 μ^2 of inner membrane surface or room for $27.4/2.3 \times 10^{-4} = 1.2 \times 10^5$ assemblies and a similar calculation for a heart mitochondrion yields a figure of 2.2×10^5 assemblies. This is a maximal number; since a minimal area was assumed and depending upon shape of the enzymes and packing distances, the calculation is at best only an order of magnitude estimate. I will point out later what the number of assemblies are in a heart mitochondrion as calculated from enzyme content.

From the first I have assumed that there is a fixed stoichiometry between the enzymes of the Krebs cycle; that is, we would expect a molar ratio of each of the enzymes. I have made such a calculation in two ways. In one I have used the analyses of Bachmann et al. (19) on the enzyme activities of pig heart mitochondria, and in the other I have used data from the literature where the individual enzymes have been purified from pig or beef heart (Table 4). In these latter cases, I have spoken with the investigators to get an impression from them as to whether a good yield of activity was obtained. From both calculations the molar ratio from high to low is about 20. One thing seems clear: the ICDH (NAD⁺) is just not active enough (there are too few molecules) to be considered part of the cycle in heart mitochondria. It must be remembered that the Q_{O_2} of heart tissue expressed in the same units is about 14.

If we assume that the concentration of one Krebs cycle assembly is about 5×10^{-5} moles/kg of mitochondria, then one heart mitochondrion would contain $(0.55 \times 10^{-15})(5 \times 10^{-5})(6 \times 10^{23})$ or about 17,000 assemblies. You may recall that I calculated earlier that a maximum of 220,000 assemblies could fit on the surface of the cristae of a heart mitochondrion. These calculations, therefore, are not inconsistent with the postulate that the Krebs cycle enzymes are on the matrix side of the inner membrane in some stoichiometric arrangement or assembly and are bound to the membrane.

Interactions between Krebs cycle components

Is there any evidence that binding occurs either between the Krebs cycle enzymes or between them and the inner membrane? One of the first observations which caused me to think along these lines was the fact that in the purification of citrate synthase one of the most difficult contaminants to remove is malate dehydrogenase. We have tried in two ways to see interactions between malate dehydrogenase and citrate synthase. First, we have performed sedimentation velocity experiments with mixtures of the two enzymes both in the presence and absence of their substrate. The results indicated that each sedimented principally with their characteristic sedimentation coefficient. Secondly, we have run Hummel and Dreyer experiments (suggested by Dr. I. Rose) with a Sephadex 100 column saturated with MDH. When citrate synthase was then placed on the column, no evidence of binding was observed with this procedure. We are at present attempting experiments of this type using sedimentation equilibrium techniques which should enable a more sensitive measurement of protein-protein interaction.

Secondly, what evidence is there for interaction between the Krebs cycle enzymes and the inner membrane? Bachmann et al. (19) found that PDC and KGDC of phospholipase-disrupted fraction sedimented with a membrane fraction and not with other Krebs cycle enzymes. STK also sedimented with a membrane fraction. Ernster and Kuylenstierna (1) have found that MDC and ICDH ($NADP^+$) in broken mitochondrial fractions appeared not only as the soluble enzyme but partially bound to inner membrane fractions. Garland et al. (20) have described experiments where 59% of MDH and 63% of CS appear in the inner membrane fraction of broken mitochondria. It does not seem unreasonable to assume, on the basis of these reports, that some affinity exists between the so-called "soluble" Krebs cycle enzymes and the inner membrane of the mitochondria.

Regulatory aspects of organization

It may be fairly asked at this point whether we have been engaged in nothing but a "Talmudic argument." The possibilities are that either the Krebs cycle enzymes in the matrix are organized or they are not. I adopted the view that an organization existed and tried to support it with a number of arguments and calculations. I feel certain, however, that the same calculations might be used to prove a random arrangement of the Krebs cycle enzyme within the matrix. Only with more precise enzyme concentration data will it be possible to make an accurate assessment on this basis. If a random non-stoichiometric arrangement does exist, then additional explanations must be forthcoming concerning (a) the discontinuous membrane-matrix location of the enzymes, (b) the apparent relation between the enzyme content and area of inner membrane, (c) the constant proportion aspect of enzyme activities, and (d) the bio-organizational aspects of mitochondrial biosynthesis.

I think it is also pertinent to ask what difference it makes if an organized or disorganized Krebs cycle exists. The consequences of order versus non-order would be pertinent to the question of the mechanism of biosynthesis of mitochondria as well as to the problem of regulation of Krebs cycle activity. Reed (21) has discussed the advantages of multi-enzyme complexes and has listed a number of these, including the α-keto acid dehydrogenase complex, fatty acid synthetase, and the tryptophan synthetase system.

Recent studies on enzymes which have been trapped within or linked to solid supports have also indicated altered kinetic behavior (22). It seems plausible to me that ordered enzyme arrays may be able to achieve higher through-puts at low substrate concentrations than would a random array of enzymes at the same substrate concentrations. A recent experiment by Mosbach and Mattiasson (23) supports such a contention. These workers attached or embedded both hexokinase and glucose-6-phosphate dehydrogenase to various gels. In every case higher rates of NADPH production were achieved with the matrix-bound system than with free enzymes in solution. These workers postulate a higher local concentration of glucose-6-phosphate in the microenvironment of the dehydrogenase than can be achieved in solution.

An example of a Krebs cycle system that may operate better as an organized system is the malate dehydrogenase-citrate synthase coupled reaction. The malate dehydrogenase reaction has a K_{eq} at pH 7.0 of 2×10^{-5}. Since Krebs and Veech (24) have shown that the $NAD^+/NADH$ ratio in rat liver mitochondria is 7 and that the malate concentration is 0.3 mM, then the free oxaloacetate concentration can be calculated to be about 4×10^{-8} M. Since the K_m for oxaloacetate of rat liver citrate synthase is about 4×10^{-6} M, then, if no other factors are involved, citrate synthase could express less than one percent of its activity, a figure much too low to account for observed respiration rates of rat liver mitochondria. It would appear that some other mechanism such as an organized system is at work so that efficient operation of the cycle is possible.

In summary, I have tried to show that it is likely that the enzymes of the Krebs cycle exist within the mitochondria in an ordered array bound to the matrix side of the inner membrane. Such an arrangement would be advantageous for the mechanism mitochondriogenesis as well as for efficient operation and regulation of the Krebs cycle.

I am indebted to Mrs. Katy Miller, Department of Anatomy, University of Texas Southwestern Medical School, for the electron micrographs (U.S.P.H.S. 10105). My colleagues were extremely helpful in formulating these ideas. The work reported in this paper was supported by U.S.P.H.S. Grant AM11313-05.

References

1. Ernster, L. and B. Kuylenstierna. FEBS Symposium 17:5(1969).
2. Greville, G.D. In: J.M. Lowenstein (Editor), Intracellular compartmentation and the citric acid cycle in citric acid cycle: control and compartmentation, Marcel Dekker, New York (1969), p. 1.
3. MacLennan, D.H. In: F. Bronner and A. Kleinzeller (Editors), Molecular architecture of the mitochondrion in current topics in membranes and transport, Academic Press, New York (1970), p. 177.
4. Lehninger, A.L. The mitochondrion, W.A. Benjamin, Inc., New York (1964).
5. Smoly, J.M., B. Kuylenstierna and L. Ernster. Topological and functional organization of the mitochondrion. Proc. Nat. Acad. Sci. 66:125-131(1970).
6. Macleod, A.G. A scope monograph on cytology, The Upjohn Co., Kalamazoo, Michigan (1957).
7. Schnaitman, C. and J.W. Greenawalt. Enzymatic properties of the inner and outer membranes of rat liver mitochondria. J. Cell Biol. 38:158-175(1968).
8. Loud, A.V. A quantitative stereological description of the ultrastructure of normal rat liver parenchymal cells. J. Cell Biol. 37:27-46(1968).
9. Weibel, E.R., W. Stäubli, H.R. Gnägi and F.A. Hess. Correlated morphometric and biochemical studies on the liver cell I. Morphometric model, stereological methods, and normal morphometric data for rat liver. J. Cell Biol. 42:68-91(1969).
10. Laguens, R. Morphometric study of myocardial mitochondria in the rat. J. Cell Biol. 48:673-676(1971).
11. Plattner, H. Personal communication, (1971).
12. Work, T.S., J.L. Coote and M. Ashwell. Biogenesis of mitochondria. Fed. Proc. 27:1174-1179(1968).
13. Roodyn, D.B. and L.A. Grivell. FEBS symposium 17:161(1969).
14. Kadenbach, B. Synthesis of mitochondrial proteins: demonstration of a transfer of proteins from microsomes into mitochondria. Biochim. Biophys. Acta 134:430-442(1966).
15. Kleitke, B. and A. Wollenberger. On the site of synthesis of enzymes tightly bound to mitochondrial structure in rat liver. FEBS Lett. 1:187-190(1968).
16. Pette, D., M. Klingenberg and T. Bücher. Comparable and specific proportions in the mitochondrial enzyme activity pattern. Biochem. Biophys. Res. Commun. 7:425-429(1962).
17. Moriyama, T. and P.A. Srere. J. Biol. Chem., in press.
18. Colowick, S.P. and N.O. Kaplan. In: J.M. Lowenstein (Editor), Methods in enzymology, Vol. 13, Academic Press, New York (1969).
19. Bachmann, E., D.W. Allmann and D.E. Green. The membrane systems of the mitochondrion I. The s fraction of the outer membrane of beef heart mitochondria. Arch. Biochem. Biophys. 115:153-164(1966).
20. Garland, P.B., B.A. Haddock and D.W. Yates. FEBS Symposium 17:111(1969).
21. Reed, L.J. Pyruvate dehydrogenase complex. In: E.R. Stadtman and B.L. Horecker (Editors), Current topics in cellular regulation, Vol. I, Academic Press, New York (1969), pp. 233-251.

22. Goldman, R., O. Kedem and E. Katchalski. Kinetic behavior of alkaline phosphatase–collodion membranes. Biochemistry 10:165-172(1971).
23. Mosbach, K. and B. Mattiasson. Matrix-bound enzymes Part II: studies on a matrix-bound two-enzyme-system. Acta Chem. Scand. 24:2093-2100(1970).
24. Krebs, H.A. and R.L. Veech. Regulation of the redox state of the pyridine nucleotides in rat liver. In: H. Sund (Editor), Pyridine nucleotide-dependent dehydrogenases, Springer-Verlag, Berlin (1970), pp. 413-438.

TABLE 1

ENZYMES OF THE KREBS CYCLE

Enzyme	Abbreviation
Citrate synthase	CS
Aconitase	A
Isocitrate dehydrogenase	ICDH
α-Ketoglutarate dehydrogenase complex	KGDC
Succinate thiokinase	STK
Succinate dehydrogenase	SD
Fumarase	F
Malate dehydrogenase	MDH

TABLE 2

KREBS CYCLE ENZYMES

Enzyme	Extramitochondrial Locations
CS	Glyoxisomes
A	Cytosol, Glyoxisome
ICDH (NAD$^+$)	None
ICDH (NADP$^+$)	Cytosol
TH	None
KGDC	None
SD	None
F	None
MDH	Cytosol, Glyoxisome

TABLE 3

MOLECULAR DIMENSIONS OF HEART KREBS CYCLE ENZYMES

Enzyme	Molecular weight	Subunits	Diameter
	daltons		Å
CS	1×10^5	2	60
A	0.9×10^5	2 ?	60
ICDH (NAD)	3×10^5	8	86
ICDH (NADP)	0.6×10^5	1	52
KGDC	4.5×10^6	24 ?	250 ?
STK	0.7×10^5	2 ?	54
SD	2×10^5	1	76
F	1.9×10^5	4	76
MDH	0.6×10^5	1	54

TABLE 4

ESTIMATES OF KREBS CYCLE ENZYMES CONCENTRATIONS

Enzyme	Estimate of Bachmann *et al.* (19)		Literature (18)	
	Activity	Molarity	Activity	Molarity
	milliunits/mg mito. protein	10^{-5} *moles/kg*	*units/g tissue*	10^{-5} *moles/kg*
CS	479	2.6	73	6.5
A	499	12	4	7.5
ICDH (NAD$^+$)	---		0.032	0.0008
ICDH (NADP$^+$)	1615	9.2		
KGDC	173	0.6	2	0.013 (0.3)
STK	220	0.5	5	0.18
SD	200	4	10	1.0
F	1745	0.8	63	0.3
MDH	3315	5.0	100	3.0

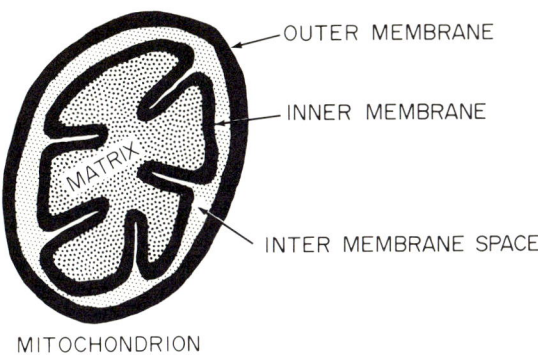

Fig. 1. *Schematic representation of a mitochondrion.*

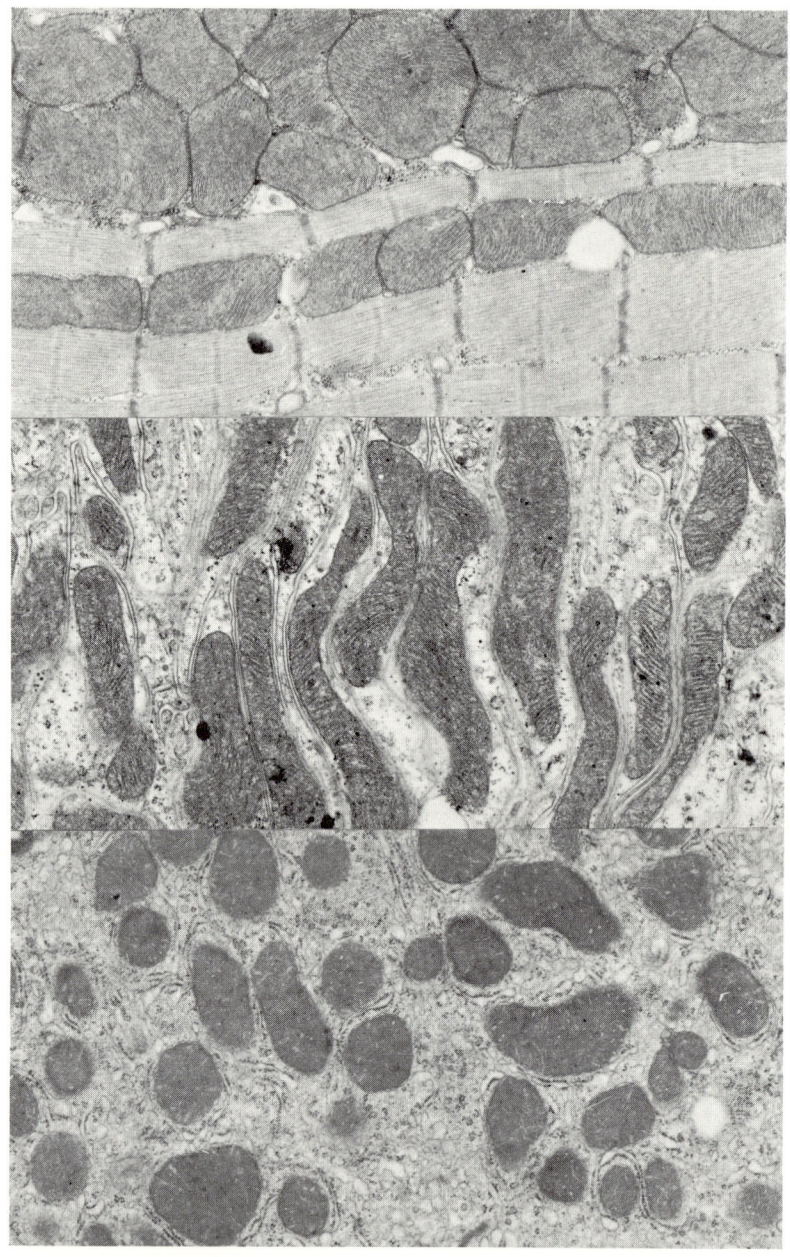

Fig. 2. *Electron photomicrographs of (from left to right) rat heart, rat kidney and rat liver.* Tissues were fixed in glutaralde-hydephosphate buffer and post fixed in 1% osmium tetroxide. The sections were stained with lead citrate and uranyl acetate. Magnification: 18,750.

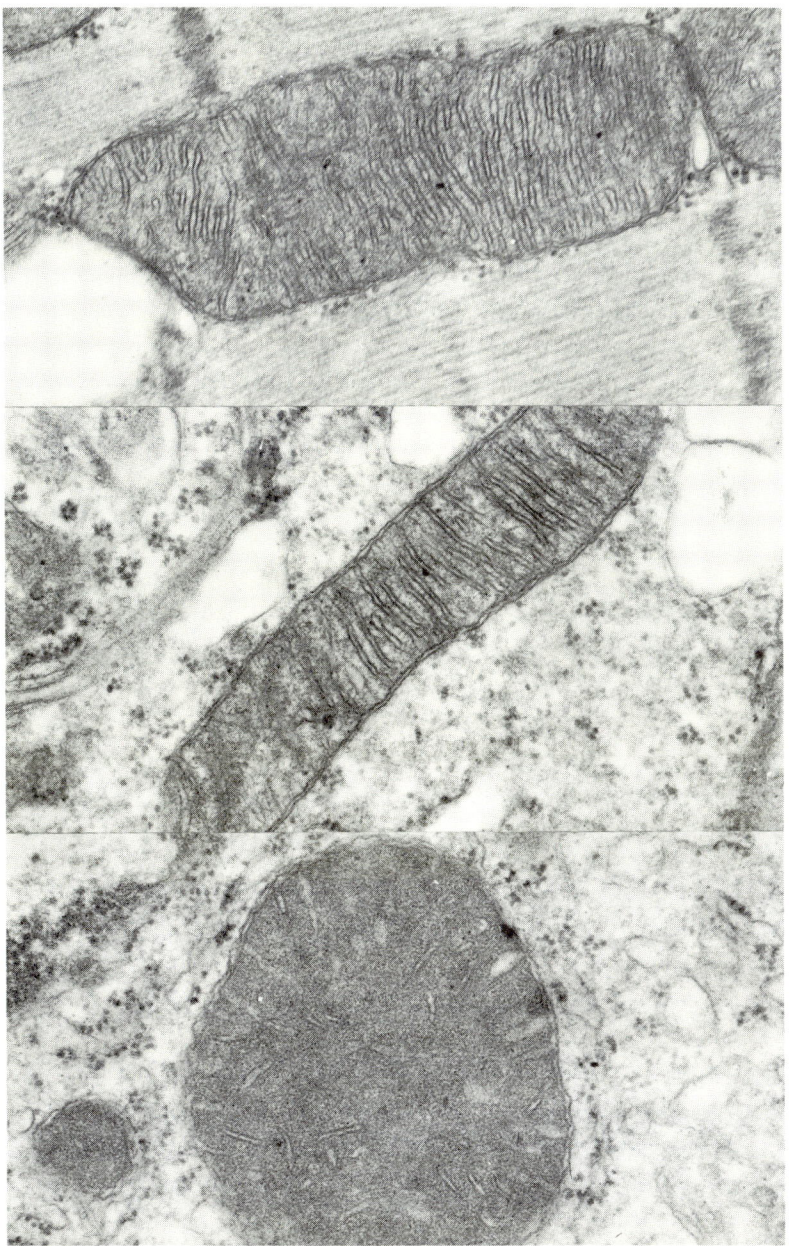

Fig. 3. *Individual mitochondria from Fig. 2.* Left to right: heart, kidney, liver. Magnification: 57,000.

CITRIC ACID CYCLE DYNAMICS AND SUBSTRATE COMPARTMENTATION IN MITOCHONDRIA FROM RABBIT HEART AND BRAIN

R.W. Von Korff

Introduction

Until recently biochemists have devoted a greater share of their attention to the isolation, purification and study of the kinetics, composition and active sites of single enzymes. They have studied mitochondria, investigated metabolic pathways, mechanisms of electron transport and of energy conservation. Intensive research has led to remarkable progress in the fields of nucleic acid metabolism, coding for protein synthesis and of genetic disorders. However, an understanding of, and perhaps solutions to, many medical problems will require more research in several areas:

1) The dynamics of multienzyme systems and of interactions between several such systems.

2) Compartmentation of enzymes, coenzymes and substrates.

3) Determination of the activities of reactants in compartmentalized systems. These are not measured by analyses of a homogenate, and concentrations do not reflect activities because of complex formation (*e.g.*, MgATP, ES, SP, E-CoENZ).

The nervous system is complex and includes many cell types: pyramidal cells, astrocytes, oligodendrocytes, motor and sensory neurons, and Purkinje cells. There may be biochemical as well as morphological compartments. There is axonal plasma flow from the cell body along the axon to the nerve ending. Mitochondria may well be different, if not in origin, at least in age along the nerve cell. Finally, as is well known, mitochondria themselves are compartmentalized.

Substrate concentrations and activities may differ markedly in the extra and intracellular fluid, and mitochondrioplasm. Dr. Srere (1) has noted that the ability to continuously measure instantaneous substrate activities in each compartment of the cell would be an important technical development to facilitate our understanding of metabolic control in animal cells. This would allow changes in substrate and effector concentrations to be related to changes in rate of a given metabolic pathway.

A chemical engineer, L.E. Scriven (2), speaking at a symposium on intracellular transport analysis in 1964 said?

> From afar, a cell like a chemical factory appears as a catalytic point in non-equilibrium surroundings. Close up, both appear from tracer, or "kinetic" studies to be compartmentalized.
>
> Were a chemical engineer to capture a strange factory the last thing he would think of would be to use a steamroller and then separate the wreckage by sedimentation. Nor would he freeze the factory and saw it up for analysis. Therefore, he is impressed by the fruitfulness of the analogous stratagems of the biologist. The engineer would probe and produce transients while the factory is functioning. Of the biologist he asks, are all available methods for manipulating individual cells being exploited? Is adequate effort being devoted to developing new techniques to study vectorial flux rates and reaction rates within single living cells? Measurements of spatial-time distributions within the cells by microspectrophotometry, microdissection, histochemical techniques, radioautography etc. could provide a set of maps showing three-dimensional distributions of of enzymes, substrates, ATP, oxygen, etc.

The low K_m of pyruvate for Krebs cycle oxidations in heart (3) and brain (4, 5) mitochondria (50 μM), the somewhat lower affinity (*i.e.*, higher K_m) of α-ketoglutarate (3, 6) for α-ketoglutarate dehydrogenase (αKG) and the far lower affinity of other substrates are of interest and importance. Permeability factors undoubtedly are crucial.* Chance and Hollunger (8) reported that intramitochondrial succinate is about 100-fold more effective in reducing intramitochondrial NAD^+ than is extramitochondrial succinate (the relative Michaelis constants were 10^{-5} and 7×10^{-4} M, respectively). They suggested that complexes (perhaps even enzyme complexes) may be involved. These markedly different activities would not be revealed by overall concentration measurements on a reaction mixture since the location of

---

*One of the reviewers objected to the fact that "the large body of work on specific mitochondrial permeases for citric acid cycle intermediates had not been mentioned." This is not an oversight. Whether permeases are involved or not is irrelevant to the points under discussion. Since 1965 we have emphasized that the customary practice of using substrates in the 5-10 mM concentration range when studying mitochondrial respiration may have little bearing on the situation which obtains within the intact cell (3). The same point has been made by Williams (7) who stated that such concentrations "may introduce artifactual interactions or may obscure physiologically important phenomena." The important facts are that a) heart mitochondria readily oxidize μM concentrations of pyruvate, α-KG, and short chain fatty acids without addition of exogenous dicarboxylic acids (3) and that b) freshly isolated brain mitochondria readily oxidize pyruvate in the μM region without addition of malate. Oxidation of succinate, and glutamate at high Q_{O_2} values requires mM concentrations of these substrates. Permeability undoubtedly is crucial, but whether permeases are involved is not pertinent to the discussion here. In addition, most studies on permeases have also been made at substrate concentration ⟩ 1 mM.

succinate or its compartmentation is as important as its average concentration. Jones and Gutfreund (9) presented evidence for the existence of enzyme-substrate or enzyme-product complexes. They concluded that the effective concentration of intramitochondrial succinate far exceeds the overall steady state concentration of this substrate. Substrate-enzyme complexes would also explain the failure of washing to remove endogenous mitochondrial substrates and the high activity of a given concentration of intra- as compared to extramitochondrial dicarboxylic acids.

Methods

Mitochondria were isolated from rabbit heart and brain as previously described (3, 4). Radioactive substrates were obtained from the New England Nuclear Corporation. Polarographic measurements were made in an Oxygraph (Gilson Medical Electronics) using a Clark type cell. The medium contained KCl, 150 mM; $MgCl_2$, 4 mM; phosphate, 5 mM and the appropriate substrate, usually in the range of 150-250 μM.

To determine the partition of substrates between intra- and extramitochondrial water, 1 ml of reaction mixture from the polarographic cell was rapidly filtered on a Millipore filter as described by McElroy et al. (10). The medium was collected in $HClO_4$ (0.05 ml of 70% per ml of reaction mixture). As rapidly as possible the Millipore filters were transferred to 1 ml of 0.6 M $HClO_4$ for each ml of reaction mixture filtered. Citric acid cycle components were separated by column chromatography on Dowex 1 X-8 (Cl⁻) columns as previously described (11) and counted in a Packard Tri-Carb scintillation counter using 0.5 ml of a given fraction added to 15 ml dioxane (scintillation grade) containing 67 ml/l of a solution of 52.5 g of 2(4-*tert*-butylphenyl)-5-(4-biphenylyl)-1,3,4-oxadiazole [Butyl PBD] in 1 l of toluene. Results are expressed as dpm/ml of medium or per ml of mitochondrial water. In the latter case the results were corrected for a 5% occlusion of medium on the Millipore filter.† Calculation of dpm/ml of mitochondrial water was made using a value of 5 μl water per mg mitochondrial protein which is larger than reported values; 3 μl/mg (12) and 1.5 μl/mg (13). Mitochondrial water content will vary with conditions; the value of 5 μl/mg is likely to yield concentration values for [^{14}C] intermediates which are too low rather than too high.

---

†Average of duplicate experiments for each of four different substrates run as controls using the procedure described but without mitochondria. McElroy et al (10) reported an occlusion factor of 2.3% for extramitochondrial fluid.

Oxidation of biogenic amines (dopamine, serotonin) and related compounds (tyramine, tryptamine) was measured using [^{14}C]-labeled compounds and separating the oxidation products. Dowex 50 (H$^+$) was used for dopamine and tyramine while extraction of the acidic solution with ether was used for dopamine, tyramine, and serotonin. For tryptamine, toluene extraction was used. Kynuramine oxidation was measured using the fluorometric procedure of Kraml (15).

Results and discussion

Requirement for exogenous malate by isolated mitochondria from brain

When mitochondria respire in a medium containing [2-or 3-^{14}C]-pyruvate (100-250 µM) or [1-^{14}C]-acetate (250 µM), radioactive citric acid cycle substrates are found in the medium and in the mitochondria (16). In the case of brain, malate usually is required for a maximum Q_{O_2} and respiratory control.

Fig. 1 illustrates a series of consecutive experiments with a single suspension of rabbit brain mitochondria showing how rapidly the respiratory control ratio and state 3 Q_{O_2} may decline with the time of storage at 0° when pyruvate is used as sole substrate. The mitochondrial suspension was kept in ice and the total elapsed time for the experiments was less than 2 hours. The effect is similar to, but far more rapid than that reported previously (17) for heart mitochondria stored at 0°.

Rapid loss of the state 3 rate and respiratory control ratio of brain mitochondria with pyruvate but without malate is also shown by the polarographic tracings of Figs. 2 and 3. The initial trace (Fig. 2A) for brain mitochondria gave a respiratory control ratio (RCR) of 3.9 using 267 µM [3-^{14}C]-pyruvate without malate and of 4.3 with added malate, 160 µM (Fig. 2B). Heart mitochondria prepared at the same time yielded a RCR of 10 without malate (Fig. 3C). However, 75 min after the first experiment, the same suspension of brain mitochondria gave a respiratory control ratio without malate of only 2.8 while with malate the RCR was 6.0 (Fig. 3D). The Q_{O_2} values for state 3 were (2A) 1.24, (2B) 1.40, (3C) 2.16, (3D) 0.80 and 1.30 µM O$_2$ sec^{-1} g protein $^{-1}$ liter^{-1} respectively.

For mitochondria from the brain, there appears to be a rough correlation between Q_{O_2} and the amount of α-ketoglutarate which accumulates (Fig. 4). Points most displaced above the line were obtained soon after isolation of the mitochondria when they are highest in endogenous substrate. Note that the data from mitochondria of heart (triangles) fall in a completely different area. As shown by LaNoue et al. (18) higher α-KG values tend to be found in reaction medium to which a higher malate concentration was added.

Table 1 shows that malate increases the absolute amount of α-KG as well as its [^{14}C] content. Although these increases occur with mitochondria from heart, with mitochondria from brain there is a greater increase in both the amount and radioactivity.

Baláz et al. (19) in 1962 reported that α-KG accumulates during oxidation of pyruvate plus malate by mitochondria from rat, pigeon and goldfish brain. Accumulation of α-KG by mitochondria from other tissues was much smaller. The K_m for pyruvate was found to be 10-fold lower than that for α-KG, leading to an apparent competition between the two substrates. They proposed that *in vivo* α-KG is converted to glutamate via transaminases. We have suggested (4) that the citric acid cycle of brain mitochondria may operate faster between malate and α-KG than between α-KG and malate.

If transamination is the route of glutamate formation from α-KG, a source of aspartate and/or alanine is required. A reductive amination of α-KG coupled to oxidation of NADH formed during glycolysis would help to explain the close association of glycolysis and respiration and the high content of glutamate in brain. As is well known, the brain is very dependent upon glucose and oxygen. Twenty percent of the basal oxygen requirement of a human being is used by the brain, while 15% of the total protein of a whole rat brain and 25% of the cerebral cortical protein is mitochondrial (20).

Fig. 5 illustrates possible modes of coupling of glycolysis and respiration in brain. NADH formed in the glyceraldehyde-3-phosphate dehydrogenase reaction may be reoxidized by reduction of pyruvate to lactate or by reduction of oxalacetate (formed via pyruvate carboxylase) to malate. The sum of the concentrations (21) of glutamate plus glutamine plus GABA (16 mM) in brain far exceeds the concentration of lactate (4 mM) or of malate plus aspartate (4 mM). Since glutamate dehydrogenase is entirely or almost entirely mitochondrial (22), reductive amination of α-ketoglutarate to glutamate would seem to be less likely. If reductive amination occurs within mitochondria, (a) oxidation of intramitochondrial NADH (without coupling of glycolysis and respiration) could occur, (b) there must be a route for transfer of electrons from extramitochondrial NADH to intramitochondrial NADH (or NADPH) or (c) mitochondrial glutamate dehydrogenase must be able to react with cytoplasmic NADH. Preliminary experiments suggest that the latter is unlikely. Further study of the factors controlling glutamate formation in brain is warranted. It should be noted that only two compartments are shown in Fig. 5, while Garfinkel (21) has found it necessary to utilize four. In addition the conversion of glutamate to aspartate via GABA is not considered here.

Intramitochondrial citric acid cycle substrates

We have been studying substrate compartmentation in both heart and brain mitochondria. In contrast to those from rat and pigeon heart, mitochondria from rabbit heart contain substantial amounts of endogenous citric acid cycle substrates. To date, most mitochondrial preparations from rabbit brain we have studied have contained relatively small amounts of these compounds. In previous collaborative work with Dr. Jennings of Northwestern University, we noted that isolated dog heart mitochondria are similar to those from rabbit heart in having high endogenous citric acid cycle substrates [unpublished; *cf.* (23) and (24)].

Polarographic data for experiments on which analyses for cycle intermediates were obtained are shown in Table 2. The data of Table 3 indicate considerably lower amounts of intramitochondrial citric acid cycle components in brain than in heart. We believe these components are lost more readily from isolated brain mitochondria than from heart mitochondria.

In the case of heart, for a substrate added at a specific activity of 3.0 μCi/μmole(6.6 $\times$ 10^6 dpm/μmole), a cycle component present with an activity of 3.3 $\times$ 10^6 dpm/ml of mitochondrial water will have an intramitochondrial concentration of 500 μM (if intramitochondrial water is not compartmented and if no dilution by non-radioactive components occurs). If the medium from which the mitochondria were isolated contained 3.3 $\times$ 10^3 dpm/ml of the same component, its concentration would be 3.3 $\times$ 10^3/6.6 $\times$ 10^6 = 0.5 $\times$ 10^{-3} μmole/ml or 0.5 μM. These data indicate a 10^3-fold difference in concentration of this component in the intra- and extramitochondrial water. Obviously, determination of the concentration on the entire reaction mixture would lead to an erroneously low prediction of the reaction rate at the time the sample was taken.

Nonequilibrium malate:fumarate ratios

We have observed both in brain and heart mitochondrial systems (4) a high malate to fumarate ratio (between 5:1 and 20:1). Similar findings have been reported by LaNoue *et al.* (18) with rat heart mitochondria and by Goldberg *et al.* (25) for intact mouse brain. Penner and Cohen (26) have reported the inhibition of fumarase by ATP and proposed that this inhibition may regulate substrate levels and thus serve as a control site in the citric acid cycle. Although ATP inhibits the rate of the fumarase reaction, as expected, we find no change in the equilibrium value. Since in the case of heart mitochondria, malate must arise from fumarate, it is difficult to see how ATP inhibition of fumarase could yield high malate/fumarate ratios. Krebs and

Eggleston (27) reported a malate:fumarate ratio of 1.1 in minced sheep heart while at approximately physiological conditions, the equilibrium value is 4.3 (28). When fumarate (100 μM) was incubated with fresh heart or brain mitochondria for 25 minutes, ratios of 4.5 and 4.1, respectively, were obtained. Likewise, when crystalline fumarase was incubated at pH 7.4 with radioactive malate, a malate:fumarate ratio of 4.35 was observed in confirmation of the data reported by Bock and Alberty (28). In experiments with [2,3-^{3}H]-succinate (580 μM) and rotenone (5 μM), the malate:fumarate ratios observed by fluorometric assay were 8, 10, 7.5, and 8 for 2 and 4 min in state 4 followed by 2 and 4 min in state 3, respectively. The reason for these high ratios under various conditions requires further study.

Monoamine oxidase (MAO) of intact mitochondria

Biogenic amines are important chemical neuro-transmitters at the synapses of the central nervous system. They are released from storage granules or synaptic vesicles and removed either by re-uptake mechanisms or by oxidative removal catalyzed by MAO present almost entirely in the mitochondria.

We are interested in this process since the K_m for oxygen of this enzyme has been reported to have values ranging from 125-700 μM (29-31), while that for Krebs cycle oxidations (32) is far lower (1 μM). Our goal is to determine if there is (a) any change in MAO activity as a result of changes in metabolic state or (b) if the nature of the products change with metabolic state. A teleological reason for location of monoamine oxidase in the mitochondria is not obvious. Data on control processes for this enzyme may aid in elucidating areas important in the control of central nervous system activity.

In spite of the fact that oxygen may be a rate limiting component for monoamine oxidase, many investigators have used air as the gas phase in the determination of the activity of this enzyme. Table 4 shows the effect of oxygen concentration on reaction rate. Although the P_{O_2} does not affect the velocity of kynuramine oxidation, pure oxygen increases the rate of oxidation for serotonin, dopamine, tyramine and tryptamine by factors of 1.3-2.1 (33, 34).

Using a single preparation of mitochondria from rabbit brain and [^{14}C]-labeled substrates and measuring the rate at which neutral and acidic radioactive products are formed from substrates at a concentration of 0.5 mM, we have observed activity ratios for tyramine, dopamine, tryptamine and serotonin of 1.0, 0.7, 0.5 and 0.3, respectively. Using dopamine as a substrate, this preparation had an activity of 3.4 mU per mg mitochondrial protein.

Current studies are being directed towards the chemical synthesis of the aldehydes of high purity corresponding to the MAO oxidation products of the biogenic amines. Development of more suitable methods for the study of possible interactions between citric acid cycle and biogenic amine oxidations will be greatly facilitated by the availability of these compounds.

It is important to remember that isolated mitochondria have properties that change with time and do not (except for very short time intervals) behave as open steady state systems. Feedback processes *in vivo* result in interactions between multienzyme systems and maintain metabolic control. The investigator is usually on the horns of a dilemma: when too many systems are present and interacting, adequate interpretation of analytical data is difficult and often impossible; when the systems are separated to permit easier interpretations of data, important interaction sites may be lost, leading to behavior not characteristic of that which exists within living cells, tissues or organs.

Summary

Citric acid cycle substrate compartmentation has been studied in mitochondria isolated from rabbit heart and brain. While mitochondria from heart contain adequate endogenous substrates to support a maximum respiratory rate without added dicarboxylic acid, mitochondria from brain rapidly develop a requirement for malate. This requirement by mitochondria from brain may be due to α-ketoglutarate formation which enters the medium. Heart mitochondria retain larger amounts of citric acid cycle intermediates during metabolism of pyruvate than do mitochondria from brain. In some cases the concentration of an intramitochondrial substrate is in the mM range, while that of the same substrate in the medium is in the μM range. Production of α-ketoglutarate by brain mitochondria (possibly by coupling of glycolysis and respiration via reductive amination of α-ketoglutarate with cytoplasmic NADH) is suggested as a source of the high glutamate content of brain. Ratios of malate to fumarate are observed to be much higher than equilibrium values. Observations on the monoamine oxidase activity of brain mitochondria and possible competition for available oxygen by the electron transport chain due to citric acid cycle operation is discussed.

References

1. Srere, P.A. Some complexities of metabolic regulation. Biochem. Med. 3:61(1969).
2. Scriven, L.E. Intracellular transport analysis. Eleventh Internatl. Cong. for Cell Biol., Brown Univ., Providence, R.I. (Sept., 1964).
3. Von Korff, R.W. Metabolic characteristics of isolated rabbit heart mitochondria. J. Biol. Chem. 240:1351(1965).

4. Von Korff, R.W., S. Steinman and A.S. Welch. Metabolic characteristics of mitochondria isolated from rabbit brain. J. Neurochem. 18:1577(1971).
5. Nicklas, W.J., J.B. Clark and J.R. Williamson. The metabolism of rat brain mitochondria: Studies on the K^+-stimulated oxidation of pyruvate. Biochem. J. 123:83(1971).
6. Chance, B. Quantitative aspects of the control of oxygen utilization. In: G.E.W. Wolstenholme and C.M. O'Connor (Editors), Regulation of cell metabolism, Ciba Foundation Symposium, Little Brown & Co., Boston (1959), p. 91.
7. Williams, G.R. Dynamic aspects of the tricarboxylic acid cycle in isolated mitochondria. Can. J. Biochem. 43:603(1965).
8. Chance, B. and J. Hollunger. The interaction of energy and electron transfer reactions in mitochondria III. Substrate requirements for pyridine nucleotide reduction in mitochondria. J. Biol. Chem. 236:1555(1961).
9. Jones, E.A. and H. Gutfreund. The kinetic behavior of enzymes in organized systems: The effective concentrations of succinate in mitochondria. Biochem. J. 91:1c(1964).
10. McElroy, F.A., G.S. Wong and G.R. Williams. Distribution of malate across the mitochondrial membrane as a significant factor in respiratory control. Arch. Biochem. Biophys. 128:563(1968).
11. Von Korff, R.W. Ion-exchange chromatography of citric acid cycle components and related compounds. In: J.M. Lowenstein (Editor), Methods of enzymology. Academic Press, New York (1969), p. 425.
12. Harris, E.J. and K. Van Dam. Changes of total water and sucrose space accompanying induced ion uptake or phosphate swelling of rat liver mitochondria. Biochem. J. 106:759(1968).
13. O'Brien, R.L. and G. Brierley. Compartmentation of heart mitochondria I. Permeability of isolated beef heart mitochondria. J. Biol. Chem. 240:4527(1965).
14. Von Korff, R.W., M. Jain and F.L. Sands. Some characteristics of monoamine oxidase (MAO) of intact brain mitochondria. Trans. Am. Soc. Neurochem. 2:116(1971).
15. Kraml, M. A rapid fluorimetric determination of monoamine oxidase. Biochem. Pharmacol. 14:1684(1965).
16. Olson, M.S. and R.W. Von Korff. Changes in endogenous substrates of isolated rabbit heart mitochondria during storage. J. Biol. Chem. 242:325(1967).
17. Olson, M.S. and R.W. Von Korff. The effect of depletion of endogenous substrates on the metabolic behavior of isolated rabbit heart mitochondria. J. Biol. Chem. 242:333(1967).
18. LaNoue, K., W.J. Nicklas and J.R. Williamson. Control of citric acid cycle activity in rat heart mitochondria. J. Biol. Chem. 245:102(1970).
19. Balázs, R., K. Magyar and D. Richter. The operation of the tricarboxylic acid cycle in brain mitochondrial preparations. Comparative neurochemistry (Proc. 5th Int. Neurochem. Symp.), Pergamon Press (1964), p. 225.
20. Abood, L.G. Brain mitochondria. In: Abel Lajtha (Editor), Handbook of neurochemistry, Vol. II: Structural neurochemistry, Plenum Press (1969), p. 303.
21. Garfinkel, D. A simulation study of brain compartments I. Fuel sources and GABA metabolism. Brain Res. 23:333(1967).
22. Balázs, R., D. Dahl and J.R. Harwood. Subcellular distribution of enzymes of glutamate metabolism in rat brain. J. Neurochem. 13:897(1966).
23. Jennings, R.B., P.B. Herdson and M.L. Hill. Pyruvate metabolism in mitochondria isolated from dog myocardium. Lab. Invest. 20:537(1969).

24. Jennings, R.B., P.B. Herdson and H.M. Sommers. Structural and functional abnormalities in mitochondria isolated from ischemic dog myocardium. Lab. Invest. 20:548(1969).
25. Goldberg, N.D., J.V. Passonneau and O.H. Lowry. Effect of changes in brain metabolism on the levels of citric acid cycle intermediates. J. Biol. Chem. 241:3997(1966).
26. Penner, P.E. and L.H. Cohen. Effect of adenosine triphosphate and magnesium ions on the fumarase reaction. J. Biol. Chem. 244:1070(1969).
27. Krebs, H.A. and L.V. Eggleston. Metabolism of acetoacetate in animal tissue. Biochem. J. 39:408(1945).
28. Bock, R.M. and R.A. Alberty. Studies of the enzyme fumarase I. Kinetics and equilibrium. J. Am. Chem. Soc. 75:1921(1953).
29. Tipton, K.F. The reaction pathways of pig brain monoamine oxidase. Eur. J. Biochem. 5:316(1968).
30. Oi, S., K. Shimada, M. Inamasu and K.T. Yasunobo. Mechanistic studies of beef liver mitochondrial amine oxidase. Arch. Biochem. Biophys. 139:28(1970).
31. Zubrzycki, Z. and H. Staudinger. Kinetik, intrazellulare Lokalisation and Induzierbarkeit der Monoamineoxydase. Hoppe-Seyler's Z. physiol. Chem. 348:639 (1967)
32. Chance, B. Cellular oxygen requirements. Fed. Proc. 16:617(1957).
33. Von Korff, R.W., M. Jain and F.L. Sands. Some characteristics of monoamine oxidase of intact brain mitochondria. Thrans. Am. Soc. Neurochem. 2:116(1971).
34. Von Korff, R.W., M.L. Jain, F.L. Sands and S. Steinman. Problems in the assay of monoamine oxidase activity. Abs. 3rd Meeting Int. Soc. Neurochem., Budapest, Hungary (1971), p. 64.

This study was supported by a grant from NINDS (Administered by Friends Medical Science Research Center) and by the Maryland Department of Health and Mental Hygiene. The collaboration and help of Dr. M. Jain, Sondra Steinman, Freeman Sands, Jane Von Korff, Tom Mayer and the lettering of figures by Mr. William Lewis are gratefully acknowledged.

TABLE 1

EFFECT OF MALATE ON α-KG FORMATION FROM PYRUVATE BY MITOCHONDRIA FROM RABBIT BRAIN AND HEART

	Mitochondria from			
	Brain		Heart	
$[2-^{14}C]$-Pyruvate (μM)	180	180	180	180
Malate (μM)	0	100	0	100
Q_{O_2}	0.21	1.2	1.8	1.9
α-KG (nmole/ml)	7	22	8	14
10^{-3} x dpm/ml	12	111	8	42
Specific radioactivity ratio: α-KG isolated/pyruvate added	0.30	0.84	0.13	0.53

See text for specific experimental conditions.

TABLE 2

POLAROGRAPHIC DATA FOR RABBIT HEART AND BRAIN MITOCHONDRIA RESPIRATIONS PRIOR TO SEPARATION OF MITOCHONDRIA AND MEDIA FOR COMPARTMENTATION MEASUREMENTS

Substrate	Malate added (μM)	Protein (mg/ml)	State	ΔO_2 (μM)	ΔO_2 ($\mu M\ O_2\ sec^{-1}\ g\ prot^{-1}\ liter^{-1}$)	Time (min)	Respiratory Control Ratio
Heart							
$3-[^{14}C]$ pyruvate	0.0	1.2	3	200	2.84	1.3	
	0.0	1.2	4	48	0.25	1.8	11.4
	0.0	1.4	3+4	180	2.04^a 0.24^b	2.0	8.4
$1-[^{14}C]$ acetate	0.0	1.5	3	200	1.50	3.0	7.1
	0.0	1.5	4	72	0.21	2.7	
Brain							
$3-[^{14}C]$ pyruvate	400	1.4	3+4	192	1.44^a 0.58^b	5.0	2.8

[a] State 3
[b] State 4

TABLE 3

COMPARTMENTATION OF TCA CYCLE SUBSTRATES BETWEEN MITOCHONDRIA AND MEDIUM

	Medium							Mitochondria[a]							
	Heart						Brain	Heart							Brain
	Pyruvate-3		Acetate-1			Pyr-3 +Malate	Pyruvate-3			Acetate-1			Pyr-3 +Malate
^{14}C-Substrate															
State	3[b]	4[b]	3+4[c]	3[d]	4[d]	3+4[c]	3[b]	4[b]	3+4[c]	3[d]	4[d]	3+4[c]
	$10^{-3} \times dpm/ml$							$10^{-3} \times dpm/ml\ H_2O$							
TCA Cycle Component															
ALA	16	25	43	0	0	9	2600	3900	2400	0	0	70
GLU	22	18	52	25	18	30	3300	3700	1200	3100	2700	360
ASP	3	0	4	0	0	3	3300	0	3500		160	120
SUC	26	9	26	14	12	68	1400	1050	800	120	140	90
MAL	25	18	63	10	25	180	6700	1300	2800	260	160	1700
CIT	n.d	n.d	28	0	0	87	1100	4300	145	0	180	900
α-KG	n.d	n.d	38	-	-	230	- -	- -	- -	- -	- -	- -
FUM	n.d	n.d	8	-	-	5	- -	- -	- -	- -	- -	- -
ACE	38	40	61	320	470	94	150	1100	- -	- -	0	- -
BOH	20	48	30	22	64	-	2400			800	40	

n.d = not determined

[a] $10^{-3} \times$ dpm measured/ml sample = 5 (mg protein/ml) $\times$ dpm/ml mitochondrial H_2O) + 0.05 dpm/ml medium
Specific activities of added substrate, μCi/μ Mole; [b], 3.1; [c], 2.1; [d], 3.0

TABLE 4

EFFECT OF [O_2] ON MAO ACTIVITY OF INTACT MITOCHONDRIA

Rabbit	Substrate	(mM)	Oxygen (mM)		Activity Ratio
			1.15	0.23	
			MAO Activity		(O_2/Air)
			mU/mg Protein		
Liver	Kynuramine	0.075	3.4	3.5	1.0
Brain	Kynuramine	0.075	0.5	0.5	1.0
	Serotonin	0.5	1.0	0.8	1.3
Liver	Dopamine	0.5	9.6	6.4	1.5
Brain	Dopamine	0.5	3.5	2.4	1.5
	Tyramine	0.5	3.8	2.3	1.7
	Tryptamine	0.5	1.9	0.9	2.1

See text for specific experimental conditions.

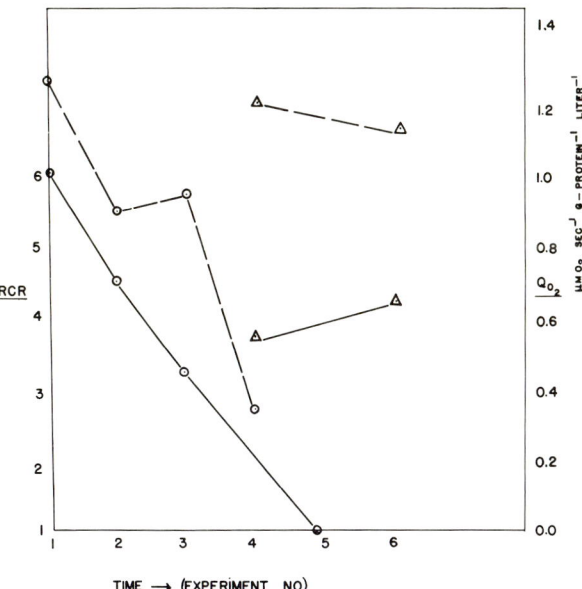

Fig. 1. *Loss of respiratory control ratio and state 3 Q_{O_2} of rabbit brain mitochondria during storage.* Solid lines, respiratory control ratios; dashed lines, Q_{O_2}; circles, pyruvate (267 μM), the sole added substrate; triangles, pyruvate (267 μM) plus malate (400 μM). Respiratory control and state 3 Q_{O_2} rapidly decreased unless malate was added as a source of dicarboxylic acid. The abcissa represents time after isolation and the numbers represent successive polarographic experiments, each requiring about 15 min. Experiment No. 6 is equivalent to 90 min.

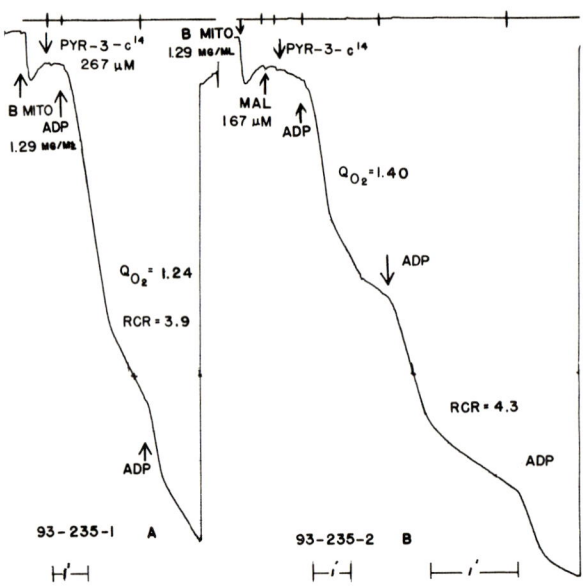

Fig. 2. *Polarographic tracings for rabbit brain mitochondria.* A: first experiment after isolation of mitochondria, pyruvate as substrate. B: second experiment after isolation of mitochondria, pyruvate plus malate as substrates.

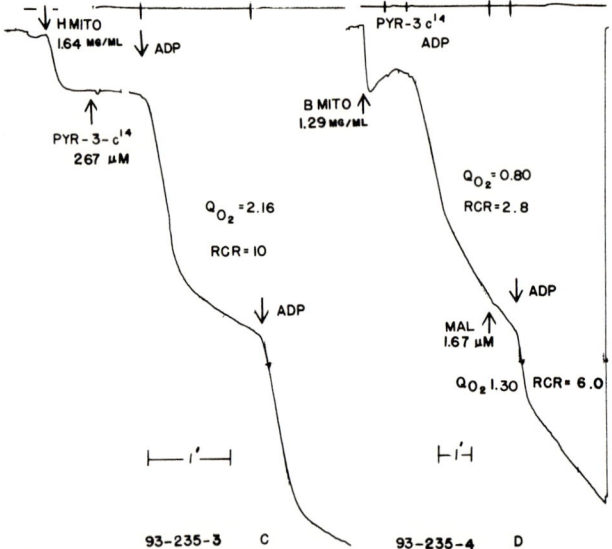

Fig. 3. *Polarographic tracings for mitochondria from rabbit heart.* (C) pyruvate as substrate; (D) rabbit brain, third experiment after isolation (1 hour), pyruvate initially the sole substrate, malate added 200 sec later.

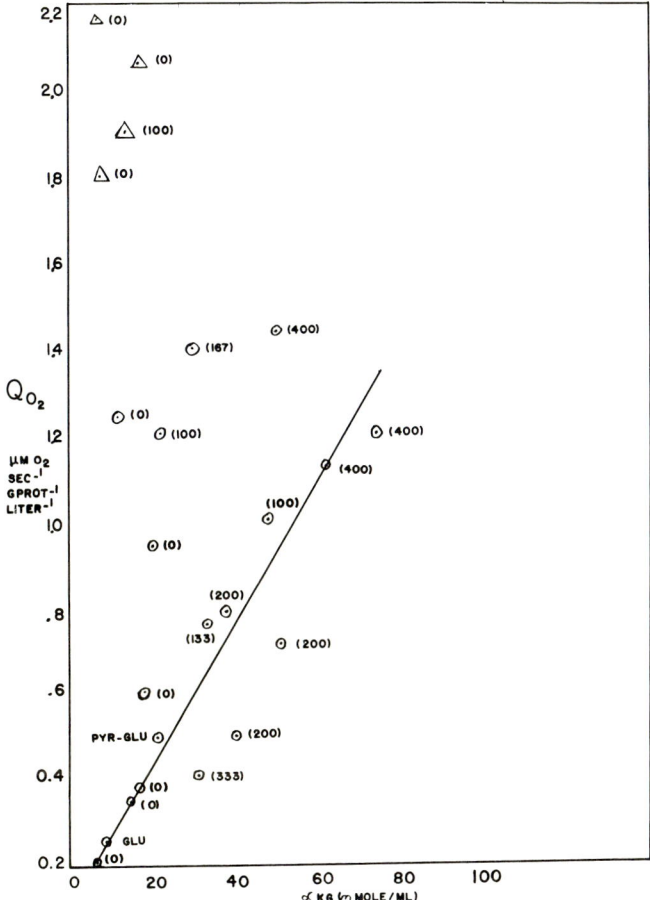

Fig. 4. *Effect of malate additions on state 3 Q_{O_2} and α-ketoglutarate formation by mitochondria from rabbit brain (circles) and heart (triangles).* Numbers in parentheses represent µM malate added.

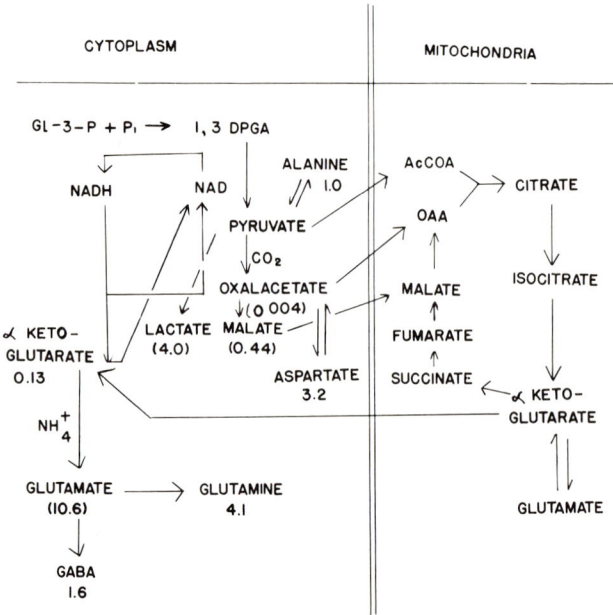

Fig. 5. *Possible interactions of cytoplasmic and mitochondrial processes in brain.*

THE SIGNIFICANCE OF MITOCHONDRIAL PHOSPHOENOLPYRUVATE FORMATION IN THE REGULATION OF GLUCONEOGENESIS IN GUINEA PIG LIVER

Alan J. Garber, F.J. Ballard and Richard W. Hanson

Introduction

Gluconeogenesis in mammalian liver involves an integrated series of reactions requiring the participation of enzymes located in both the cytosol and the mitochondria. Although the characteristic enzymatic reactions which constitute the overall process are known, the detailed regulation of carbon flow over this pathway has remained elusive. Most of the studies in this area have been carried out using the isolated, perfused rat liver or rat kidney cortex slices as model systems and have been based on the extensive literature of the properties of isolated mitochondria and purified enzymes from rat tissues. Far less information is available, however, concerning the regulation of gluconeogenesis in species other than the rat in spite of the fact that in many species, certain basic variations in enzyme distribution and in metabolic responses to dietary and hormonal alterations have been observed. In this paper we will present our own studies on the mitochondrial formation of precursors for gluconeogenesis in guinea pig liver and attempt to formulate from them a model for the regulation of gluconeogenesis in this species which may be applicable to other species having a mitochondrial form of P-enolpyruvate carboxykinase. This model will be supported by certain preliminary studies with the isolated perfused guinea pig liver.

Results

The pathway of gluconeogenesis in rat liver.

The proposed pathway of gluconeogenesis in rat liver is shown in Fig. 1. This abbreviated scheme stresses the importance of those mitochondrial reactions which may provide intermediates for cytosolic oxalacetate formation. In the rat, P-enolpyruvate carboxykinase is found almost totally in the cytosol

(Table 1), whereas pyruvate carboxylase is a mitochondrial enzyme (1, 2). Gluconeogenic substrates such as alanine and lactate must first be converted in the cytosol to pyruvate which is then carboxylated to oxalacetate within the mitochondria. Since oxalacetate itself does not appear to translocate rapidly from the mitochondrial matrix (3), it has been proposed that either malate or aspartate, or both, may function as translocating anions (4) which transport the oxalacetate carbon skeleton from the matrix space in order to continue the gluconeogenic sequence in the cytosol. Aspartate may then be reconverted to oxalacetate via cytosolic aspartate aminotransferase, whereas malate regenerates oxalacetate in the cytosol by the action of NAD^+-malate dehydrogenase. The latter reaction has the advantage of supplying cytosolic NADH, if necessary, for the reversal of the triosephosphate dehydrogenase step during gluconeogenesis. Furthermore, the intramitochondrial formation of malate from pyruvate proceeds by the direct reversal of NAD^+-malate dehydrogenase rather than by forward flow through the citric acid cycle. This hypothesis is based upon studies involving the incubation of isolated rat liver mitochondria with pyruvate and radioactive bicarbonate, which resulted in the formation of large quantities of labeled malate (5, 6). Also, if either ammonia or glutamate is included in the incubation medium, aspartate formation is greatly increased, whereas malate synthesis correspondingly decreases (5). Recent studies on the mechanism of gluconeogenesis have used the inhibitors aminoxyacetic acid and D-malate (7). Aminoxyacetic acid inhibits aspartate aminotransferase (7, 8) and was shown to block gluconeogenesis from lactate but not pyruvate. D-malate, on the other hand, inhibited gluconeogenesis from pyruvate by rat kidney cortex slices, presumably by blocking the cytosolic reoxidation of malate. These studies suggest that aspartate and malate do function as carriers for oxalacetate, but they are not conclusive since inhibitors such as aminoxyacetate usually have more than one site of action.

The intracellular distribution of P-enolpyruvate carboxykinase in livers of various species.

The intracellular distribution of hepatic P-enolpyruvate carboxykinase varies widely in mammalian species (Table 1). Of all the animals studied to date, the rat, hamster, and mouse are the only species having a predominately cytosolic enzyme (9). The livers of rabbits and several species of birds have an almost exclusively mitochondrial P-enolpyruvate carboxykinase, whereas widely diverse species such as the guinea pig, cow, cat, and man have significant activities of this enzyme in both the cytosol and mitochondria. The cytosolic enzyme is adaptive, increasing 2 to 5-fold after 48 hrs of starvation, but the activity of the mitochondrial form of P-enolpyruvate carboxykinase is unaltered under these conditions in any species studied. Immunochemical studies using purified antibodies to rat liver cytosolic P-enolpyruvate carboxykinase indicate major antigenic differences between the cytosolic and mitochondrial forms of the enzyme (10). Based simply on the intracellular distribution of P-enolpyruvate carboxykinase in the liver of various species, it is probable that the mechanisms of gluconeogenesis developed from studies of the rat may not be applicable to other animals. It is likely that in avian liver, for example, all of the P-enolpyruvate generated for glucose synthesis is formed in the mitochondria and then translocated to the cytosol. The situation is less clear in species such as man and the guinea pig, which contain P-enolpyruvate carboxykinase in both compartments of the liver cell. The following discussion will deal primarily with the metabolic role of the mitochondrial form of this enzyme in the guinea pig liver, with special emphasis on the integration of P-enolpyruvate formation with the functioning of the citric acid cycle.

The substrate and energetic requirement for mitochondrial P-enolpyruvate formation.

Previous studies from our laboratory (11) have shown that substantial rates of P-enolpyruvate synthesis by isolated guinea pig liver mitochondria can be observed with oxalacetate, pyruvate, malate, fumarate and α-ketoglutarate as substrates when either ATP or ADP is also added. With oxalacetate or pyruvate, P-enolpyruvate formation was greater in the presence of added ATP, whereas ADP caused a marked stimulation of P-enolpyruvate synthesis from malate and fumarate. Both ATP and ADP were equally effective in increasing the rate of P-enolpyruvate formation from α-ketoglutarate. These studies also demonstrated that P-enolpyruvate formed by isolated guinea pig liver mitochondria can rapidly translocate across the mitochondrial membrane. In fact, using a Millipore filtration system, Garber and Ballard (11) reported a nearly quantitative transfer of P-enolpyruvate, with only a slight amount remaining within the mitochondria.

In species such as the guinea pig and rabbit, 2,4-dinitrophenol or oleic acid uncoupling was found to be a requisite for substantial rates of P-enolpyruvate synthesis by isolated mitochondria (12, 13, 14). Also, under these conditions α-ketoglutarate alone or in combination with other intermediates was the preferred mitochondrial substrate. These observations have been explained on the basis of the GTP requirement for P-enolpyruvate carboxykinase, and it was proposed that P-enolpyruvate formation was stoichiometrically linked to the intramitochondrial metabolism of α-ketoglutarate (12). Our own studies (11, 15) have indicated that GTP may be formed by guinea pig liver mitochondria by the transphosphorylation of

GDP from the ATP pool by the action of nucleoside diphosphokinase as well as by the substrate level phosphorylation of GDP to GTP which accompanies the oxidation of α-ketoglutarate. As shown in Table 2, the addition of 2,4-dinitrophenol, an uncoupler which induces a marked ATPase activity in the mitochondria, depressed the synthesis of P-enolpyruvate from oxalacetate, malate and α-ketoglutarate. The addition of α-ketoglutarate to mitochondrial suspensions incubated with ADP and with oxalacetate or malate produced an increase in the amount of P-enolpyruvate formed. This effect was not evident with oxalacetate when ATP was the added nucleotide. In this case, dinitrophenol inhibited both ATP synthesis and P-enolpyruvate formation, an effect which may be ascribed to the reduced rate of nucleoside diphosphokinase-catalyzed formation of GTP. In no instance did the simultaneous oxidation of α-ketoglutarate relieve the dinitrophenol-induced inhibition of P-enolpyruvate formation. In all of these experiments, a good correlation existed between ATP generation and P-enolpyruvate synthesis. The kinetics of this relationship are illustrated in Fig. 2. When guinea pig liver mitochondria were incubated with malate, ATP and dinitrophenol, there was a linear, increased rate of malate oxidation and a rapid decline in the ATP concentration. As the ATP levels fell during the first 10 min. of incubation there was a decline followed by a complete cessation of P-enolpyruvate synthesis after 30 min. Throughout the course of this incubation the concentration of AMP increased due to the combined actions of adenylate kinase and the dinitrophenol-stimulated adenosine triphosphatase. A correlation between the mitochondrial levels of ATP and the observed concentration of GTP during a parallel series of incubations has also been demonstrated (15). In these experiments, dinitrophenol reduced the steady state level of GTP, and this reduction could not be overcome by the addition of α-ketoglutarate.

From these observations we have concluded that the pools of the various adenine and guanine nucleotides in guinea pig liver mitochondria rapidly equilibrate with one another; furthermore, the rate of reaction between nucleotides may be more rapid than the rate of GTP utilization by P-enolpyruvate carboxykinase. We have assumed that the formation of GTP is catalyzed by nucleoside diphosphokinase which has an activity of 5.4 units per g guinea pig liver and would be sufficiently active to account for the observed rates of P-enolpyruvate synthesis. Recent studies in our laboratory have shown that rabbit liver mitochondria are more dependent on substrate level phosphorylation of GDP than are guinea pig liver mitochondria. These studies are in essential agreement with those of other groups (13, 16) and indicate that uncoupling levels of dinitrophenol are necessary for maximal rates of P-enolpyruvate formation when α-ketoglutarate is present. We have attributed this difference between guinea pig and rabbit liver mitochondria to a lower activity of nucleoside diphosphokinase (1.4 units/g liver) in the rabbit. It is probable, therefore, that differences in the regulation of P-enolpyruvate synthesis exist even in species such as the rabbit and guinea pig, both of which have a substantial portion of the total hepatic P-enolpyruvate carboxykinase activity within the mitochondria.

The influence of oxidation-reduction potential on P-enolpyruvate formation.

Mitochondrial incubations with substrates such as malate and fumarate, which must be oxidized to form oxalacetate, yield greater rates of P-enolpyruvate formation when ADP rather than ATP is the added adenine nucleotide. A similar result was noted with dinitrophenol and hexokinase plus glucose (Table 3). This finding is somewhat paradoxical since compounds which provide the greatest rate of malate utilization yield the greatest initial rate of P-enolpyruvate synthesis regardless of nucleotide triphosphate availability. Uncoupled mitochondria had markedly increased adenosine triphosphatase activity and should have low intramitochondrial levels of ATP. We therefore expected decreased rates rather than the increased rates of P-enolpyruvate formation observed in these experiments. However, when non-oxidizable substrates such as oxalacetate and pyruvate were used in the incubation studies with isolated guinea pig liver mitochondria, there was an increased rate of P-enolpyruvate formation with ATP as compared to ADP addition (11). This finding suggests a dependence of P-enolpyruvate formation upon the rate of malate oxidation, in addition to the previously noted dependence of GTP formation. This point was clarified in the experiment shown in Fig. 3, in which succinate was added to guinea pig liver mitochondria incubated with both malate and ADP to produce NADH by reverse electron flow (17). In this system, P-enolpyruvate formation and malate oxidation were inhibited by increasing concentrations of succinate although the ATP concentration is maintained. It is therefore likely that the generation of increased levels of NADH reduced the rate of P-enolpyruvate synthesis by guinea pig liver mitochondria.

In order to study the control of P-enolpyruvate formation by guinea pig liver mitochondria an attempt was made to re-create *in vitro* the oxidation-reduction potentials noted in the *in vivo* condition. We first determined the alterations in the NAD^+ to NADH ratio occuring in the liver *in vivo* during the transition from the fed to the fasted state. Such a determination should provide insight into the changes in the oxidation-reduction potential which occur when the animal experiences the accelerated rate of gluconeogenesis characteristic of fasting (18). These measurements, using freeze clamped liver (Tables 4 and 5), suggest that certain fundamental differences in response to fasting exist between the guinea pig and the more widely studied rat liver system. The ATP:ADP ratio in the livers of fed guinea pigs was 5.52 and it declined only slightly during prolonged fasting (96 hrs). In addition, no significant change in the phosphate potential $[ATP]/[ADP] \cdot [P_i]$ of the livers was noted in animals fasted for either 48 or 96 hrs. Although the levels of lactate in the livers of fed and fasted guinea pigs were similar to those reported in the rat (19, 20), the concentration of pyruvate in guinea pig liver was lower, and as a consequence the lactate to pyruvate ratio was higher than in rat liver. The NAD^+ to NADH ratio in guinea pig liver is therefore considerably more reduced than in rat liver, and in the guinea pig no significant variation was noted upon fasting.

The oxidation-reduction potential of the mitochondrial nicotinamide adenine dinucleotides was calculated from both the β-hydroxybutyrate dehydrogenase and the glutamate dehydrogenase equilibria (19). The calculated mitochondrial oxidation-reduction potential of livers from both fed and fasted animals using either couple resulted in ratios of NAD^+ to NADH which approximately agree (Table 5). The most striking observation was the shift toward oxidation of the intramitochondrial nucleotides noted in the livers of fasted as compared to fed guinea pigs, which resulted in an increase in the mitochondrial ratio of NAD^+ to NADH. This contrasts with the reported finding for rat liver in which the oxidation-reduction potential shifts toward reduction during the transition from the fed to the fasted state (19). Mitochondria from rabbit liver also demonstrate an increased NAD^+ to NADH ratio after 48 hrs of fasting (Table 6). Since these two species both have a mitochondrial form of P-enolpyruvate carboxykinase, this may be a significant factor in the regulation of gluconeogenesis in these animals.

A number of metabolic and enzymatic factors may interact to produce this shift toward oxidation in the mitochondria of species containing a mitochondrial form of P-enolpyruvate carboxykinase. Although the exact causes are poorly understood at present, preliminary evidence suggests that, in addition to variations in the activity of the respiratory chain, alterations in the nature and relative amounts of the various hepatic substrates such as glutamate and pyruvate may act to shift the intramitochondrial oxidation-reduction potential. Furthermore, the presence of P-enolpyruvate carboxykinase within the matrix space allows a rapid regeneration of **ADP** in the mitochondria as well as providing a ready means of removing mitochondrial oxalacetate. The regeneration of **ADP** by the action of P-enolpyruvate carboxykinase and nucleoside diphosphokinase, together with an observed failure to remove oxalacetate when **ATP** levels are low (1) may cause a marked shift toward oxidation in the free pool of adenine nucleotides.

Unfortunately, these observations are not completely understood at present, and further work must be done, particularly on the possibility of multiple pools of oxalacetate within the matrix space.

A degree of competition for oxalacetate must exist in the mitochondria since the total activity of those enzymes removing oxalacetate (aspartate aminotransferase, NAD^+-malate dehydrogenase and P-enolpyruvate carboxykinase) is almost 100 times greater than the maximal activity of pyruvate carboxylase. This implies that there are effective controls on the removal of intramitochondrial oxalacetate which are as important to the mechanism of gluconeogenesis as factors regulating oxalacetate formation. In order to evaluate the metabolic significance of mitochondrial P-enolpyruvate formation, mitochondrial incubations were carried out using substrate concentrations as close to physiological as is possible experimentally. In these experiments with guinea pig liver mitochondria, the oxidation-reduction potential was established using varying amounts of β-hydroxybutyrate and acetoacetate and was altered to correspond to observed changes in the NAD^+:NADH ratio noted after fasting. The importance of this approach in any attempt to understand the regulation of gluconeogenesis in guinea pig liver mitochondria is well illustrated by the results shown in Fig. 4. Mitochondria were incubated for 10 min with 1 mM pyruvate at varying oxidation-reduction states and the formation of P-enolpyruvate, malate, aspartate and citrate was measured. At an NAD^+:NADH ratio close to that calculated for mitochondria of liver from fed guinea pigs (about 12.5) the synthesis of malate and aspartate predominated over that of P-enolpyruvate. However, if the oxidation-reduction potential was adjusted to that noted in mitochondria of freeze-clamped liver from fasted animals (about 33), the synthesis of P-enolpyruvate is greater than either malate or aspartate formation. The key to this approach is the use of either β-hydroxybutyrate or acetoacetate to shift the mitochondrial NAD^+ to NADH ratio without altering the level of acetyl CoA within the mitochondria. It is important to point out that the concentration of substrate chosen for these experiments can critically affect the intramitochondrial oxidation-reduction potential. If the pyruvate concentration in the incubation medium is raised to 5 mM, the rates of both malate and aspartate formation exceed those of P-enolpyruvate, presumably due to a decline in the intramitochondrial oxidation-reduction potential (21).

Experiments with a single substrate such as pyruvate are not of themselves sufficient for a careful analysis of many of the citric acid cycle interactions. For example, mitochondrial aspartate formation is a potentially important process for cytosolic oxalacetate generation to support gluconeogenesis in rat liver. Since aspartate aminotransferase is an equilibrium enzyme, the ratio of glutamate to α-ketoglutarate within the cell may determine the rate of oxalacetate transamination. The concentration of aspartate in freeze-clamped fed guinea pig liver was found to be about 0.94 mM (21). Using this concentration together with 1 mM α-ketoglutarate and 4 mM glutamate, the rates of P-enolpyruvate, malate and citrate formation from 1 mM pyruvate were studied at various ratios of NAD^+ to NADH (Fig. 5). In this, as in all other arrangements of substrates, only the production of P-enolpyruvate increased at more oxidized ratios of NAD^+ to NADH, whereas malate synthesis and pyruvate utilization and carboxylation were decreased.

Recent studies on the mechanism of gluconeogenesis in rat liver have suggested that with lactate as a substrate, aspartate may function as the translocating anion (4). Perfusion studies with isolated rat liver have indicated that both the cytosolic and mitochondrial oxidation-reduction potentials are more reduced with lactate as substrate as compared to pyruvate. In Figs. 4 and 5 we have shown that aspartate formation increased with decreased NAD^+:NADH ratios. Aspartate aminotransferase, which generates aspartate within the mitochondria, has a K_m for oxaloacetate that is greater than the steady state level of that intermediate (22). It is therefore unexpected that the formation of aspartate is increased at these NAD^+:NADH ratios. Rather, the oxalacetate concentration should decline owing to the NAD^+-malate dehydrogenase equilibrium which appears to regulate the steady state level of oxalacetate. This indicates that some other factor related to the oxidation-reduction potential must have controlling significance for aspartate formation. We have previously demonstrated that the ratio of glutamate to α-ketoglutarate varies inversely with the NAD^+ to NADH ratio (21). It is therefore likely that alterations in this ratio may in turn regulate the rate of aspartate formation. Further experimental proof for this suggestion is shown in Fig. 6, in which mitochondria from guinea pig liver were incubated with 1 mM pyruvate and varying ratios of glutamate to α-ketoglutarate. In these incubations the intramitochondrial oxidation-reduction potential was controlled by the addition of β-hydroxybutyrate. By this method two isopotential NAD^+:NADH ratios, 25±3 and 9±1, were established. It can be seen from Fig. 6 that the production of aspartate, but not malate, by guinea pig liver mitochondria was directly related to the ratio of glutamate to α-ketoglutarate. Furthermore, at any given ratio of glutamate to α-ketoglutarate the production of aspartate is greater at the more oxidized ratios of NAD^+:NADH, whereas malate production is always less. This suggests that aspartate aminotransferase and NAD^+-malate dehydrogenase compete for the same pool of oxalacetate. From these studies one may conclude that the relative rates of formation of the various gluconeogenic precursors are largely determined by alterations of the intramitochondrial oxidation-reduction potential.

Regulation of gluconeogenesis in guinea pig liver.

These observations using isolated guinea pig liver mitochondria allow us to formulate a model for the regulation of gluconeogenesis in this species which may then be tested in the isolated, perfused liver. It is readily apparent that in guinea pig liver, any factor which reduces the intramitochondrial oxidation-reduction potential should also reduce mitochondrial P-enolpyruvate synthesis and possibly lower the rate of gluconeogenesis. This suggestion is in striking contrast to observations derived from studies with rat liver mitochondria which show that a shift in the NAD^+ to NADH ratio toward reduction, as would be observed during fasting, favors the formation of malate and aspartate for gluconeogenesis. Fatty acids stimulate gluconeogenesis from lactate in perfused rat liver (23, 24). This stimulation has been attributed to either an acetyl-CoA dependent activation of pyruvate carboxylase (24) or to an increased supply of mitochondrial reducing equivalents (23), or both. Fatty acids have been shown to significantly inhibit glucose synthesis from lactate in perfused guinea pig liver (18). Since there are no apparent kinetic differences in pyruvate carboxylase from rat and guinea pig liver (18) or in the concentration of free CoA and acetyl-CoA in either species (18), the divergent effects of fatty acids must then be explained by alterations in the NAD^+ to NADH ratio within the mitochondria.

These observations on the divergent effects of fatty acids have been confirmed as noted in Figures 7 and 8. Using a constant, non-recycling perfusion of 2 mM lactate, a rate of glucose synthesis of 0.42 μmoles per min per g liver were observed (Fig. 7). The infusion of low levels of octanoate (0.1 and 0.2 mM) caused a marked reduction in gluconeogenesis. Upon cessation of fatty acid infusion, the rate of gluconeogenesis returned to normal control values. Livers from fasted rats, on the other hand, had an enhanced rate of gluconeogenesis after infusion with octanoate (Fig. 8).

In order to determine whether this inhibition of gluconeogenesis by fatty acids in the perfused guinea pig liver results from a shift in the mitochondrial oxidation-reduction potential, infusion studies were carried out using β-hydroxybutyrate. Since mammalian liver can only convert β-hydroxybutyrate to acetoacetate (25), thereby decreasing the mitochondrial NAD^+ to NADH ratio, this experiment should directly indicate whether a shift of the hepatic oxidation-reduction potential alters gluconeogenesis in the absence of other effects. As noted in Fig. 9, glucose synthesis from lactate in guinea pig liver is markedly decreased by simultaneous β-hydroxybutyrate infusion. In sharp contrast, glucose synthesis is stimulated by β-hydroxybutyrate infusion in the perfused rat liver (Fig. 10). Qualitatively the effects of fatty acids in both guinea pig and rat liver may be explained by an increased intramitochondrial availability of NADH as has been previously suggested for rat liver (23).

Conclusions

In this review we have shown that it is possible to predict, based on studies of enzyme distribution and mitochondrial function, potential regulatory interactions for gluconeogenesis in species containing a mitochondrial form of P-enolpyruvate carboxykinase. This model offers a basis for understanding the observations of Willms *et al.* (26) that glucagon does not stimulate gluconeogenesis from lactate in perfused guinea pig liver as it does in rat liver. This would suggest that in guinea pig liver the effects of glucagon may be mediated by a direct effect on fatty acid release, a mechanism already proposed for rat liver (23). Another example of the divergent mechanism of regulation in different species is the inhibition of hepatic gluconeogenesis by phenethylbiguanide (Phenformin®), an oral hypoglycemic agent. It has been known for some time that this compound is effective in lowering blood glucose in humans at far lower doses than in rats. Recently, Haeckel and Haeckel (27) have shown an inhibition of gluconeogenesis from lactate by phenethylbiguanide in guinea pig liver at concentrations significantly lower than that needed for inhibition of this process in rat liver. Biguanides are known inhibitors of the second site of oxidative phosphorylation (28) and as such could easily cause an elevation in mitochondrial NADH. If such a mechanism is related to the hypoglycemic action of this compound *in vivo*, it may be possible to understand this effect on the basis of the previously demonstrated controlling mechanisms of P-enolpyruvate formation in isolated mitochondria. We may therefore predict that phenethylbiguanide should reduce gluconeogenesis from lactate in guinea pig liver and presumably in human liver, by two possible mechanisms, both consistent with the model described previously. A drug-induced fall in the mitochondria ratio of NAD^+:NADH would act to depress P-enolpyruvate formation, and an associated decline in the intramitochondrial ATP and GTP pools would further inhibit gluconeogenesis.

A number of important questions remain unanswered from work on the regulation of metabolic processes in guinea pig liver. The physiological relationship between fatty acid metabolism and gluconeogenesis in guinea pig liver remains obscure. However, it may be helpful to note that the rate of ketogenesis in starving guinea pigs is only one-fourth the rate noted in starving rats (18). It is possible that fatty acid metabolism in the liver of the guinea pig is not as intimately related to gluconeogenesis as it is in rat liver. However, it should be emphasized that further studies are necessary before any definitive answers to this question are possible.

Another intriguing difference between the guinea pig and rat is the strikingly reduced oxidation-reduction potential of the guinea pig liver cytosol, which is unchanged by prolonged fasting (Table 4). This contrasts with the more oxidized cytosol of the rat liver. Calculated ratios of NAD^+ to NADH in the fed state give values of 725 (19). However, this declines after 48 hrs of fasting to 508 (19). This comparison suggests that the generation of NADH in the cytosol may not be limiting for gluconeogenesis in guinea pig liver.

Presented by Alan J. Garber. The experimental work reported in this paper has been supported by U.S.P.H.S. Grants CA-10916, HD-02758, AM-11279, F2-AM-40,463, CA-12227 and American Cancer Society Grant P202.

References

1. Bottger, I., O. Wieland, D. Brdiczka and D. Pette. Intracellular localization of pyruvate carboxylase and phosphoenolpyruvate carboxykinase in rat liver. Europ. J. Biochem. 8:113-119(1969).
2. Ballard, F.J., R.W. Hanson and L. Reshef. Immunochemical studies with soluble and mitochondrial pyruvate carboxylase activities from rat tissues. Biochem. J. 119:735-742(1970).
3. Haslem, J.M. and H.A. Krebs. The permeability of mitochondria to oxaloacetate and malate. Biochem. J. 107:659-667(1968).
4. Walter, P., V. Paetkau and H.A. Lardy. Paths of carbon in gluconeogenesis and lipogenesis III. The role and regulation of mitochondrial processes involved in supplying precursors of phosphoenolpyruvate. J. Biol. Chem. 241:2523-2532(1966).
5. Mehlman, M.A., P. Walter and H.A. Lardy. Paths of carbon in gluconeogenesis and lipogenesis VII. The synthesis of precursors for gluconeogenesis from pyruvate and bicarbonate by rat kidney mitochondria. J. Biol. Chem. 242:4594-4602(1967).
6. Haynes, R.C., Jr. The fixation of carbon dioxide by rat liver mitochondria and its relation to gluconeogenesis. J. Biol. Chem. 240:4103-4106(1965).
7. Rognstad, R. and J. Katz. Gluconeogenesis in the kidney cortex. Effects of D-malate and amino-oxyacetate. Biochem. J. 116:483-491(1970).
8. Hopper, S. and H.L. Segal. Kinetic studies of rat liver glutamic-alanine transaminase. J. Biol. Chem. 237:3189-3195(1962).
9. Nordlie, R.C. and H.A. Lardy. Mammalian liver phosphoenolpyruvate carboxykinase activities. J. Biol. Chem. 238:2259-2263(1963).
10. Ballard, F.J. and R.W. Hanson. Purification of phosphoenolpyruvate carboxykinase from the cytosol fraction of rat liver and the immunochemical demonstration of differences between this enzyme and the mitochondrial phosphoenolpyruvate carboxykinase. J. Biol. Chem. 244:5625-5630(1969).
11. Garber, A.J. and F.J. Ballard. Phosphoenolpyruvate synthesis and release by mitochondria from guinea pig liver. J. Biol. Chem. 244:4696-4703(1969).

12. Ishihara, N. and G. Kikuchi. Studies on the functional relationship between the phosphopyruvate synthesis and the substrate level phosphorylation in guinea-pig liver mitochondria. Biochim. Biophys. Acta 153:733-748(1968).
13. Davis, E.J. and D.M. Gibson. Regulation of the metabolism of rat liver mitochondria by long chain fatty acids and other uncouplers of oxidative phosphorylation. J. Biol. Chem. 244:161-170(1969).
14. Nordlie, R.C. and H.A. Lardy. The synthesis of phosphoenolpyruvate in liver mitochondria. Biochem. Z. 338:356-363(1963).
15. Garber, A.J. and F.J. Ballard. Regulation of phosphoenolpyruvate metabolism in mitochondria from guinea pig liver. J. Biol. Chem. 245:2229-2240(1970).
16. Gamble, S.L., Jr. and J.A. Mazar. Intramitochondrial metabolism of phosphoenolpyruvate. J. Biol. Chem. 242:67-72(1967).
17. Chance, B. and G. Hollunger. The interaction of energy and electron transfer reactions in mitochondria I. General properties and nature of the products of succinate-linked reduction of pyridine nucleotide. J. Biol. Chem. 236:1534-1543(1961).
18. Soling, H.D., B. Willms, J. Kleineke and M. Gehlhoff. Regulation of gluconeogenesis in the guinea pig liver. Europ. J. Biochem. 16:289-302(1970).
19. Williamson, D.H., P. Lund and H.A. Krebs. The redox state of free nicotinamide-adenine dinucleotide in the cytoplasm and mitochondria of rat liver. Biochem. J. 103:514-527(1967).
20. Lardy, H.A. On the direction of pyridine nucleotide oxidation-reduction reactions in gluconeogenesis and lipogenesis. In: B. Chance, R. Estabrook and J. Williamson (Editors), Control of energy metabolism, Academic Press, New York, pp. 245-248(1965).
21. Garber, A.J. and R.W. Hanson. The interrelationships of the various pathways forming gluconeogenic precursors in guinea pig liver mitochondria. The influence of the oxidation-reduction state of nicotinamide adenine dinucleotides on phosphoenolpyruvate, malate, and aspartate formation. J. Biol. Chem. 246:589-598(1971).
22. Henson, C.P. and W.W. Cleland. Kinetic studies of glutamic-oxalacetic transaminase isozymes. Biochem. 3:338-345(1964).
23. Struck, E., J. Ashmore and O. Wieland. Effects of glucagon and long chain fatty acids on glucose production by isolated perfused rat liver. Adv. Enz. Reg. 4:219-224(1966).
24. Williamson, J.R., R. Scholz and E.T. Browning. Control mechanisms of gluconeogenesis and ketosis II. Interactions between fatty acid oxidation and the citric acid cycle in perfused rat liver. J. Biol. Chem. 244:4617-4627(1969).
25. Krebs, H.A. The physiological role of the ketone bodies. Biochem. J. 80:225-233(1961).
26. Willms, B., J. Kleineke and H.D. Soling. The redox state of $NAD^+/NADH$ systems in guinea pig liver during increased fatty acid oxidation. Biochim. Biophys. Acta 215:438-448(1970).
27. Haeckel, R. and H. Haeckel. Interference of ethanol oxidation with gluconeogenesis in the perfused guinea pig liver. Biochem. 7:3803-3810(1963).
28. Chappell, J.B. The effect of alkylguanidines on mitochondrial metabolism. J. Biol. Chem. 238:410-417(1963).

29. Ballard, F.J. and R.W. Hanson. Phosphoenolpyruvate carboxykinase and pyruvate carboxylase in developing rat liver. Biochem. J. 104:866-871(1967).

TABLE 1

THE INTRACELLULAR DISTRIBUTION OF PHOSPHOENOLYPYRUVATE CARBOXYKINASE IN THE LIVERS OF VARIOUS SPECIES

Values are the mean ± standard error for the number of animals shown in brackets. Rats were fasted 24 hrs, guinea pigs and rabbits for 48 hrs, lactating dairy cows for 96 hrs and man overnight. Assay conditions have been described in detail previously (29).

Species	Mitochondria		Cytosol	
	Fed	Fasted	Fed	Fasted
		units/g liver		
Man [1]	–	8.62	–	4.4
Rat [12]	0.110 ±0.03	0.114 ±0.02	2.49 ±0.13	6.73 ±0.13
Guinea pig [6]	5.10 ±0.32	5.53 ±0.09	1.84 ±0.21	4.10 ±0.20
Rabbit [6]	3.49 ±0.17	5.43 ±0.39	0.42 ±0.004	2.37 ±0.30
Cow [3]	5.46 ±0.83	4.97 ±1.02	4.73 ±0.69	4.25 ±0.33
Cat [4]	6.07 ±0.76	–	4.78 ±1.01	–

TABLE 2

EFFECTS OF DINITROPHENOL AND α-KETOGLUTARATE ON PHOSPHOENOPYRUVATE SYNTHESIS FROM OXALACETATE AND MALATE BY GUINEA PIG LIVER MITOCHONDRIA

Approximately 20 mg of mitochondria dry weight were incubated for 30 min at 30° with 75 μmoles of malate or α-ketoglutarate, or with 30 μmoles of oxalacetate. Nine μmoles of either ATP or ADP were added together with 500 μmoles of sucrose, 6 μmoles of Tris (pH 7.5), 60 μmoles of potassium phosphate and 7.5 μmoles of $MgCl_2$ in a final volume of 3.0 ml 2,4-Dinitrophenol, 120 nmoles, or 75 μmoles of α-ketoglutarate (or both) were added where indicated. Values are the mean ± standard error for four experiments and expressed as μmoles/mg mitochondria. Asterisks indicate malate uptake. Experimental details are given in (15).

Substrate	Addition	P-enolpyruvate	Malate	ATP
Oxalacetate ATP	none	4.62 ± 0.23	1.89 ± 0.23	8.05 ± 0.09
	2,4-dinitrophenol	1.48 ± 0.04	0.51 ± 0.07	0.58 ± 0.06
	α-ketoglutarate	4.29 ± 0.07	8.33 ± 0.37	8.49 ± 0.05
	2,4-dinitrophenol + α-ketoglutarate	1.62 ± 0.08	7.91 ± 0.16	1.19 ± 0.05
Malate ADP	none	2.89 ± 0.04	* 7.00 ± 0.96	8.97 ± 0.20
	2,4-dinitrophenol	0.38 ± 0.03	*11.97 ± 0.19	0.50 ± 0.15
	α-ketoglutarate	4.06 ± 0.04	* 4.81 ± 0.14	8.67 ± 0.14
	2,4-dinitrophenol + α-ketoglutarate	0.64 ± 0.05	*10.02 ± 0.20	1.59 ± 0.12
α-Ketoglutarate ADP	none	1.83 ± 0.24	2.54 ± 0.35	8.81 ± 0.10
	2,4-dinitrophenol	0.51 ± 0.02	3.10 ± 0.12	0.56 ± 0.05

TABLE 3

INITIAL RATE OF PHOSHOENOLPYRUVATE SYNTHESIS AND MALATE UPTAKE DURING VARYING PHASES OF MITOCHONDRIAL RESPIRATION

Approximately 20 mg dry wt of mitochondria were incubated for 20 min at 30° with 75 μmoles of malate and 9 μmoles of either ATP or ADP or 10 units of crystalline hexokinase and 100 μmoles glucose or 120 nmoles of 2,4-dinitrophenol. All other conditions are as outlined in Table 2. Values are the means ± standard error for 6 incubations and are expressed in nmoles/min/mg dry wt. All other experimental details are given in (15).

Respiratory stimulant	Initial rate	
	P-enolpyruvate	Malate uptake
Control	0.18 ± 0.04	0.51 ± 0.02
ATP	2.61 ± 0.18	3.80 ± 0.14
Hexokinase, glucose + ATP	4.95 ± 0.15	6.34 ± 0.18
2,4-Dinitrophenol + ATP	6.15 ± 0.24	8.61 ± 0.35
ADP	7.85 ± 0.32	11.80 ± 0.14

TABLE 4

LEVELS OF ATP, ADP, AMP, PHOSPHATE, LACTATE, AND PYRUVATE AND THE CALCULATED CYTOSOLIC OXIDATION REDUCTION POTENTIALS OF THE NICOTINAMIDE ADENINE DINUCLOETIDES IN FREEZE-CLAMPED GUINEA PIG LIVER

The livers of fed and fasted guinea pigs were freeze clamped and intermediates were determined enzymatically on the neutralized tissue extracts. Values given are the means ± the standard error for 11 to 14 animals. The oxidation reduction potential of the cytosolic pyridine nucleotides was calculated with the $K_{eq} = 1.11 \times 10^{-4}$ for lactate dehydrogenase. Further experimental details are given in (21).

Dietary status	ATP	ADP	Ratio of ATP/ADP	AMP	Phosphate	Lactate	Pyruvate	Ratio of lactate to pyruvate	Ratio of NAD^+/NADH
				μmoles/g tissue (wet wt)					
Fed	2.32±0.06	0.42±0.05	5.52	0.19±0.04	4.31±0.62	1.38±0.05	0.045±0.010	30.7	293
Fasted 48 hrs	2.12±0.10	0.41±0.07	5.17	0.14±0.05	4.97±0.43	0.34±0.03	0.013±0.004	26.2	344
Fasted 96 hrs	2.05±0.09	0.39±0.04	5.25	0.14±0.04		0.41±0.06	0.015±0.003	27.3	330

TABLE 5

LEVELS OF INTERMEDIATES OF GLUTAMATE DEHYDROGENASE AND β-HYDROXYBUTYRATE DEHYDROGENASE EQUILIBRIA IN FREEZE-CLAMPED GUINEA PIG LIVER

The livers of fed and fasted guinea pigs were freeze clamped and intermediates were determined enzymatically on the neutralized tissue extracts. Values given were micromoles per g of liver, wet wt, and are the means ± the standard error for 11 to 14 animals. The oxidation-reduction potential were calculated using a $K_{eq} = 3.87 \times 10^{-3}$ for glutamate dehydrogenase, and $K_{eq} = 4.93 \times 10^{-2}$ for β-hydroxybutyrate dehydrogenase. Further experimental details are given in (2).

Dietary status	α-Ketoglutarate	Glutamate	Ammonia	Ratio of $NAD^+/NADH$	β-Hydroxybutyrate	Acetoacetate	Ratio of $NAD^+/NADH$
			μmoles/g liver (wet wt)				
Fed	0.31±0.03	3.57±0.21	0.37±0.06	8.30	0.08±0.03	0.05±0.02	12.71
Fasted 48 hrs	0.19±0.03	2.76±0.08	0.63±0.08	28.23	0.18±0.03	0.30±0.04	33.80
Fasted 96 hrs	0.17±0.02	1.47±0.15	0.73±0.09	40.94	0.40±0.03	0.82±0.05	41.60

TABLE 6

THE LEVELS OF INTERMEDIATES OF THE GLUTAMATE DEHYDROGENASE AND β-HYDROXYBUTYRATE DEHYDROGENASE EQUILIBRIA IN FREEZE-CLAMPED RABBIT LIVER

The livers of fed and fasted rabbits were freeze clamped and the intermediates determined enzymatically on neutralized liver extracts. Values are the means ± the standard error for 6 animals. The oxidation-reduction potentials were calculated using a $K_{eq} = 3.87 \times 10^{-3}$ for glutamate dehydrogenase, and $K_{eq} = 4.93 \times 10^{-2}$ for β-hydroxybutyrate dehydrogenase. Experimental details in (21).

Dietary	α-Ketoglutarate	Ammonia	Glutamate	Ratio of $NAD^+/NADH$	β-Hydroxybutyrate	Acetoacetate	Ratio of $NAD^+/NADH$
			μmoles/g liver				
Fed	0.17 ± 0.02	0.29 ± 0.04	2.85 ± 0.14	4.5	0.13 ± 0.02	0.06 ± 0.01	9.3
Fasted 96 hrs	0.20 ± 0.02	0.71 ± 0.09	2.11 ± 0.15	17.4	0.26 ± 0.04	0.27 ± 0.02	21.1

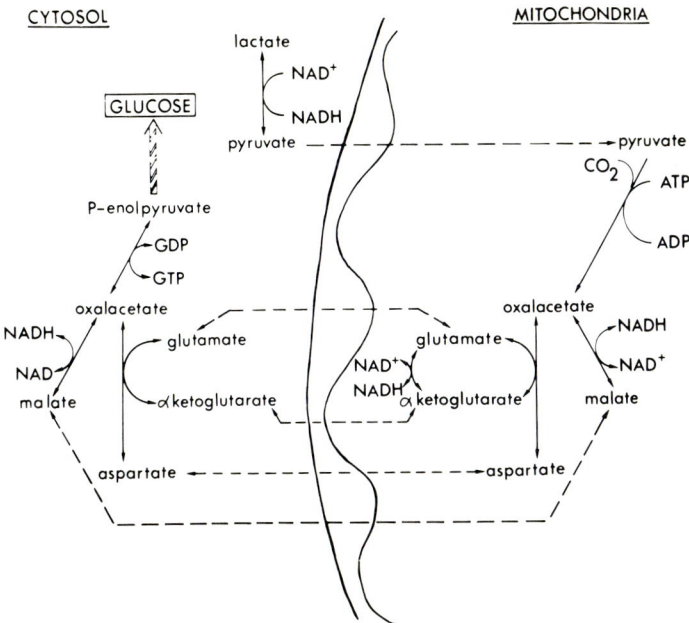

Fig. 1. *The mechanism of P-enolpyruvate formation for gluconeogenesis as proposed for rat liver.*

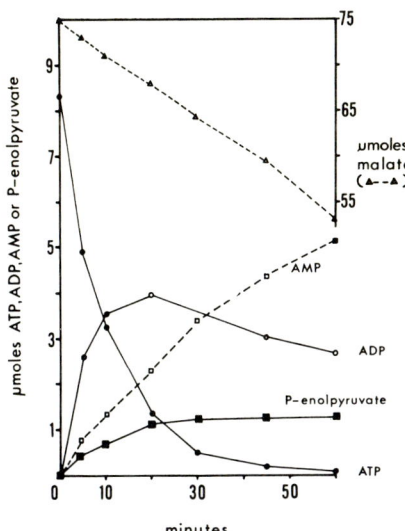

Fig. 2. *Time-course of P-enolpyruvate formation, malate uptake, and the concentrations of ATP, ADP, and AMP using 19.6 mg mitochondria incubated with 120 nmoles of 2,4-dinitrophenol, 9μmoles of malate.* Mitochondria were incubated at 30° in 0.25 M sucrose, 20 mM potassium phosphate, 10 mM $MgCl_2$ and 0.01 M Tris, pH 7.5. Intermediates were determined enzymatically on the neutralized perchloric acid extracts. Values are the means ± SEM, in μmoles.

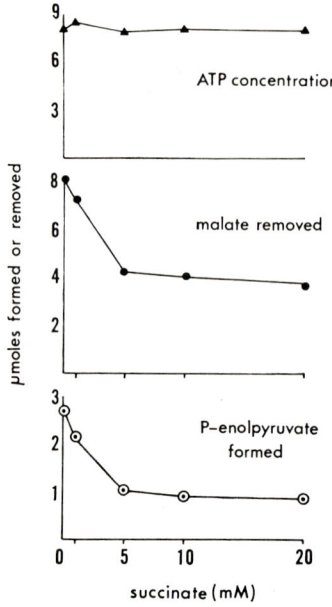

Fig. 3. *The effects of succinate addition upon the mitochondrial formation of P-enolpyruvate, malate uptake, and ATP synthesis.* Mitochondrial incubations contained 21.3 mg mitochondria with between zero and 60 μmoles of succinate added. Other details as in Fig. 2.

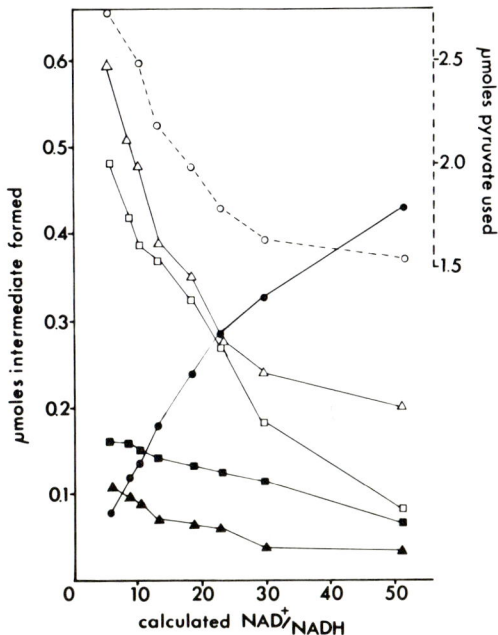

Fig. 4. *The net change in P-enolpyruvate (●), malate (□), aspartate (▲) and citrate formation (■) and pyruvate carboxylation (△) and utilization (○) as a function of the intramitochondrial oxidation-reduction potential.* Guinea pig liver mitochondria (0.97 mg nitrogen) were incubated with 1.03 mM pyruvate, and the resultant ratio of NAD^+ to NADH was calculated from the intermediates of the β-hydroxybutyrate dehydrogenase equilibrium using $K_{eq} = 4.93 \times 10^{-2}$.

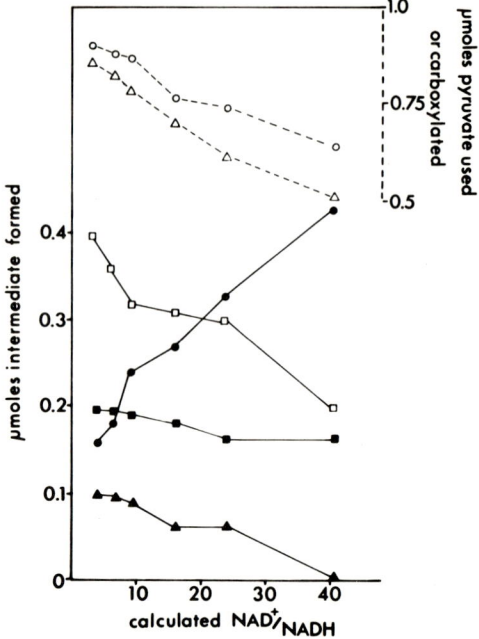

Fig. 5. *The formation of P-enolpyruvate (●), malate (□), citrate (■), and the net formation of aspartate (▲), and pyruvate carboxylation (△), and utilization (○) as a function of the intramitochondrial oxidation-reduction potential.* Mitochondria (0.60 mg nitrogen) were incubated with 1 mM pyruvate, 4 mM glutamate, 1 mM α-ketoglutarate, and 1 mM aspartate. Other details as in Fig. 4 and Table 1.

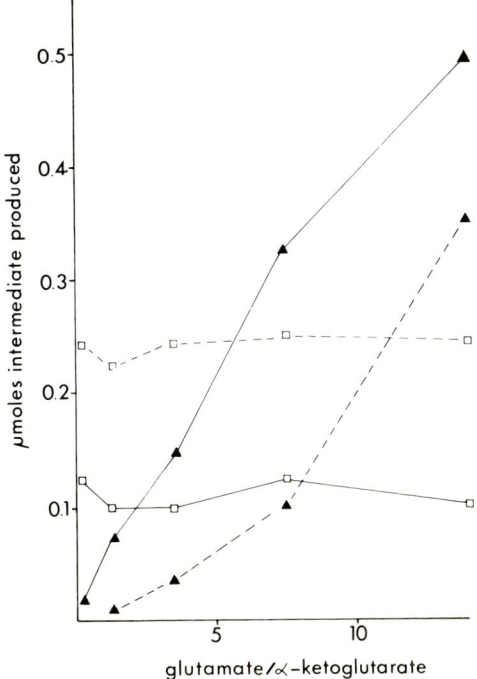

Fig. 6. *Aspartate (▲) and malate (□) formation as a function of the ratio of glutamate to α-ketoglutarate.* Guinea pig liver mitochondria (0.90 mg nitrogen) were incubated with varying ratios of glutamate and α-ketoglutarate and 1 mM pyruvate. Intramitochondrial ratios of NAD^+:NADH were adjusted to equilibrium with varying ratios of β-hydroxybutyrate to acetoacetate. Two isopotential ratios of NAD^+:NADH are shown 25±3 (—) and 9±1 (---). Other details as in Fig. 4.

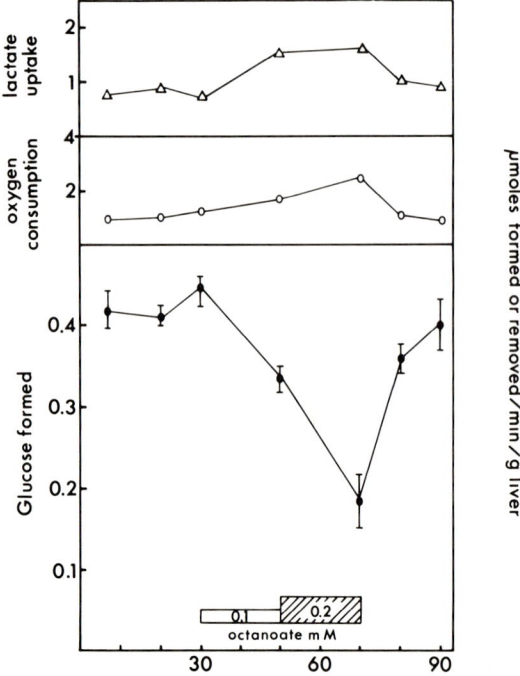

Fig. 7. *The rates of glucose formation (●), lactate uptake (▲), and oxygen consumption in the isolated perfused guinea pig liver.* Liver from guinea pigs fasted 48 hrs was perfused intraportally with Krebs-Henseleit buffer containing 2 mM lactate in a non-recirculating system. Oxygen consumption was determined on the venous effluent obtained by canulation of the inferior vena cava. At the thirtieth minute of lactate perfusion, an intraportal infusion of sodium octanoate was begun, and the rate of infusion adjusted to produce the concentrations in the arterial input as noted in the figure. All values are given as μmoles formed or removed per min per gram liver wet weight.

METABOLIC REGULATION

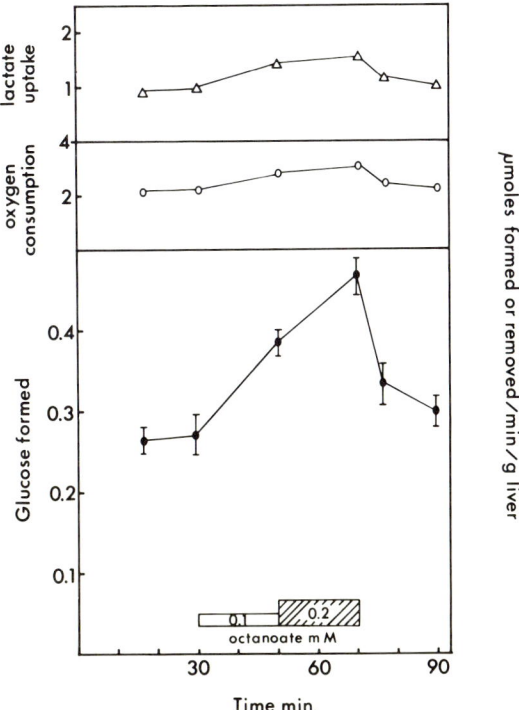

Fig. 8. *The rates of glucose formation (●), lactate uptake (▲), and oxygen consumption in the isolated perfused rat liver.* Other details as given in Fig. 6.

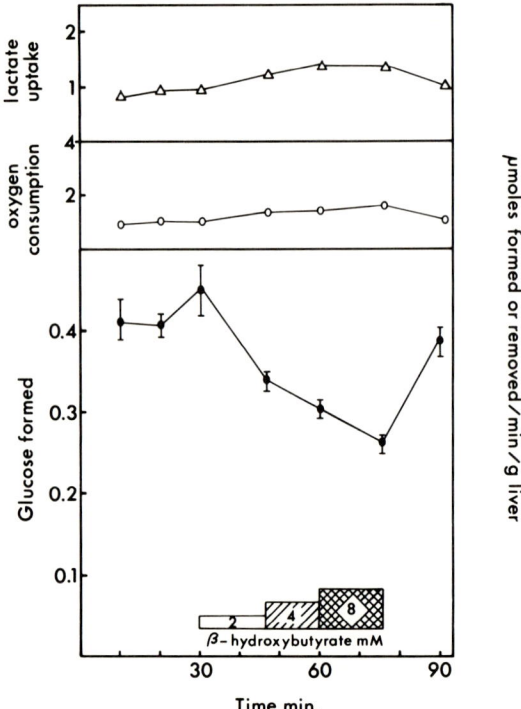

Fig. 9. *The rates of glucose formation (●), lactate uptake (▲), and oxygen consumption (○), in the isolated perfused guinea pig liver.* At the thirtieth minute of lactate perfusion, an intraportal infusion of sodium β-hydroxybutyrate was begun and was adjusted to produce the arterial concentrations as indicated in the figure.

METABOLIC REGULATION

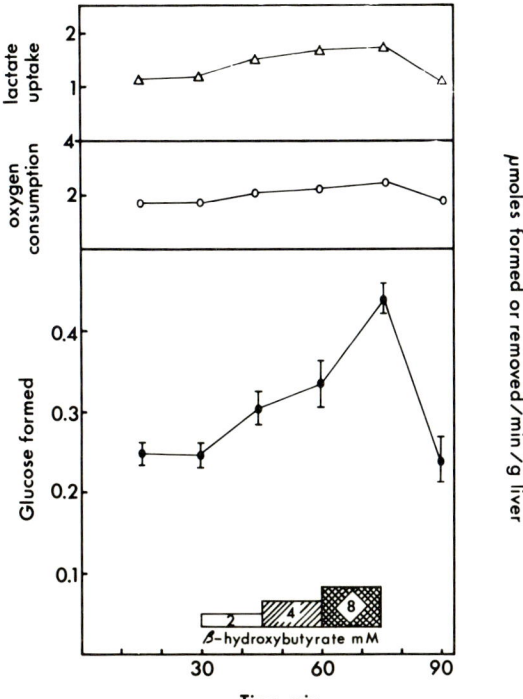

Fig. 10. *The rate of glucose formation (●), lactate uptake (▲), and oxygen consumption (○) in the isolated perfused rat liver.* Other procedures and details as described in Figs. 6 and 9.

PRODUCTION AND UTILIZATION OF ATP AND THE REGULATION OF LIPOGENESIS IN ADIPOSE TISSUE

J.P. Flatt

Introduction

Because of its high fat content (greater than 85%) and the high caloric value of lipids, adipose tissue provides the most compact means for storing energy reserves. In warm blooded animals, whose energy expenditure is high even at rest, the ability to carry sufficient caloric reserves is crucial, and adipose tissue plays a very important role in energy metabolism (1, 2).

The fatty acids stored in the form of triglycerides in adipose tissue can be derived from dietary fats or synthesized *de novo* from carbohydrate, depending on the relative proportion of carbohydrate and fat supplied in the diet. *De novo* lipogenesis is carried out in adipose tissue itself or in liver followed by transport to adipose tissue [for review, *cf.* (2)].

The entry of glucose into adipose tissue is regulated by the concentration of insulin (3, 4, 5). In the fed state, with a diet containing limited amounts of fat, adipose tissue promptly converts to fat the glucose taken up under the influence of insulin (3, 6). Thus, insulin is a primary regulator of adipose tissue lipogenesis. However, when on a high fat diet, or during starvation or diabetes, the response of adipose tissue to insulin is considerably impaired, even with high concentrations of this hormone (2, 7, 8). In these metabolic situations intracellular control phenomena, rather than the control of glucose entry, appear to play a major role in controlling adipose tissue lipogenesis (9).

Because of the variety of metabolic pathways and enzymes cooperatively involved in the conversion of glucose to fat (2) and because of the number of enzymes which are known to possess regulatory properties, there are many possible intracellular control sites (2, 8 10). The discussion presented here will serve to identify the rate-limiting phenomena in the conversion of glucose to fat in intact rat adipose tissue, as deduced from *in vitro* experiments performed with rat epididymal fat pads (11).

Results

Acetyl-CoA formation

The rate of fatty acid (FA) synthesis in adipose tissue can be determined by the incorporation of ^{14}C from ^{14}C-labeled precursors. To test whether the pathway from acetyl-CoA to FA is rate-limiting or not, epididymal fat pads were incubated with increasing concentrations of insulin in the presence of either glucose alone or both glucose and acetate. So that the fate of the carbon from each of these substrates could be determined, the later experiments included parallel incubations with either glucose-U-^{14}C and unlabeled acetate or unlabeled glucose and acetate-U-^{14}C.

The results (Fig. 1) show that in the presence of insulin FA synthesis proceeds at a higher rate when acetate is available in addition to glucose. The capacity to convert acetyl-CoA into FA is therefore always greater than required during *de novo* lipogenesis from glucose. This implies that the enzymes for FA synthesis from acetyl-CoA as well as those involved in supplying NADPH and ATP for this synthesis are not rate-limiting (9). FA synthesis must consequently be limited by the rate of acetyl-CoA formation from glucose, itself governed by the inflow of glucose at low concentrations of insulin, and/or by the flow of intermediates through the Embden-Meyerhof pathway and the control phenomena influencing this pathway. As shown by Fig. 1, this conclusion is valid in adipose tissue from rats maintained on a high carbohydrate diet, regular lab chow, or a "mixed" diet containing 20% fat, as well as in rats fasted for two days. It is of interest to note that under all the above conditions the activity of the lipogenic enzymes remains above that required when lipogenesis from glucose is maximally stimulated by insulin. In view of its complex regulatory properties acetyl-CoA carboxylase has often been regarded as the most important pace-maker of FA synthesis (2, 8, 12). The experiments described above demonstrate that this enzyme is not in a position to control FA synthesis during *de novo* lipogenesis in adipose tissue, since this conversion is limited by the rate of acetyl-CoA formation from glucose (9, 13, 14).

Pathway flow

Detailed quantitative data on the flow of intermediates through the major metabolic pathways involved in lipogenesis in adipose tissue have been available for some time. Very similar pathway flow patterns have been obtained by Flatt and Ball (15) and Katz and Rognstad (14), using approaches designed to exploit radioactive data gathered from various combinations of specifically labeled ^{14}C-glucose. The rates of production and of utilization of reducing equivalents can be calculated from such flow data (Fig. 2). The reducing equivalents not used for reductive synthesis are seen to sustain oxidative phosphorylation, in agreement with the measured rates of oxygen consumption (Fig. 2).

During insulin induced lipogenesis, the pentose cycle provides only 50% to 60% of the NADPH required for FA synthesis (15, 16), with the remainder of the NADPH now known to be produced by malic enzyme (17, 18, 19). The ability of adipose tissue to produce NADPH by other reactions than those of the pentose cycle is demonstrated by the finding that pyruvate can sustain an appreciable rate of fatty acid synthesis (3, 20, 21). As shown in Fig. 2, the tissue's ability to produce NADPH from pyruvate closely matches the amount of NADPH needed during fatty acid synthesis from glucose in addition to that supplied by the pentose cycle.

Fig. 3 illustrates the pathways involved in lipogenesis from glucose. Acetyl-CoA is shown to be carried out of the mitochondria in the form of citrate after condensation with oxaloacetate (22), with acetyl-CoA and oxaloacetate regenerated in the cytosol by the action of citrate cleavage enzyme (23). Oxaloacetate is then reduced to malate using the reducing equivalent generated at the glyceraldehyde-3-P dehydrogenase step during the formation of acetyl-CoA from glucose in the Embden-Meyerhof pathway (23). Malate can be oxidatively decarboxylated to pyruvate by malic enzyme with the production of NADPH, followed by resynthesis of oxaloacetate in the mitochondria by pyruvate carboxylase (Fig. 3A). Thus this scheme described a situation where one of the NADPH required for the reductive incorporation of acetyl-CoA into FA is produced by the pentose cycle and the other by the "malate cycle" (double line arrows of Fig. 3A). Alternatively malate diffuses back into the mitochondria (24) as represented in Fig. 3B. In this case the reducing equivalent generated in the Embden-Meyerhof pathway is transferred to the mitochondria and the pentose cycle provides all the NADPH for FA synthesis. Since the pentose cycle is found to produce more than half of the NADPH required for FA synthesis, the two mechanisms illustrated in Fig. 3 are involved in the conversion of glucose to fat.

Energy metabolism during de novo *lipogenesis*

Figure 3 describes the energy balance during the conversion of glucose to fat. If one includes the high energy bonds generated at the substrate level and those produced from the NADH not used for reductive synthesis, the energy requirement for fat synthesis is more than covered. Thus 1 ATP (or one high energy bond) is gained per acetyl-CoA derived from glucose and converted to FA according to scheme A and 5 ATP according to scheme B. Therefore, the conversion of glucose to fat is an energy generating process.

This conclusion was in fact already made obvious by the data given in Fig. 2, which show that most of the reducing equivalents used for oxidative phosphorylation are derived from the catabolism of glucose when lipogenesis is stimulated by insulin. Concomitantly the oxidation of endogenous substrates which furnishes the tissue with energy when glucose entry is limited by lack of insulin is markedly suppressed (15). Since the glycogen content of adipose tissue is very limited (25), the endogenous substrate for energy production is essentially FA (15).

In order to evaluate more directly the energy metabolism associated with lipogenesis from glucose, the difference in the catabolism of glucose and of endogenous FA catabolism was calculated from the pathway flow values obtained in the presence or absence of insulin-induced lipogenesis (15). These differences are given in Table 1 for all the steps in which either high energy bonds or reducing equivalents are involved. When lipogenesis is induced there is an increase of 2.52 (μmoles/100 mg/h) in the amount of reducing equivalents gained from the catabolism of glucose, while 1.37 fewer reducing equivalents are formed by endogenous FA oxidation, leading to an overall increase of 2.52 − 1.27 = 1.15 in the reducing equivalents used for oxidative phosphorylation, or of 0.58 μmoles of O_2/100 mg/h. The observed increase in oxygen consumption was 0.83; the difference between these values is 0.25, corresponding to a discrepancy of

$$\frac{(0.25)\ (100)}{1.98} = 3\%$$

relative to the O_2 consumption of the tissue incubated with insulin.

Assuming a P:O ratio of 3 for the reoxidation of NADH, 3.47 more μmoles of ATP (per 100 mg/h) are derived from the catabolism of glucose in the presence of insulin than in its absence. This is almost exactly offset by a decrease of 3.83 μmoles in the ATP obtained by oxidation of endogenous FA. Relative to the total amount of ATP (11.76 μmoles/100 mg/h) entering into this balance, the discrepancy amounts to but

$$\frac{(3.83-3.47)\ (100)}{11.76} = 3\%$$

With the present knowledge of the pathways involved in lipogenesis, the energy metabolism implied by the pathway flow data can be accounted for rather precisely! This close agreement supports the conclusions drawn from the metabolic schemes given in Fig. 3, and, furthermore, it indicates that the basic energy expenditure in adipose tissue is not appreciably altered upon induction of lipogenesis.

Pathway capacity

Adipose tissue has been incubated *in vitro* under a variety of conditions, resulting in different rates of flow through its major metabolic pathways, as displayed in Fig. 4. It can be seen that conditions exist which permit any of the pathways involved in the conversion of glucose to fat to operate faster than during maximally stimulated fat synthesis from glucose, except the malate cycle. In three different situations of high FA synthesis, where glucose and insulin, or glucose plus acetate and insulin, or pyruvate, were available, the malate cycle is seen to operate at approximately the same rate. This can be explained if each time the maximal capacity of this cycle is attained, making it the only pathway to reach saturation during conversion of glucose to fat. However, this does not in itself set a limit to this conversion, since it can proceed also without the participation of the malate cycle, as shown in Fig. 3B. The limitation in lipogenesis cannot therefore be attributed to any particular segment of the metabolic pathway reaching saturation. Lipogenesis must consequently be limited by some constraint which results from the interplay of all the pathways involved.

It should be remembered at this point that glucose conversion to fat is an energy yielding process. As such it must reach a limit when it supplies enough energy to sustain the total energy expenditure of adipose tissue, as no evidence of uncoupling was found in white adipose tissue incubated with glucose, even if the tissue was obtained from hyperthyroid rats and exposed to lipolytic agents (26).

To examine whether the dissipation of the energy produced during *de novo* lipogenesis could become a rate-limiting factor, the ATP gain as a function of the rate of FA synthesis is graphically represented in Fig. 5. As long as the malate cycle participates, the line giving the ATP yield has a slope of 1 (Fig. 3A), which changes to a slope of 5 (ATP generated per acetyl-CoA converted from glucose to FA) (Fig. 3B) beyond the point where the malate cycle has reached its maximal capacity. This maximal capacity can be evaluated a) from pathway flow studies (15) using the NADPH requirement for FA synthesis less the NADPH production by the pentose cycle: 4.45 µg of glucose-C incorporated into FA/100mg/h x 0.875 µmoles NADPH/µg-at of C incorporated into FA (11) − 2.35 µmoles NADPH from pentose cycle = 1.55 µmoles/100mg/h, or b) on the basis of the tissue's ability to synthesize FA from pyruvate (20): 1.86 µg-at of pyr-C incorporated into FA/100mg/h x 0.875 µmoles NADPH/µg-at of C incorporated into

FA = 1.63 μmoles/100mg/h, which is the value used in the construction of Fig. 5.

The basic energy expenditure of adipose tissue in the absence of FA synthesis corresponds to the consumption of 1 μmole of O_2/100 mg/h (11). Assuming a P:O ratio of 3, this corresponds to the utilization of 6 μmoles of ATP/100 mg/h. As shown in Fig. 5, this rate of ATP production is achieved when the incorporation of glucose into FA reaches a rate of 2.5 μmoles of acetyl-CoA/100 mg/h. However, this value will lead to an overestimate since adipose tissue in addition to adipose cells contains also stromal-vascular cells, whose energy expenditure obviously cannot be covered through lipogenesis. A better estimation of the tissue's ability to utilize the high energy bonds generated by lipogenesis is possible using as an index the decrease in the energy produced by oxidation of endogenous FA (15), which must be directly related to the generation of ATP by lipogenesis. The observed decrease in endogenous CO_2 production was 0.48 μmoles/100 mg/h (15), corresponding to the consumption of 0.48/0.7 = 0.68 μmoles of O_2, and to the utilization of 0.68 x 2 x 3 = 4.1 μmoles of ATP/100 mg/h. This value indicates that 2/3 of the adipose tissue's energy requirement can be covered through lipogenesis. Since about one-half of the adipose tissue protein is associated with other cells than the adipocytes (27), it is clear that the oxidation of endogenous substrates in the adipocytes themselves is very extensively suppressed under conditions of high lipogenesis. Using this figure to evaluate the rate at which ATP generated in lipogenesis can be utilized, or produced, the maximal possible rate of FA synthesis from glucose can be estimated at 2.1 μmoles of acetyl-CoA incorporated into FA/100 mg/h. If instead of a P:O ratio of 3, the P:O ratio of 2.42, reported by Fisher and Ball (26) for intact adipose tissue, is used for this estimation, the limiting rate of FA synthesis would be similar, corresponding to 2.3 μmoles of acetyl-CoA incorporated into FA/100 mg/h. The approach is therefore not dependent on a precise knowledge of the P:O ratio. These estimates are in close agreement with the rate of 2.2 μmoles of acetyl-CoA/100 mg/h determined experimentally by the ^{14}C incorporation (15). Thus it becomes evident that *the conversion of glucose to fat in intact rat adipose tissue is limited by the tissue's ability to utilize the high energy bonds generated during the process.* This conclusion explains why the limitation in *de novo* lipogenesis cannot be attributed to the saturation of a particular pathway or enzyme involved in converting acetyl-CoA into FA or in supplying the required cofactors.

In the construction of Fig. 5 it is assumed that the malate cycle does reach its highest capacity during fat synthesis from glucose. This pathway requires $NADP^+$, since in the presence of acetate this cycle is capable of operating nearly twice as rapidly as in the presence of glucose alone (13, 14).

A study of the three NADP-linked dehydrogenases involved in these processes shows that in adipose tissue of fed rats malic enzyme has the highest activity as well as the highest affinity for $NADP^+$ (Fig. 6). Malic enzyme should thus be able to react with enough $NADP^+$ for the malate cycle to reach its full capacity. Fig. 7 shows that in regard to the competition for $NADP^+$ the same situation prevails in adipose tissue of rats maintained on a high carbohydrate, regular lab chow, or a high fat diet, as well as during fasting.

Impairment of lipogenesis in fasting

The maximal ability of adipose tissue from fed rats to convert glucose to fat is determined by the ATP yield with which this conversion proceeds and by the rate of utilization of the ATP generated by lipogenesis. The former is related to the capacity of the malate cycle, the latter to the tissue's ability to curtail the oxidation of endogenous FA so that a maximal amount of ATP generated by glucose catabolism may be utilized. These two determining factors are influenced by fasting. FA synthesis from pyruvate, which reflects the capacity of the malate cycle, decreases with fasting (Fig. 4, extreme right). As shown by the results of Table 2, the production of endogenous CO_2, after induction of lipogenesis with a maximal dose of insulin, is not nearly as well curtailed in adipose tissue from fasted as in tissue from fed rats.

The effect of fasting on the parameters which determine the maximal possible rate of lipogenesis are shown in Fig. 8. The effects of changes in glycolysis and in reesterification are also shown in order to demonstrate that their effect on the energy balance is insignificant in comparison to the effects of endogenous FA oxidation and malate cycle capacity. It can be seen that the effects of fasting, when considered at the level of the energy metabolism, provide an explanation for the decrease in the ability of adipose tissue to synthesize fat from glucose, an impairment for which so far no satisfactory explanation has been advanced (28, 29).

The loss in the ability to decrease FA oxidation would appear to be related to the high levels of FFA which prevail in starvation. However, FFA or their CoA derivatives are seen here to inhibit adipose tissue lipogenesis by virtue of being excellent substrates for energy production rather than by inhibiting a particular enzyme, such as acetyl-CoA carboxylase, generally considered to be primarily responsible for the control of FA synthesis (8, 12).

Lactate/pyruvate ratios during lipogenesis

The ratio of the amounts of lactate to pyruvate in the incubation medium is related to the cytoplasmic redox potential (30). As the rate of glucose conversion to fat increases, and the amount of energy generated by this process becomes greater, the cytoplasmic redox potential increases (Fig. 9). This is most striking in the case of tissue from fasting animals where the utilization of the energy generated by lipogenesis becomes more rapidly limiting due to the tissue's apparent inability to stop oxidizing endogenous FFA. This supports the view that the continued oxidation of endogenous FFA creates an obstacle to lipogenesis and argues against the reverse interpretation, namely that endogenous FA oxidation continues because lipogenesis cannot reach a rate sufficient to provide all the ATP needed to cover the tissue's basic energy expenditure.

It is of interest to note that acetate has an effect on the lactate/pyruvate ratio. As measured by the incorporation of ^{14}C, the greatest part of the utilized acetate is incorporated into FA (9, 13). Thus, the utilization of acetate results in an increased demand for NADPH, which is entirely met by a greater flow of glucose through the pentose cycle, and in an enhanced utilization of ATP. The concomitant decrease in the lactate/pyruvate ratio indicates that the cytoplasmic redox potential is related to the availability of energy in the tissue.

Fig. 9 also shows that the proportion of pyruvate converted to lactate is related to the cytoplasmic redox potential. A mechanism important for the conservation of carbohydrate in starvation becomes evident. In starvation, as opposed to diabetes, significant concentrations of insulin are maintained, which at the level of adipose tissue serve to control the rate of FFA mobilization. This is achieved through the direct antilipolytic action of insulin (31) and by it permitting some glucose uptake to provide glycerol-P for FFA reesterification (32). Because of its energy-yielding nature, lipogenesis leads to a very sharp increase of the cytoplasmic redox potential in the FFA loaded adipose tissue of starving animals. This sets up a metabolic feedback effect causing increased diversion of pyruvate to lactate. Since lactate leaves the tissue, a three carbon fragment derived from glucose, but not used as glycerol-P, is thereby preserved for glucose resynthesis in the liver.

Effect of acetate on lipolysis

Since adipose tissue contains no significant glycerolkinase activity (33, 34), the release of glycerol into the incubation medium reflects the rate of lipolysis, whereas the release of FFA is the result of the rates of lipolysis as well as of FFA reesterification. Acetate causes not only a decrease in lactic production but also a decrease in glycerol release (9). In the absence of glucose, the acetate effect on glycerol release is comparable to that exerted by of glucose to fat, one would expect this conversion to be enhanced by uncoupling agents. But these would have to be added at concentrations sufficiently low to leave enough ATP and acetyl-CoA available for the synthesis of FA. Thus, for example, addition of a high dose of epinephrine induces lipolysis and increases ATP expenditure for the reesterification of FFA to such an extent that acetyl-CoA is primarily oxidized for energy production and lipogenesis is greatly reduced (32). The failure to demonstrate an increase in lipogenesis by addition of dinitrophenol (24) is probably related to the difficulty of achieving the right concentration of this lipophilic substance in a tissue containing 85% fat.

However, when the ATP utilization is only moderately increased, either by addition of a rather small dose of epinephrine (0.2 μg/ml) (15) or by supplying acetate in the medium (cf. Fig. 1, tissue from fasted rats), an enhancement of glucose conversion to fat can be demonstrated. Assuming that an increase in ATP utilization of 1 μmole permits an increase of FA synthesis of 0.2 μmole of acetyl-CoA derived from glucose (cf. slope = 5 in Fig. 5), the effect of epinephrine or of acetate on FA synthesis from glucose can be calculated from the increases in ATP utilization due to the metabolism of acetate and to changes in FFA reesterification and lactate production. The effects predicted by these calculations have been used in the construction of Fig. 12, where the ratios of predicted to observed changes are plotted as a function of the rate of FA synthesis from glucose. At low rates of FA synthesis, where ATP generation by lipogenesis is small in comparison to the tissue's energy requirement, these ratios are, as one would expect, far from the value of 1 which would indicate a perfect prediction. However, with high rates of FA synthesis, the rationale used gives very satisfactory results, supporting the concept that the limitation in lipogenesis finds its causes at the level of the energy metabolism. The discrepancy seen in the case of the highest rate of FA synthesis indicates that the conversion of glucose to fat is not limited by the energy metabolism in one of the two incubations which are compared. Apparently, in the case of the incubation with glucose plus acetate (but not in that with glucose alone) some pathway has reached saturation (e.g., FA synthesis from acetyl-CoA, NADPH production in the pentose cycle, or any other).

Among the possible limiting factors for *de novo* lipogenesis it has been considered that with the saturation of the malate cycle the reoxidation of the cytoplasmic NADH formed by glyceraldehyde-P dehydrogenase becomes impossible, preventing further formation of pyruvate and acetyl-CoA from glucose (39). Such a view, however, fails to account for the increase caused by acetate or epinephrine in glucose-C incorporation into FA, an increase which

60 μunits/ml of insulin, and a further decrease in lipolysis is seen when both acetate and insulin are present (Fig. 10).

In order to analyze separately the effects of acetate and those of insulin, the results described by the vertical bars on the left side of Fig. 10 are presented in another manner on the right side of Fig. 10, where glycerol release is shown as a function of lactate production. Two paired values are connected by arrows using full or broken lines to indicate respectively the presence or absence of glucose. The origins and heads of the arrows correspond to the releases of lactate and glycerol and their direction describes the effect of insulin addition (top graph) or of acetate addition (lower graph). It can be noted first that the points determined by the two ends of the arrows fall into a zone whose several shapes indicate a correlation between lactate and glycerol releases. While the arrows showing the effect of acetate are always pointing in the same general direction, those showing the effects of insulin point in nearly opposite directions depending upon whether or not glucose is present.

It has long been puzzling why the antilipolytic effect of insulin (31) is counteracted by the presence of glucose (35, 36). Since insulin increases the production of lactate only in the presence of glucose, it is of interest to note an apparent correlation between the rate of lipolysis and the lactate production, which is itself closely related to the cytoplasmic redox potential (Fig. 9) and possibly the relative energy charge within the adipose cells. To obtain further evidence on this correlation adipose tissue was incubated with glucose and either pyruvate or lactate, two substrates which would be expected to differ only by their effect on the cytoplasmic redox potential. The experiments reported in Fig. 11 show indeed that pyruvate decreases, whereas lactate increases the release of glycerol, even though the total amount of substrate metabolized is greater with pyruvate than with lactate or with glucose alone.

An antilipolytic effect of acetate is also manifest *in vivo*. Abramson and Arky (37) found that infusion of sodium acetate in man lowered plasma FFA and plasma glycerol levels by some 30%. The antilipolytic effect of acetate is able to counteract epinephrine stimulated lipolysis in man (37), as well as in isolated adipose tissue (38). In man the effects of acetate were similar to those of ethanol (37), and one wonders whether the frequent use of alcoholic beverages in the diet of diabetic patients before insulin was available could owe its empirical justification to a beneficial effect associated with a decreased rate of lipolysis.

Discussion

If the utilization of ATP is the rate-limiting factor for the conversion is easily understood when the limitation of the process is attributed to the utilization of the ATP generated by lipogenesis (Fig. 12).

It is a basically accepted notion that the oxidative degradation of substrates is controlled and limited by the expenditure of high energy bonds. In the case of adipose tissue, this notion can now be extended to a synthetic pathway, the energy-generating conversion of glucose to fat. Since the synthesis of ketone bodies in the liver is also an energy-generating process (40, 41), ketogenesis could conceivably be limited as well by the rate of utilization of the energy which it generates. Taking oleic acid to be representative of the FFA mobilized from adipose tissue and using a ratio of 2:1 in the proportion of β-hydroxybutyrate to acetoacetate released by the liver (42), the following equation can summarize hepatic ketogenesis:

$$2 \text{ oleate} + 12 \text{ O}_2 + 54 \text{ ATP} + 54 P_i \longrightarrow$$
$$6 \text{ } \beta\text{-hydroxybutyrate} + 3 \text{ acetoacetate} + 54 \text{ ATP} + 55 H_2O$$

Thus 1.33 moles of O_2 are consumed in the formation of one mole of ketoacid. In man the consumption of oxygen by the splanchnic bed is approximately 1/5 of the total basal oxygen consumption, with the liver itself accounting for 3/4 of the splanchnic O_2 consumption (43). With a basal rate of O_2 utilization of 0.25 L/min, the consumption of oxygen by the liver is of some 2.4 moles/day. Ketogenesis must therefore be limited to the formation of 2.4/1.33 = 1.8 moles, or 185 g of ketoacid per day. A situation in which 90% of the hepatic oxygen consumption is due to ketone body production has in fact been described by Krebs and Hems (45), using livers from fat-fed ketotic rats. According to Wieland (42) the oxygen consumption by the liver can be expected to be some 30% higher in diabetes, so that 240 g/day would be the highest possible rate of ketoacid production in acutely decompensated diabetes.

Table 3 presents data which indicate that the production of ketones in man during acutely decompensated diabetes can be estimated to reach 200 g per day. The operation of the citric acid cycle must be very extensively reduced to permit the utilization of the energy generated by such rates of ketogenesis. This then suggests an analogy between adipose tissue and liver where FA synthesis and ketoacid formation provide mechanisms for recovering CoA from acetyl-CoA. Given an excess of glucose or FFA, respectively, the maximal rates of lipogenesis or ketogenesis are determined by the energy yields of these processes, the effectiveness of the mechanisms by which the generation of energy by the citric acid cycle can be curtailed, and by the energy expenditure of the tissue.

It has been long noted that high rates of ketogenesis are associated with high rates of gluconeogenesis (40, 41, 45, 46). It has been proposed that a reduction in mitochondrial oxaloacetate concentration (47) and/or inhibition by ATP of citrate synthetase (40) prevent the normal operation of the citric acid cycle, thus forcing the hepatic metabolism to obtain its energy through conversion of FFA to ketoacids (40, 41, 48). If the balance between energy production and utilization were to play a role in controlling ketogenesis, gluconeogenesis could exert a permissive action on ketogenesis by raising the energy expenditure in the liver, thus allowing for a higher rate of energy generation through ketogenesis.

One of the most significant differences between the metabolic state of diabetes and that of starvation lies in the rates of gluconeogenesis from protein, the synthesis of glucose from lactate and glycerol presumably proceeding at comparable rates. With urinary nitrogen excretion of 25 g/day in decompensated diabetes (49, 50) but only 4-5 g/day in prolonged starvation (42, 51) the difference in hepatic gluconeogenesis from protein is of some 75 g per day, since under conditions of high demand for gluconeogenesis 3.65 g of glucose are formed per g of urinary N (52). Alanine is the amino acid most efficiently removed by the liver (53). It can be calculated that 10 moles of ATP are required for the synthesis of one mole of glucose and one mole of urea from 2 moles of alanine. Since the equation describing ketogenesis from FFA shows that 6 moles of ATP are gained per mole of ketoacid formed from oleate, the synthesis of 75 g of glucose from alanine uses the energy generated in the synthesis of $(75/180)(10/6) = 0.7$ moles, or 70 g of ketoacid. If one assumes that the dissipation of the energy generated is the rate-limiting factor for ketogenesis, one would expect the production of ketoacids in starvation to be some 70 g/day less than the 200 g formed in acute diabetes, or some 130 g/day. This rate is not far from that observed in prolonged starvation, where ketoacid production approaches 100 g/day (42).

In obese human subjects undergoing complete starvation the concentrations of glucose, FFA, ketoacids, glucagon and growth hormone in blood reach levels after 24 days of starvation which remain constant for the continuation of the fast (54). In spite of a small decline in insulin levels (from 22 μunits/ml on day 24 to 17 μunits/ml on day 36), the excretion of urinary ketoacids decreases. In view of the stability of all the substrate concentrations, this decrease probably indicates a reduction in ketogenesis. This may be related to the continuing decline after day 24 of gluconeogenesis from protein evidenced by the decreasing urinary N loss. The calculation developed in Table 4 shows that the corresponding reduction in the hepatic energy expenditure for gluconeogenesis is adequate to explain the apparent decrease in ketogenesis beyond the 24th day of starvation.

Summary

The experimental data on the energy metabolism of adipose tissue during the conversion of glucose to fat are in excellent agreement with present knowledge of the metabolic reactions involved in *de novo* lipogenesis. The conversion of glucose to fat is an energy-yielding process; as such it avoids being stimulated by FFA, in contrast to hepatic gluconeogenesis which is promoted by high FFA concentrations. Given sufficient insulin to facilitate glucose entry, lipogenesis reaches a rate which is limited by the tissue's ability to utilize the energy thus generated. It is crucial that the flow of intermediates through the citric acid cycle be halted if high rates of *de novo* lipogenesis are to be achieved. The impairment of lipogenesis in metabolic situations characterized by elevated FFA levels is attributed to the tissue's inability to shut-off energy generation in the citric acid cycle.

An analogy is made between lipogenesis in adipose tissue and ketogenesis in liver, which is also an energy-generating process. Data are presented which suggest that the dissipation of the energy generated in the conversion of FFA to ketoacids may become a rate-limiting factor for ketogenesis. High rates of gluconeogenesis may contribute to the establishment of pathological rates of ketogenesis by raising the energy utilization in the liver. Instead of being limited by a particular rate-limiting enzyme, lipogenesis and possibly ketogenesis can reach rates at which they are controlled by restrictions which originate at the level of the cellular energy metabolism.

Presented by J.P. Flatt. The experimental work reported in this paper was supported by National Institute of Health Grant AM 14161.

References

1. Renold, A.E. and G.F. Cahill, Jr. Preface. In: A.E. Renold and G.F. Cahill, Jr. (Editors), Handbook of Physiology, Section 5, Adipose Tissue. American Physiological Society, Washington, D.C. (1965), pp. 1-5.
2. Jeanrenaud, B. Adipose tissue dynamics and regulation, revisited. Ergeb. Physiol. Biol. Chem. Exp. Pharmakol. 60:57-140(1968).
3. Winegrad, A.I. and A.E. Renold. Studies on rat adipose tissue *in vitro* I. Effects of insulin on the metabolism of glucose, pyruvate, and acetate. J. Biol. Chem. 233:267-272(1958).
4. Crofford, O.B. and A.E. Renold. Glucose uptake by incubated rat epididymal adipose tissue. Rate-limiting steps and site of insulin action. J. Biol. Chem. 240:14-21(1965).
5. Crofford, O.B. and A.E. Renold. Glucose uptake by incubated rat epididymal adipose tissue. Characteristics of the glucose transport system and action of insulin. J. Biol. Chem. 240:3237-3244(1965).

6. Ball, E.G., D.B. Martin and O. Cooper. Studies on the metabolism of adipose tissue I. The effect of insulin on glucose utilization as measured by the manometric determination of carbon dioxide output. J. Biol. Chem. 234:774-780(1959).
7. Herrera, M.G., G.R. Phillips and A.E. Renold. Stimulation of metabolic activity of adipose tissue from fasted rats by prolonged incubation *in vitro*. Biochim. Biophys. Acta 106:221-233(1965).
8. Wieland, O. Der intermediäre Stoffwechsel des Fettgewebes in Hinblick auf die Koordination des Energiehaushaltes. Symp. deut. Ges. Endokrinol. 12:138-153(1967).
9. Delboca, J. and J.P. Flatt. Fatty acid synthesis from glucose and acetate and the control of lipogenesis in adipose tissue. Europ. J. Biochem. 11:127-134(1969).
10. Masoro, E.J. Biochemical mechanisms related to the homeostatic regulation of lipogenesis in animals. J. Lipid Res. 3:149-164(1962).
11. Flatt, J.P. Conversion of carbohydrate to fat in adipose tissue: an energy-yielding and, therefore, self-limiting process. J. Lipid Res. 11:131-143(1970).
12. Numa, S., W.M. Bortz and F. Lynen. Regulation of fatty acid synthesis at the acetyl-CoA carboxylation step. Advanc. Enzyme Reg. 3:407-423(1965).
13. Flatt, J.P. and E.G. Ball. Studies on the metabolism of adipose tissue XIX. An evaluation of the major pathways of glucose catabolism as influenced by acetate in the presence of insulin. J. Biol. Chem. 241:2862-2869(1966).
14. Katz, J. and R. Rognstad. The metabolism of tritiated glucose by rat adipose tissue. J. Biol. Chem. 241:3600-3610(1966).
15. Flatt, J.P. and E.G. Ball. Studies on the metabolism of adipose tissue XV. An evaluation of the major pathways of glucose catabolism as influenced by insulin and epinephrine. J. Biol. Chem. 239:675-685(1964).
16. Rognstad, R. and J. Katz. The balance of pyridine nucleotides and ATP in adipose tissue. Proc. Nat. Acad. Sci. (Washington) 55:1148-1156(1966).
17. Wise, E.M., Jr. and E.G. Ball. Malic enzyme and lipogenesis. Proc. Nat. Acad. Sci. (Washington) 52:1255-1263(1964).
18. Ball, E.G. Regulation of fatty acid synthesis in adipose tissue. Advanc. Enzyme Reg. 4:3-18(1966).
19. Rognstad, R. and J. Katz. Acetyl group transfer in lipogenesis II. Fatty acid synthesis from intra- and extramitochondrial acetyl-CoA. Arch. Biochem. Biophys. 127:437-444(1968).
20. Kneer, P. and E.G. Ball. Studies on the metabolism of adipose tissue XXI. An evaluation of the major pathways of pyruvate metabolism. J. Biol. Chem. 243:2863-2870(1968).
21. Schmidt, K. and J. Katz. Metabolism of pyruvate and L-lactate by rat adipose tissue. J. Biol. Chem. 244:2125-2131(1968).
22. Spencer, A.F. and J.M. Lowenstein. The supply of precursors for the synthesis of fatty acids. J. Biol. Chem. 237:3640-3648(1962).
23. Kornacker, M.S. and E.G. Ball. Citrate cleavage in adipose tissue. Proc. Nat. Acad. Sci. (Washington) 54:899-904(1965).
24. Rognstad, R. and J. Katz. The effect of 2,4-dinitrophenol on adipose tissue metabolism. Biochem. J. 111:431-444(1969).
25. Frerichs, H. and E.G. Ball. Studies on the metabolism of adipose tissue XI. Activation of phosphorylase by agents which stimulate lipolysis. Biochemistry 1:501-509(1962).
26. Fisher, J.N. and E.G. Ball. Studies on the metabolism of adipose tissue XX. The effect of thyroid status upon oxygen consumption and lipolysis. Biochemistry 6:637-647(1967).

27. Rodbell, M. Localization of lipoprotein lipase in fat cells of rat adipose tissue. J. Biol. Chem. 239:753-755(1964).
28. Jeanrenaud, B. and A.E. Renold. Studies on rat adipose tissue *in vitro* VII. Effects of adrenal cortical hormones. J. Biol. Chem. 235:2217-2223(1960).
29. Denton, R.M. and M.L. Halperin. The control of fatty acid and triglyceride synthesis in rat epididymal adipose tissue. Roles of coenzyme-A derivatives, citrate and L-glycerol 3-phosphate. Biochem. J. 110:27-38(1968).
30. Hohorst, H.J., F.H. Kreutz and Th. Bücher. Über Metabolitgehalte und Metabolit-Konzentrationen in der Leber der Ratte. Biochem. Z. 332:18-46(1959).
31. Jungas, R.L. and E.G. Ball. Studies on the metabolism of adipose tissue XII. The effects of insulin and epinephrine on free fatty acid and glycerol production in the presence and absence of glucose. Biochemistry 2:383-391(1963).
32. Cahill, G.F., Jr., B. Leboeuf and R.B. Flinn. Studies on rat adipose tissue *in vitro* VI. Effect of epinephrine on glucose metabolism. J. Biol. Chem. 235:1246-1250(1960).
33. Wieland, O. and M. Suyter. Glycerokinase: Isolierung und Eigenschaften des Enzyms. Biochem. Z. 329:320-331(1957).
34. Margolis, S. and M. Vaughan. α-Glycerophosphate synthesis and breakdown in homogenates of adipose tissue. J. Biol. Chem. 237:44-48(1962).
35. Hall, C.L. and E.G. Ball. Factors affecting lipolysis rates in rat adipose tissue. Biochim. Biophys. Acta. 210:209-220(1970).
36. Ho, R.J., R. England and H.C. Meng. Effect of glucose on lipolysis and energy metabolism in fat cells. Life Sciences 9:137-150(1970).
37. Abramson, E.A. and R.A. Arky. Acute antilipolytic effects of ethyl alcohol and acetate in man. J. Lab. Clin. Med. 72:105-117(1968).
38. Flatt, J.P. Production de $NADPH_2$ cytoplasmique autre que par le cycle des pentoses dans le tissu adipeux. Helv. Physiol. Pharmacol. Acta 25:CR180-182(1967).
39. Halperin, M.L. and B.H. Robinson. The role of the cytoplasmic redox potential in the control of fatty acid synthesis from glucose, pyruvate and lactate in white adipose tissue. Biochem. J. 116:235-240(1970).
40. Krebs, H.A. The regulation of the release of ketone bodies by the liver. Advanc. Enzyme Reg. 4:339-354(1966).
41. Wieland, O. Ketogenesis and its regulation. Advanc. Metab. Disorders 3:1-47(1968).
42. Owen, O.E., P. Felig, A.P. Morgan, J. Wahren and G.F. Cahill, Jr. Liver and kidney metabolism during prolonged starvation. J. Clin. Invest. 48:574-583(1969).
43. Brauer, R.W. Liver circulation and function. Physiol. Rev. 43:115-213(1967).
44. Krebs, H.A. and R. Hems. Fatty acid metabolism in the perfused rat liver. Biochem. J. 119:525-533(1970).
45. Peters, J.P. and D.D. Van Slyke. Quantitative clinical chemistry: Interpretations, Vol. 1. Williams and Wilkins Co., Baltimore (1946), p. 440.
46. Renold, A.E. and G.F. Cahill, Jr. Diabetes mellitus. In: J.B. Stanbury, J.B. Wyngaarden and D.S. Fredrickson (Editors), The Metabolic Basis of Inherited Disease, McGraw-Hill, New York (1960), pp. 69-108.
47. Wieland, O. Proc. 4th Congr. Inter. Diabetes Fed. Editions Med. Hyg., Geneva (1961), pp. 131-137.
48. Shepherd, J. and P.B. Garland. ATP-controlled acetoacetate and citrate synthesis by rat liver mitochondria oxidizing palmitoyl-carnitine and the inhibition of citrate synthase by ATP. Biochem. Biophys. Res. Comm. 22:89-93(1966).
49. Benedict, F.G. and E.P. Joslin. In: A Study of Metabolism in Severe Diabetes, Carnegie Institution, Washington, D.C. (1912).

50. Shaffer, P. Antiketogenesis IV. The ketogenic-antiketogenic balance in man and its significance in diabetes. J. Biol. Chem. 54:399-441(1922).
51. Owen, O.E., A.P. Morgan, H.G. Kemp, J.M. Sullivan, M.G. Herrera and G.F. Cahill, Jr. Brain metabolism during fasting. J. Clin. Invest. 46:1589-1595(1967).
52. Mandel, A.R. and A. Lusk. Stoffwechselbeobachtungen an einem Falle von Diabetes mellitus, mit besonderer Berücksichtigung der Prognose. Deutsch. Arch. klin. Med. 81:472-492(1904).
53. Felig, P., O.E. Owen, J. Wahner and G.F. Cahill, Jr. Amino acid metabolism during prolonged starvation. J. Clin. Invest. 48:584-594(1969).
54. Cahill, G.E., Jr. and T.T. Aoki. How metabolism affects clinical problems. Medical Times 98L10:106-122(1970).
55. Flatt, J.P. and E.G. Ball. Studies on the metabolism of adipose tissue XIV. The manometric determination of total CO_2 production and oxygen consumption in bicarbonate buffer. Biochem. Z. 338:73-83(1963).
56. Bondy, P.K., W.L. Bloom, V.S. Whitner and B.W. Farrar. Studies of the role of liver in human carbmhydrate metabmlism nj tte venous catheter technic II. Patients with diabetic ketosis, before and after the administration of insulin. J. Clin. Invest. 28:1126-1133(1949).
57. Mosenthal, H.O. and D.S. Lewis. The D:N ratio in diabetes mellitus. Bull. Johns Hopkins Hospital 28:187-191(1917).
58. Joslin, E.P. Metabolism in diabetic coma, with especial reference to acid intoxication. J. Med. Res. 6:306-330(1901).
59. Flatt, J.P. Energy metabolism and the control of lipogenesis in adipose tissue. In: B. Jeanrenaud and D. Hepp (Editors), Adipose Tissue: Regulation and Metabolic Function, Academic Press, New York (1970), pp. 93-101.

TABLE 1

On the basis of the data given in Fig. 3 of Flatt and Ball (15) the effect of insulin on the intermediate flow was calculated by subtracting from the flow in the presence of insulin the flow in the control tissue incubated without insulin. The values in the Table are expressed in terms of μmoles/100 mg wet weight tissue/h. The symbol * identifies the reducing-equivalents in the form of NADPH. The flow through the malate cycle was calculated by subtracting from the NADPH required for FA synthesis (1.75/μmole acetyl-CoA incorporated into FA, [cf. (11)]) those supplied by the pentose cycle. In adipose tissue all the reducing-equivalents carried by NADPH in the cytoplasm are utilized for FA synthesis (14).

	REDUCING EQUIVALENTS				ATP	
	Production		Utilization		Production	Utilization
	From Glucose					
Cytoplasm	GAP→DPH	2.10	DHAP→Glyc-P	0.13	DPG→3-PG 2.10	G→G-6-P 1.33
			Pyr→Lact	0.12	PEP→Pyr 2.10	G-6-P→Glycogen 0.14
			OA→Mal	1.89		F-6-P→FDP 0.96
		2.10		2.14		Citr→Acetyl-CoA + OA 1.89
	G-6-P→Pent-P	1.80*	Acetyl-CoA→FA 3.30*			Acetyl-CoA→Malonyl-CoA 1.65
	Mal→Pyr	1.50*				FA (new)→FAcyl-CoA 0.47
		3.30*		3.30*		FA(endog.)→FAcyl-CoA 0.35
Mitochrondria	Pyr→Acetyl-CoA	1.98			Oxid. Phosphor. 7.52	Pyr→OA 1.50
	Acetyl-CoA→CO$_2$	0.15			Succ-CoA→Succ 0.04	
	Mal→OA	0.39			11.76	8.29
		2.52			−8.29	
				Balance	3.47	

153

TABLE 1 (continued)

			From endogenous FA			
Cytoplasm					FA→FAcyl-CoA	−0.06
Mitochondria	FAcyl-CoA→ Acetyl-CoA	−0.42	Oxid. Phosphor. Succ-CoA→Succ	−3.65 −0.24		−0.06
	Acetyl-CoA→CO_2	−0.95		−3.89 −0.06		
			Balance	−3.83		

Combined

From glucose	2.52	3.47
From endogenous FA	−1.37	−3.83
	1.15	−0.36

TABLE 2

EFFECT OF FASTING ON FATTY ACID SYNTHESIS AND ON OXIDATION OF ENDOGENOUS SUBSTRATES BY ADIPOSE TISSUE

	Fed rats	Fasted rats
	μg-at. of C/100 mg tissue $\pm$ S.E.M.	
Glucose uptake	20.85 $\pm$ 2.70	13.45 $\pm$ 2.80
Incorporation of Gluc-^{14}C into FA	8.40 $\pm$ 1.40	2.35 $\pm$ 0.85
Total CO_2 production	8.25 $\pm$ 0.85	4.44 $\pm$ 0.65
Incorporation of Gluc-^{14}C into CO_2	7.55 $\pm$ 1.00	3.08 $\pm$ 0.75
Endogenous CO_2	0.70 $\pm$ 0.45	1.36 $\pm$ 0.34
	(0.2 $\rangle$ P $\rangle$ 0.1)	(P = 0.01)
Total CO_2	0.089 $\pm$ 0.059	0.326 $\pm$ 0.077
	(0.05 $\rangle$ P' $\rangle$ 0.02)	
Production of lactic acid	1.85 $\pm$ 0.35	4.75 $\pm$ 0.75
Release of glycerol	1.85 $\pm$ 0.35	2.20 $\pm$ 0.35

Incubations (120 min) were performed in Krebs-Ringer bicarbonate buffer containing glucose-U-^{14}C (10 mM) and insulin (10^3 μunits/ml). The total production of CO_2 was determined manometrically (55). The data for adipose tissue of fed rats are taken from Flatt and Ball (13), and those for tissue of fasted rats from identical experiments (unpublished), performed concomitantly with rats of the same batches but fasted for 16 hr (1 night) preceding these experiments. Since rats eat mostly at night, this is practically equivalent to a fasting period of 1 day. The fasting rats weighed an average 25 g less than the fed animals. For a detailed description of the experimental procedure cf. (13). The P values show the probabilities that the endogenous CO_2 is not significantly different from zero and the P' value, the probability that the ratio of endogenous to total CO_2 is not increased by fasting.
[Table reproduced from Flatt (11)].

TABLE 3
DATA ON KETONE BODY METABOLISM IN MAN

Metabolic State	grams/day	Reference
Diabetes production	240	(41)
	195	(56)
	385	(56)
Urinary excretion	110	(57)
	120	(50)
	150	(58)
Prolonged starvation		
Production	85	(42)
Utilization	75	(42)
Urinary excretion	10	(42)

Wieland, in a recent review (41), considered that the maximal capacity of the human liver to produce ketone bodies could be estimated at 150g/kg liver/day, or 240 g/day for an adult man's liver which weighs 1.6kg. The data of Bondy *et al.* (56) were obtained by the venous catheter technique in three patients with decompensated diabetic ketoacidosis; the production of ketoacids was calculated from hepatic blood flow and arterio-hepatic venous differences (one subject whose rate of ketone body production was only 11 g/day is not shown in the Table.)

The data on urinary ketoacid excretion were chosen to illustrate the extreme rates which were reported for acutely decompensated diabetic patients before insulin was available for treatment (body weights were respectively 55, 55, and 49 kg).

The rates of ketone body production and utilization reported by Owen *et al.* (42) are based on arterio-venous differences across the splanchnic bed, the kidney and the brain in a series of obese patients (body weight range: 99-124 kg) after 5-6 weeks of complete starvation.

TABLE 4

DECREASE IN HEPATIC GLUCONEOGENESIS AND IN URINARY KETOACID EXCRETION DURING PROLONGED STARVATION

Total urinary nitrogen excretion for day 24 was calculated by averaging the amounts excreted on days 23, 24 and 25 reported by Owen et al. (42) for an obese subject undergoing complete starvation. Similar calculations gave the values for the total nitrogen excretions on days 31 and 36. The starvation was terminated on day 37.

Urinary ketoacids were obtained by adding the excretions of β-hydroxybutyrate and acetoacetate during prolonged starvation reported by Cahill and Aoki (54).

The calculations presented in the Table are based on the difference in urinary total nitrogen and ketoacid excretion between days 24 and 31 (= Δ^{31}_{24}) and days 24 and 36 (= Δ^{36}_{24}). The relative rates of alanine extractions by the liver and the kidney are based on the data of Owen et al. (43) on hepatic and renal blood flows and arterio-hepatic venous differences of α-amino nitrogen. The rationale for the calculation is explained in the text in discussing the difference between the rates of ketogenesis observed in diabetes and in prolonged starvation.

	Observed		Predicted
	Urinary Nitrogen Excretion	Urinary Nitrogen Excretion	Ketoacid Synthesis
	gN/day	moles ketoacid/day	moles ketoacid synthesized/day
Day 31	4.4	0.094	---
Day 24	5.3	0.107	---
Δ^{31}_{24}	−0.9	−0.013	−0.014
Day 36	4.1	0.092	---
Day 24	5.3	0.107	---
Δ^{36}_{24}	−1.2	−0.015	−0.018

The following conversion factors were used in the calculation of the constant 0.015 moles ketoacid formed/g urinary nitrogen:

$0.44 \times \dfrac{\text{Liver } \alpha\text{-amino N extraction}}{\text{Liver + kidney } \alpha\text{-amino N extraction}}$, (42);

3.65g glucose/g urinary N, (52);
10 moles ATP/mole glucose formed from alanine:
6 moles ATP/mole ketoacid formed from oleic acid.

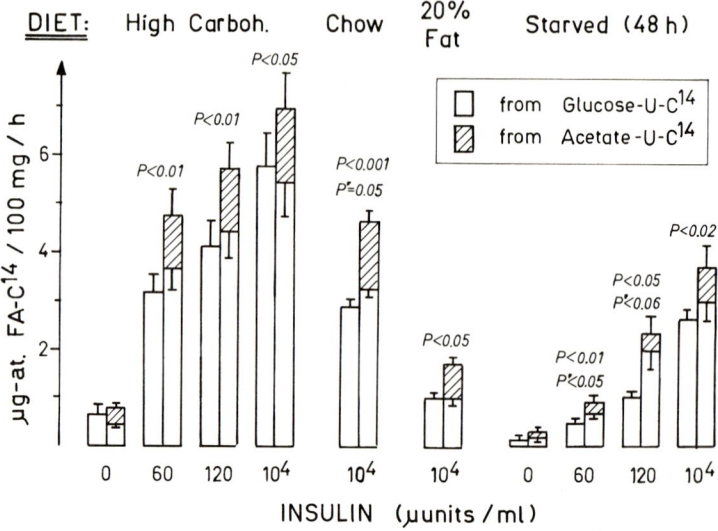

Fig. 1. *Incorporation of glucose-C and acetate-C into fatty acid by rat epididymal adipose tissue.* The data are taken from Delboca and Flatt (9) and Ackerman and Flatt (unpublished). In each pair of columns, open areas show: the incorporation of glucose-C into FA during incubation with glucose-U-^{14}C (10mM) (left column), or with glucose-U-^{14}C in the presence of unlabeled acetate (15 mM) (right column). Incubations with unlabeled glucose and acetate-U-^{14}C in the medium gave the incorporations shown by hatched areas; these are represented on the top of the right hand columns to show the total rate of fatty acid synthesis in the presence of the two substrates. The brackets show the size of the standard errors of the mean. The P values were obtained by the Student's *t* test, using the difference between fatty acid synthesis from glucose plus acetate and from glucose alone; P_r values apply to the difference between glucose-C incorporation into fatty acid with and without acetate in the incubation medium.

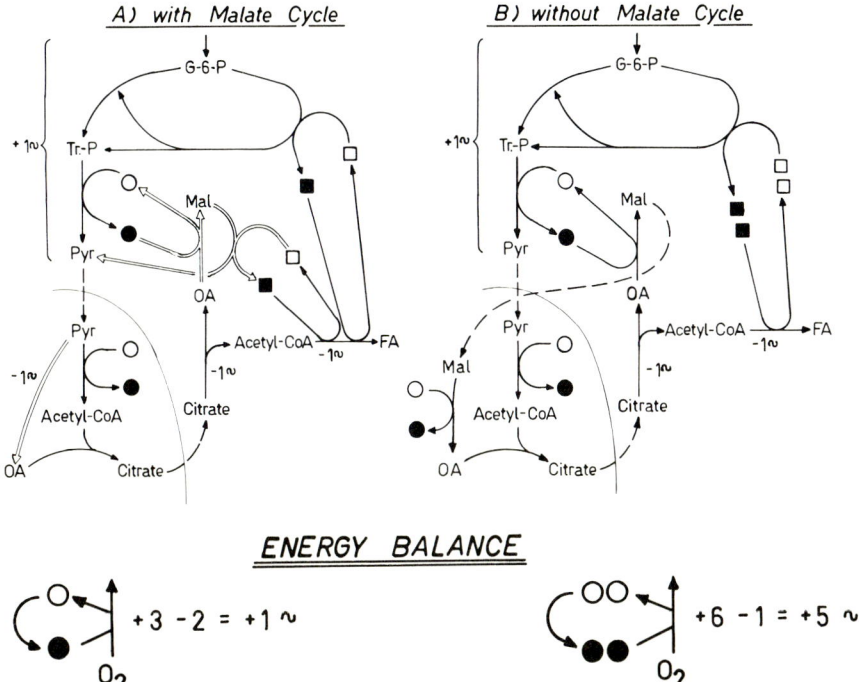

Fig. 2. *Rates of production and utilization of reducing-equivalents during lipogenesis in rat adipose tissue*. The horizontal bars show by their length the number of μmoles of reducing equivalents (2 electrons) produced or utilized in one hour in 100 mg of wet weight adipose tissue, as calculated from the pathway flow estimation of Flatt and Ball (15). The number of μg-atoms of oxygen utilized, shown by horizontal segments, was determined manometrically. The transfer of reducing equivalents from NADH to NADP$^+$ by the malate cycle, calculated from the data of Kneer and Ball (20) during FA synthesis in the presence of pyruvate but not glucose, has also been represented. The production of reducing equivalents from endogenous fatty acids is based on measurements of the production of endogenous CO_2: this was determined by subtracting from the total CO_2 produced (determined manometrically) the CO_2 formed from glucose-U-^{14}C (15); 1.75 reducing-equivalents are considered to be required (or generated) per acetyl-CoA incorporated into (or formed from) fatty acid [*cf.*(11). Figure adapted from Flatt (59)].

159

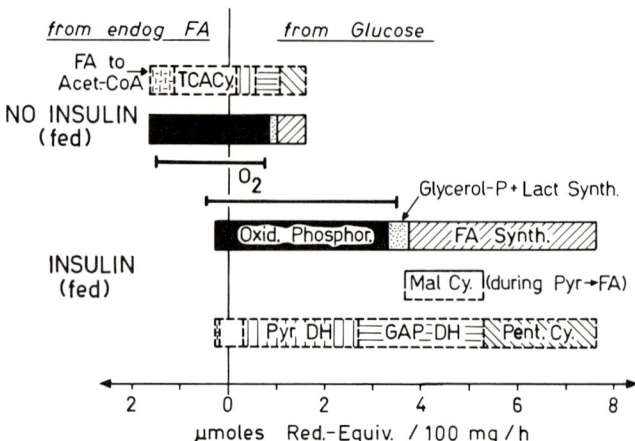

Fig. 3. *Production and utilization of reducing equivalents and high-energy bonds as one acetyl-CoA is formed from glucose and incorporated into FA.* The oxidized and reduced forms of the coenzymes which transport these reducing equivalents are: NAD^+, ○; NADH, ●; $NADP^+$, □; and NADPH, ■. In Scheme A, one reducing equivalent for the incorporation of an acetyl-CoA into FA is obtained by transfer from NADH to $NADP^+$ through the reactions of the malate cycle (shown by double-line arrows), while the other required NADPH is generated in the pentose cycle. In Scheme B all the NADPH needed for FA synthesis is generated by the pentose cycle. The simplified stoichiometry shown here neglects the fact that one acetyl-CoA per molecule of long-chain FA synthesized requires no reducing equivalent and that one reducing equivalent is required for the synthesis of glycerol-P for each molecule of triglyceride formed; the error due to this simplification is negligible [*cf.* (11) Table 1].

The net Energy Balance is also shown; it includes the high-energy bonds formed by reoxidation (with a P:O ratio of 3) of the reducing-equivalents generated in excess of those required for reductive synthesis. [Figure adapted from Flatt (11).]

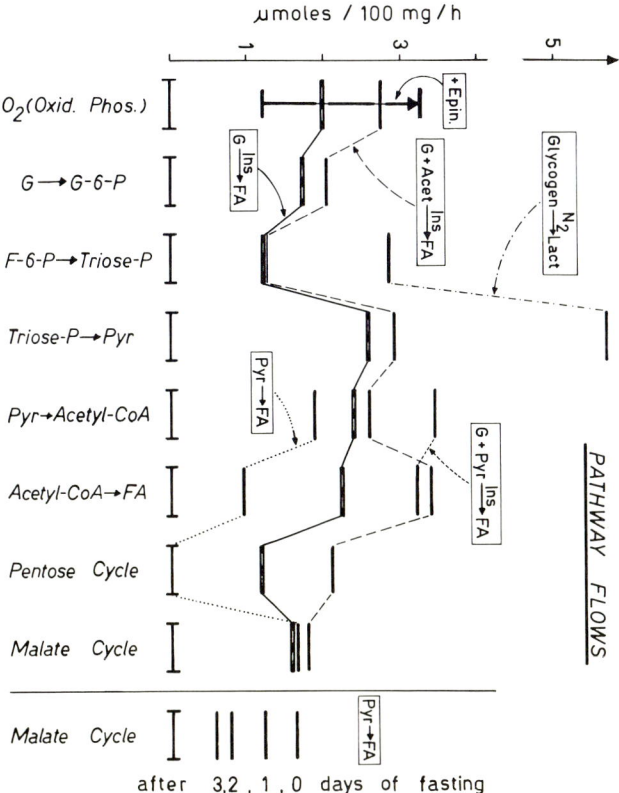

Fig. 4. *Rates of flow of intermediates through the metabolic pathways involved in converting glucose to fat in rat adipose tissue.* The horizontal segments connected by thin lines show the flow of intermediates as determined during *in vitro* incubations of rat adipose tissue under the various conditions indicated by inserts. The following data were used: glucose, in the presence of 10^3 μunits of insulin/ml: Flatt and Ball (15); glucose plus acetate, in the presence of 10^3 μunits of insulin/ml: Flatt and Ball (13); glucose plus pyruvate, in the presence of 10^4 μunits of insulin/ml: Delboca and Flatt (9); pyruvate: Kneer and Ball (20); anaerobic glycolysis: Frerichs and Ball (25). The vertical segment at left shows the tissue's ability to increase its O_2 consumption upon addition of epinephrine in the presence of glucose (31). At far right the effect of fasting on the rates of flow through the malate cycle is shown, as calculated from measurements of FA synthesis during FA synthesis from pyruvate (11). [Figure reproduced from Flatt (59).]

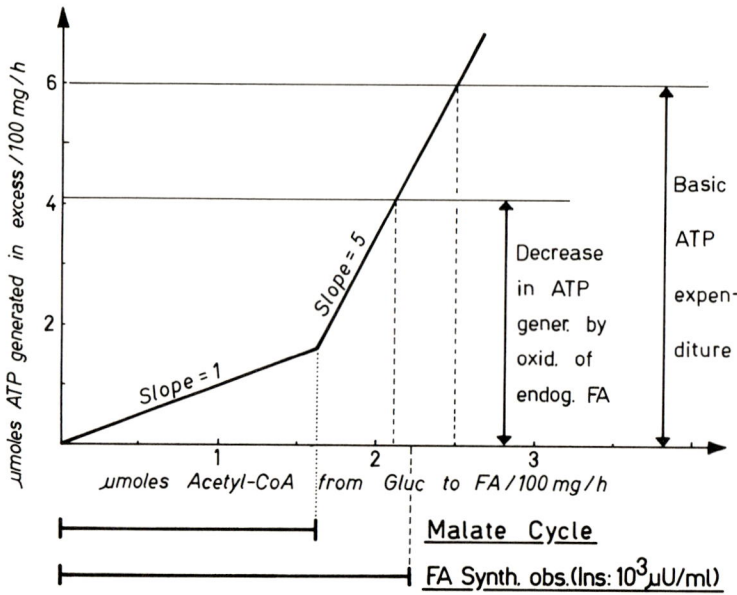

Fig. 5. *Number of high energy bonds (or ATP) formed during FA synthesis from glucose.* As long as the malate cycle participates (Fig. 3A), 1 mole of ATP is yielded per mole of acetyl-CoA derived from glucose and incorporated into FA. When the conversion of glucose to FA synthesis exceeds the capacity of the malate cycle (Fig. 1B), the ATP yield becomes 5. The maximal capacity of the malate cycle is based on the assumption that this cycle works at its full capacity during fat synthesis from pyruvate, generating 0.875 moles of NADPH per μg-atom of pyruvate-C found to be incorporated into FA in the experiments of Kneer and Ball (20). Thus, the heavy line shows the μmoles of ATP generated as a function of the number of μmoles of acetyl-CoA derived from glucose and incorporated into FA.

The distance between the abcissa and the upper horizontal line is the amount of ATP used by adipose tissue in the absence of fat synthesis, based on the consumption of 1 μmole of O_2/100 mg of tissue/h. [*cf.* (11)]. The distance between the abcissa and the lower horizontal line shows the suppressible production of ATP by oxidation of endogenous FA, determining the upper limit for the conversion of glucose to FA. For comparison, the observed rate of FA synthesis is shown, as measured by the incorporation of ^{14}C from glucose-U-^{14}C into FA by adipose tissue incubated in the presence of 1000 μunits of insulin/ml (15). [Figure reproduced from Flatt (11).]

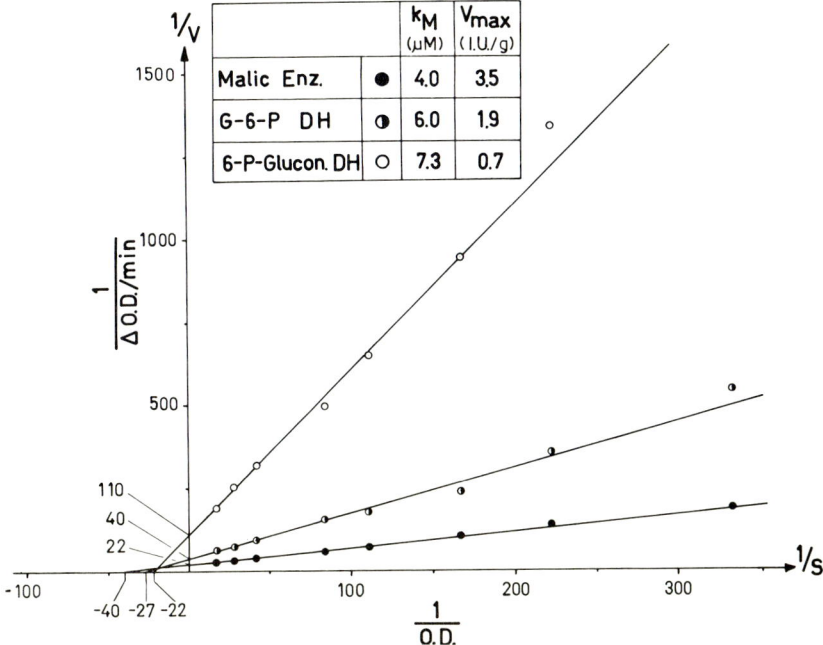

Fig. 6. *Lineweaver-Burke plot relative to $NADP^+$ for three NADP-linked dehydrogenases of rat adipose tissue.* The reaction mixture contained in a total volume of 2.3 ml: triethanolamine]HCl (pH 7.5; 112 μmoles), $MgCl_2$ (17 μmoles), EDTA (4.4 μmoles), adipose tissue homogenate (supernatant, 100,000 g × 30 min) equivalent to 4.8 mg of adipose tissue, and either Na Malate (11 μmoles), Na 6-P-gluconate (4.4 μmoles), or Na 6-P-gluconate (4.4 μmoles) plus Na glucose-6-P (2.2 μmoles). [*cf.* (11). Figure reproduced from Flatt (11)].

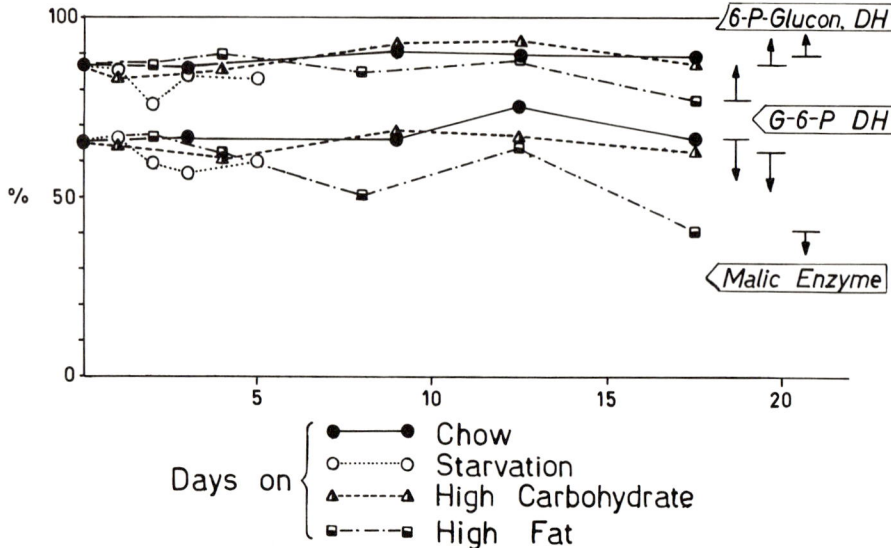

Fig. 7. *Effect of dietary status on relative proportions of NADP-linked dehydrogenases in epididymal rat adipose tissue.* The activities of malic enzyme, glucose-6-P dehydrogenase and 6-P-gluconate dehydrogenase were determined in high-speed supernatants of rat epididymal fat pad homogenates (Ellis and Flatt, unpublished). Each point represents the average of 6 determinations (or 3 determinations for the points on days 12-13 and 17-18). The activity of each enzyme is expressed in terms of percentage relative to the sum of the activities of the three measured enzymes. Assay conditions were: Malic enzyme: triethanolamine·HCl, 64 mM, pH 7.4; Mn^{++}, 4 mM; $NADP^+$, 0.2 mM; malate, 5 mM. 6-Phosphogluconate dehydrogenase: tris·HCl, 64 mM, pH 8.4; $NADP^+$, 0.2 mM; 6-P-gluconate, 2 mM. Glucose-6-P, dehydrogenase: same as previous assay system but supplemented with glucose-6-P, 1 mM. The difference between the reaction rates in the two latter assays was taken to represent glucose-6-P dehydrogenase activity.

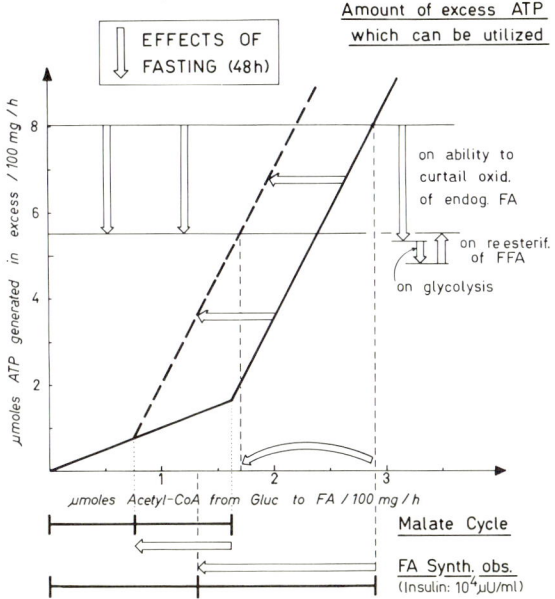

Fig. 8. *Effect of a fasting period of 48 hr on the maximal possible rate of glucose conversion to fat.* The figure is constructed on the same principle as Fig. 5, using the data shown on the extreme right of Fig. 4 to calculate the maximal capacities of the malate cycle. The amount of ATP produced during fat synthesis from glucose is obtained from the solid line which shows ATP production as a function of the rate of FA synthesis from glucose-U-^{14}C, in the presence of 10^4 μunits/ml of insulin, in adipose tissue of rats fed a high carbohydrate diet (9). It corresponds also to the amount of ATP which can be utilized in the adipocytes for sustaining their basic energy expenditure.

The changes caused by a 48-hr fasting period (shown by arrows in the figure) were evaluated in the following manner: 1) The ability to curtail the oxidation of endogenous FA during fat synthesis was found to be decreased by 0.33 μmoles of CO_2/100 mg/hr in adipose tissue of rats fasted for one day (Table 2). In the absence of other data the above value was used to estimate the reduction in the amount of ATP which can be utilized in tissue of rats fasted for 2 days: (0.33/0.7)(2)(2.85) = 2.7 μmoles/100 mg/hr (where 0.7 is the RQ for FA oxidation, 2 is the number of reducing equivalents oxidized per μmole of O_2 consumed, and 2.85 the theoretical P:O ratio for palmitate oxidation (27). 2) The increase of 0.5 μmoles/100 mg/hr in the production of lactic acid (9) and hence of ATP was considered to reduce the utilizable ATP by a like amount. 3) The increase of 0.1 μmole/100 mg/hr in the release of glycerol (9) reflects a greater ATP expenditure for reesterification [7 μmoles of ATP/1 μmole of glycerol (26)] considered to augment the utilizable ATP by 0.7 μmoles/100 mg/hr.

The estimated amount of ATP (generated by conversion of glucose to fat) which adipose tissue of rats fasted for 48 hr is able to utilize is shown by a horizontal line. Its intersection with the broken line, which shows the ATP production as a function of FA synthesis in tissue of fasting rats, determines the maximal predicted rate of glucose conversion to fat. The rate observed in the presence of 10^4 μunits/ml of insulin (9) is shown for comparison. [Figure reproduced from Flatt (11).]

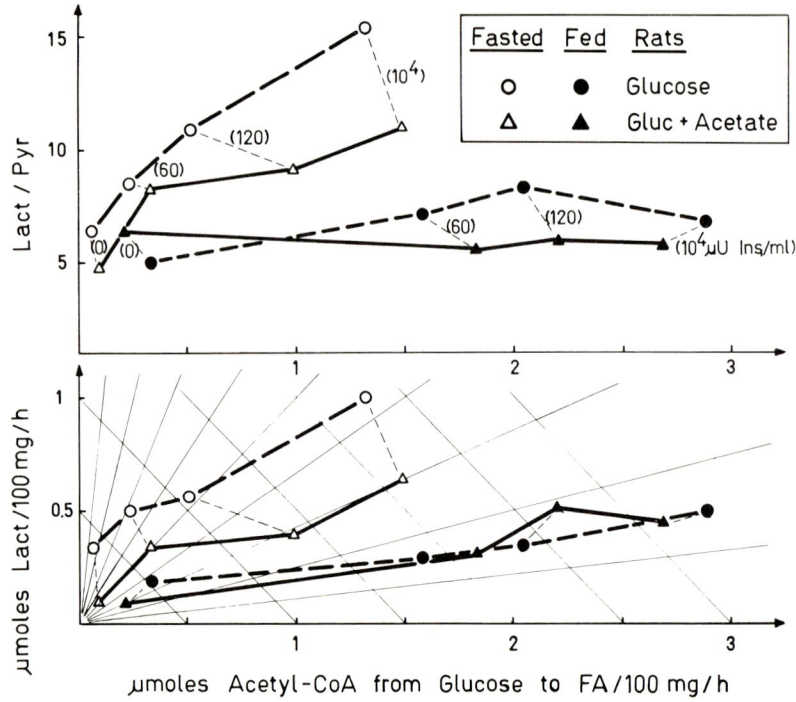

Fig. 9. *Effect of fasting and of acetate on the relative rates of lactate and FA synthesis from glucose.* The figure was constructed from data of Delboca and Flatt (9) obtained with rat adipose tissue incubated *in vitro* in Krebs-Ringer bicarbonate buffer. The experiments were performed with tissue from rats fed *ad libitum* (●, ▲) and from rats fasted for 48 hr (○, △), incubated with glucose [10 mM (●, ○)] and with glucose plus sodium acetate [15 mM (▲, △)]. The insulin concentrations were 0, 60, 120, and 10^4 μunits/ml, as indicated for each series of paired incubations.

The diagonal grid aids in showing the approximate number of μmoles of pyruvate formed from glucose; since in adipose tissue only very small amounts of glucose-C are oxidized to CO_2 in the citric acid cycle (15, 16). The radial lines assist in evaluating the percentage of pyruvate diverted to lactate. [Figure reproduced from Flatt (11).]

METABOLIC REGULATION

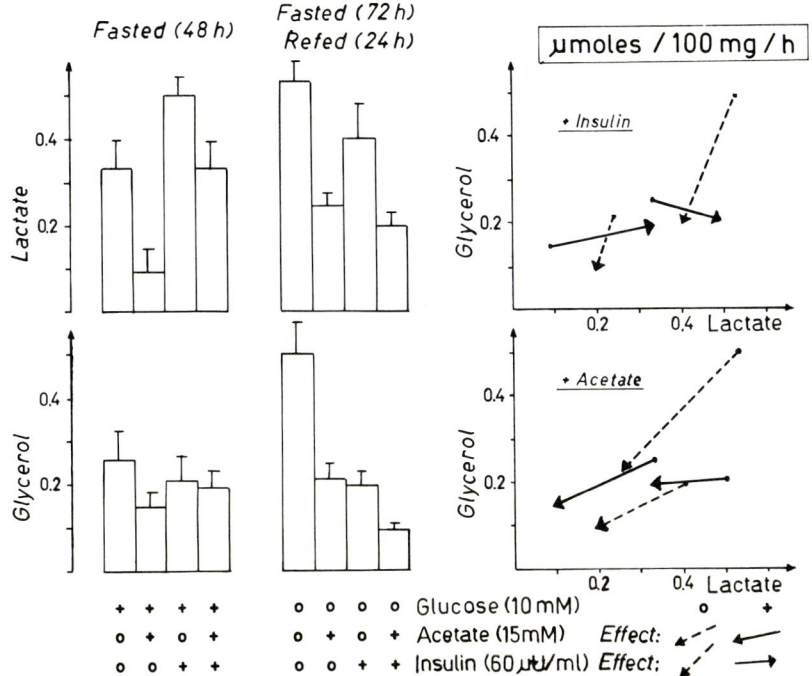

Fig. 10. *Antilipolytic effect of acetate.* Epididymal fat pads were incubated for 150 min in bicarbonate buffer supplemented with glucose, acetate and insulin as indicated. The vertical bars and brackets give the number of μmoles/100 mg/h (average of 4-9 experiments ± S.E.M.) of lactate and glycerol released into the medium (determined enzymatically). The effect of insulin addition or of acetate addition is shown by the direction and length of the arrows, whose origins and heads correspond respectively to the lactate and glycerol releases in identical experiments except for the absence and presence of insulin (upper graph) or of acetate (lower graph) in the incubation medium. [From Delboca and Flatt (9, 38).]

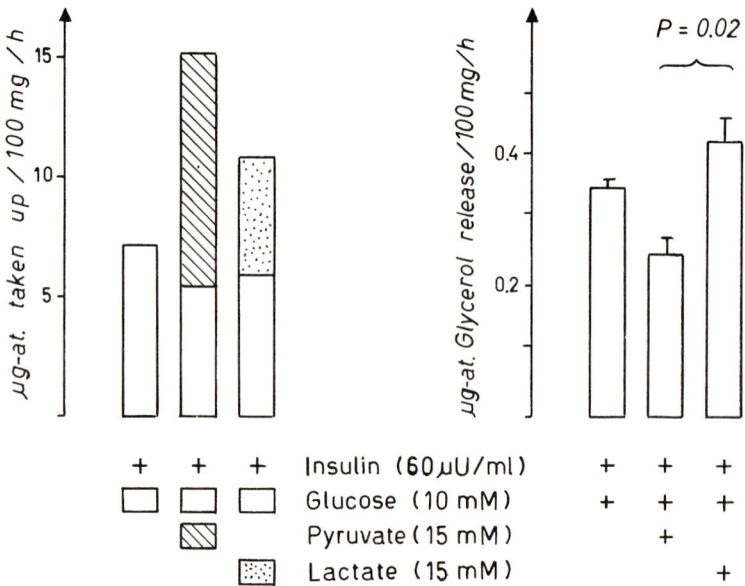

Fig. 11. *Epididymal fat pads of fed rats were incubated for 150 min in bicarbonate buffer containing 2 mg of gelatin (6) in the presence of insulin, glucose, pyruvate or lactate as indicated (Ackerman and Flatt, unpublished).* The vertical bars to the right show the average uptakes of glucose ☐, pyruvate ☐ and lactate ☐, and to the left the average releases of glycerol in terms of µg-atoms of C per 100 mg wet weight per hr. The results are the averages of 5-6 experiments with brackets to show the S.E.M. for the glycerol measurements. The P value, obtained by the Student's t test, gives the probability that the glycerol releases in the presence of pyruvate and lactate are not different.

METABOLIC REGULATION

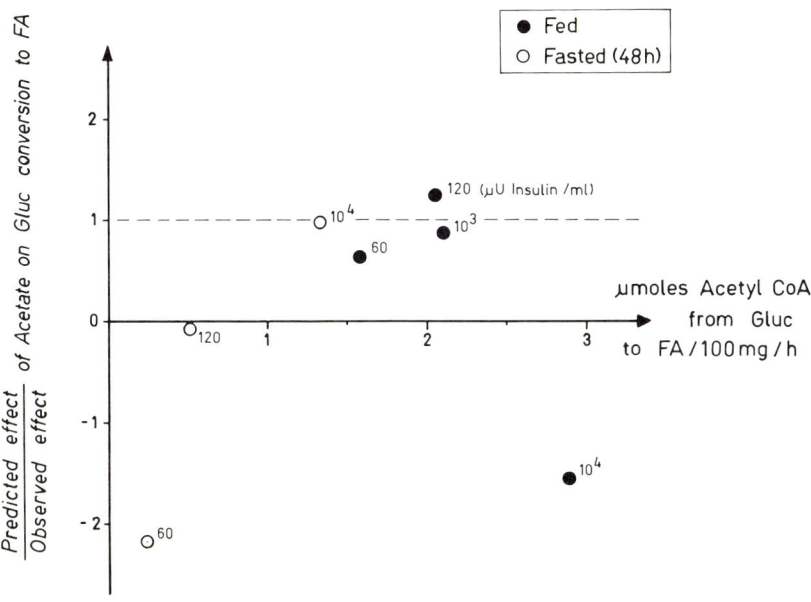

Fig. 12. *Effects of acetate or epinephrine on the conversion of glucose-C to FA.* The figure is constructed from published data (9, 13, 15). "Predicted effects" were calculated on the assumption that the conversion of glucose-C to FA is increased by 0.2 μmoles of acetyl-CoA/μmole increase in the demand for ATP resulting from the presence of acetate (15 mM) or epinephrine (0.2 μg/ml) in the incubation medium, in addition to glucose (10 mM). Solid and open symbols apply respectively to experiments performed with fed or 48-hr-fasted rats; the concentrations of insulin (μunits/ml) are shown by small numbers. a) Circles: The effect of acetate utilization on the ATP balance was computed from the data of Delboca and Flatt (9) and Flatt and Ball (13); +10 ATP per molecule of acetate oxidized to CO_2; −3 ATP per molecule of acetate converted to FA. Allowance was made for the changes caused by acetate in the production of glycerol and lactate, considering that the release of 1 μmole of glycerol indicates an expenditure of 7 μmoles of ATP for FFA reesterification (26), and that 1 μmole of ATP is gained per μmole of lactate formed from glucose. b) Triangle: The changes in pathway flow due to the addition of a small dose of epinephrine (0.2 μg/ml) in the experiments of Flatt and Ball (15) were calculated (= flow of intermediates with insulin and epinephrine minus flow of intermediates with insulin). From these values there was deducted the flow of intermediates due to the increase in FA synthesis, assuming the required NADPH to be generated in the pentose cycle. The values so obtained served to evaluate the effect of epinephrine on the ATP balance. A P:O ratio of 3 was employed to convert reducing equivalents to high energy bonds. [Figure reproduced from Flatt (11).]

LIVER METABOLITE CONTENT, REDOX AND PHOSPHORYLATION STATES IN RATS FED DIETS CONTAINING 1,3-BUTANEDIOL AND ETHANOL

R.L. Veech and M.A. Mehlman

Introduction

Calories can be supplied by many metabolizable organic compounds. We have investigated the feasibility of supplementing animal and human diets with organic compounds which are chemically synthesized. One such compound, 1,3-butanediol (BD) is readily metabolized by rats. BD containing diets have been shown to modify lipid content (1, 2) and gluconeogenic enzymes and metabolite patterns (3-8). It has been shown that BD is metabolized initially in a similar manner to ethanol, with subsequent oxidation to β-hydroxybutyrate (9).

The metabolism of alcohols produces reducing equivalents in the cytoplasm. Since alteration in the cytoplasmic redox state can cause profound effects on hepatic metabolism, the redox and phosphorylation states and metabolite levels in livers of BD fed rats were studied. The results of this study were compared with those found during fasting and after ethanol administration.

Methods

Treatment of animals

Albino male rats of Sprague-Dawley strain which weighed 116 to 124 g were housed in individual stainless steel cages with wire screen bottoms and kept at 25 ± 2°. All animals were fed *ad libitum*. Two groups of eight rats were fed either a 30% fat diet or a 30% fat diet plus 20% BD. Food consumption was measured daily. The compositions of these diets are shown in Table 1. All experiments were performed on the 43rd day after starting the diet. Early in the morning eating was permitted until immediately prior to sacrifice.

Procedures for urine collection and analysis for BD and ketones

Urine was collected in flasks kept in a dry ice-alcohol mixture for twenty-four hour periods. β-Hydroxybutyric acid and acetoacetic acid were analyzed by enzymatic procedures (10, 11). 1,3-Butanediol (BD) was analyzed by vapor phase chromatography on a Varian Aerograph Model 1200 using a flame ionization detector by the procedure developed by Dr. John L. Couvillion of Celanese Chemical Corporation. A 2 ft × 1/8 in stainless steel column packed with 25% Carbowax 20 MTPA on acid-washed dimethylchlorosilane-treated Chromosorb W (60-80 mesh) was used with nitrogen carrier at a flow rate of 40 cc/min. Samples of 0.5 µl were injected into the injection port which was operated at 195°C; the column and detector temperatures were 165° and 250° respectively. Under these conditions the retention time of BD was 4 min with a detection limit of 10 ppm.

Procedures for metabolic analysis

Livers were removed within 8 sec after cervical dislocation and pressed between aluminum blocks previously cooled in liquid N_2. The treatment of tissue has been described previously by Veech, Eggleston and Krebs (12). Tissue metabolites were measured by the following procedures: lactate, malate and α-glycerophosphate (13), pyruvate (14), dihydroxyacetone phosphate (15), α-ketoglutarate (16), acetoacetate and β-hydroxybutyrate (17), 3-phosphoglycerate and phosphoenolpyruvate (18), glucose-6-phosphate (19), citrate (20), ATP (21), ADP and AMP (22), inorganic phosphate (23), CoA (24), and acetyl-CoA (25). NH_4^+ was determined by the method of Folbergrova (26) except that the samples were treated with 0.05 g of Norite A per ml of neutralized perchlorate extract of liver.

Experiments in which known amounts of the measured metabolites were added to the tubes prior to the addition of the frozen liver showed that recovery was between 90 and 105%. The spectrophotometric measurements were carried out with a Zeiss PMQII.

Calculation of redox and phosphorylation states

For all calculation the intracellular pH was assumed to be 7.0. The concentration of isocitrate was assumed to be equal to measured citrate concentration/15. Previous measurements of citrate/isocitrate ratios yielded 15 as a mean value for rat liver (27) and which agrees well with the equilibrium ratio of citrate/isocitrate of 13.5 found by Krebs (28) using purified aconitase (EC 4.2.1.3).

The cytoplasmic [NAD⁺]/[NADH] ratio was calculated from the lactate dehydrogenase reaction according to the method of Hohorst et al. (19), using $K_{LDH} = 1.4 \times 10^{-4}$ (30).

In addition the cytoplasmic [NADP⁺]/[NADPH] ratio was also calculated from the isocitric dehydrogenase reaction according to the proposal of Veech et al. (12), assuming a constant CO_2 concentration of 1.16×10^{-3} M where:

$$K_{ICDH} = \frac{[\alpha\text{-ketoglutarate}][CO_2][NADH]}{[\text{Isocitrate}][NADP^+]} = 1.17M$$

Cytoplasmic phosphorylation state was calculated from the direct measurement of ATP, ADP and P_i. The P_i concentration was multiplied by 0.6 in order to convert total phosphate to the ionic species HPO_4^-. The mitochondrial [NAD⁺]/[NADH] ratio was calculated using the proposal of Williamson et al. (29):

$$K_{GDH} = \frac{[\alpha\text{-ketoglutarate}][NH_4^+][NADH]}{[\text{glutamate}][NAD^+]} = 3.87 \times 10^{-6}M$$

Results

The average body weight, food consumption and food efficiency (g gain/g food intake x 100) of control and animals fed 20% 1,3-butanediol containing diets for 43 days are presented in Table 2. The results showed that animals fed BD containing diets consumed less food and had significantly lower body weight gain. These results are in agreement with previous publications (3-7), which showed that addition of BD to diets resulted in decreased body weight gain and food consumption of rats fed BD for shorter periods of time than in the present study.

The excretions of 1,3 butanediol and ketone bodies in rat urine for a 24 hour period are presented in Table 3. In non-fasted rats fed dietary BD it was found that 4.67% of ingested BD was excreted in the urine. These data indicate that most of the dietary BD is readily metabolized by the rat. Urinary excretion of ketones (β-hydroxybutyrate and acetoacetate) was greatly increased by feeding BD. This increase in the levels of ketone bodies in the urine and in the blood (8) is consistent with previously reported metabolic pathways for oxidation of BD (8, 9) as depicted in Fig. 1.

Figure 2 shows metabolite levels in frozen-clamped livers for starved rats and rats administered ethanol. A comparison between control and BD fed rats showed that BD feeding caused highly significant decreases in lactate (P less than 0.001), pyruvate (P less that 0.001), 3-phosphoglycerate (P less than 0.002), dihydroxyacetone phosphate (P less than 0.002), glycerophosphate (P less than 0.05), glucose-6-phosphate (P less than 0.01) and α-ketoglutarate (P less than 0.02) levels in livers of BD fed rats; malate, citrate, isocitrate, glutamate, NH_4^+, CoA, ATP, ADP, AMP and P_i levels were unchanged, but acetoacetate, β-hydroxybutyrate and acetyl-CoA were increased.

Ethanol caused large increases in the levels of lactate (P less than 0.001), malate (P less than 0.001), α-glycerophosphate (P less than 0.001) and AMP (P less than 0.05). A larger decrease was seen with ethanol than with BD in the levels of phosphoenolpyruvate and 3-phosphoglycerate. The changes in the metabolite levels observed with BD resembled those obtained with 48 hour fasted rats.

The results in Fig. 3. show redox and phosphorylation states in frozen-clamped rat liver. The cytoplasmic $[NAD^+]/[NADH]$ ratio calculated from lactate dehydrogenase was greatly decreased by BD feeding, ethanol administration and starvation (P less than 0.001). The cytoplasmic $[NADP^+]/[NADPH]$ ratio calculated from isocitrate dehydrogenase was decreased in BD fed rats, but not in starved or ethanol treated rats. The mitochondrial $[NAD^+]/[NADH]$ ratio calculated from β-hydroxybutyrate dehydrogenase was decreased by the three treatments. The phosphorylation state in the cytosol was decreased by starvation and BD feeding, but was increased in ethanol administration.

Discussion

Rats fed 1,3-butanediol-containing diets had a metabolite pattern in frozen-clamped liver very similiar to that seen after 48 hours of starvation (Fig. 2 and Fig. 4). However, the degree of metabolite change was smaller with BD feeding than after starvation.

The pathway for metabolism of BD in animal tissues was proposed by Mehlman *et al.* (8), and the enzymatic steps and intermediates involved in this pathway were established by Tate *et al.* (9). Fig. 1 shows that 1,3-butanediol is metabolized by NAD^+-linked alcohol and aldehyde dehydrogenases to β-hydroxybutyrate. β-Hydroxybutyrate formed from BD can be partially converted to acetoacetate in the liver mitochondria. The extent of this conversion of β-hydroxybutyrate to acetoacetate is determined by the mitochondrial redox state described by Williamson *et al.* (29). Thus, the administration of BD shows that the drastic fall in the metabolites below **DHAP** in the glycolytic pathway after ethanol administration is not due to the action of alcohol dehydrogenase alone, but due to the further metabolism of ethanol. The mechanism of this fall can be understood by considering the reactions of glyceraldehyde phosphate dehydrogenase and 3-phosphoglycerate kinase, depicted below, which have been shown to be near equilibrium (30).

$$K = \frac{[\text{GAP}]\ [\text{MgATP}]\ [\text{HPO}_4^{-2}]\ [\text{NADH}]}{[\text{3-GP}]\ [\text{MgADP}]\ [\text{NAD}^+]}$$

A rise in the factors

$$\frac{[\text{MgATP}]\ [\text{HPO}_4^{-2}]}{[\text{MgADP}]} \quad \text{and} \quad \frac{[\text{NADH}]}{[\text{NAD}^+]}$$

leads to a profound fall in 3-phosphoglycerate and its isomerase partners 2-phosphoglycerate, 3-phosphoglycerate and phosphoenolpyruvate. Use of the "crossover analysis" can lead to an erroneous conclusion about changes in enzyme activity unless proper consideration is given to the cofactors involved in the reactions.

It would appear that feeding BD as a replacement for carbohydrate results in a state in liver similar to a very mild starvation. Despite the similarity of metabolic pathways, BD does not produce a metabolic pattern similar to ethanol. It seem unlikely, therefore, that BD would produce the hepatotoxic effects associated with chronic ethanol feeding.

Summary

Diets containing 1,3-butanediol (BD) as a replacement for carbohydrate were fed to rats for 43 days, and the concentrations of metabolites were measured in rat liver. In animals fed BD-containing diets there were decreases in the contents of acetate, pyruvate, 3-phosphoglycerate, dihydroxyacetone phosphate and α-ketoglutarate, while the contents of acetoacetate, β-hydroxybutyrate and acetyl-CoA were increased. The cytoplasmic [NAD$^+$]/[NADH], [NADP$^+$]/[NADPH] ratios were significantly decreased in livers of animals fed BD diets.

The effects observed on liver metabolites during ethanol administration were very different from those obtained with BD in spite of the close similarity of the metabolic pathways of BD and ethanol. The differences in the cytoplasmic [NAD$^+$]/[NADH] and [NADP$^+$]/[NADPH] ratios were even more pronounced.

Hepatotoxic changes expected from chronic ethanol administration are unlikely to result from 1,3-butanediol feeding despite their similarity in metabolic pathways.

Presented by M.A. Mehlman. The experimental work reported in this paper was supported by Grant AM-13782 from the National Institute of Arthritis and Metabolic Diseases.

References

1. Schlüssel, H. Beitrag Zur Verwertung mehr wertiger Alkohole in der Ernährung. Arch. Exp. Pathol. Pharmakol. 221:67-75(1954).
2. Miller, S.A. and H.A. Dymsza. Utilization by the rat of 1,3-butanediol as a synthetic source of dietary energy. J. Nutrition 91:79-88(1967).
3. Stoewsand, G.S., H.A. Dymsza, M.A. Mehlman and D.G. Therriault. Influence of 1,3-butanediol on tissue lipids of cold exposed rats. J. Nutrition 87:464-468(1965).
3. Stoewsand, G.S., H.A. Dymsza, M.A. Mehlman and D.G. Therriault. Influence of 1,3-butanediol on tissue lipids of cold exposed rats. J. Nutrition 87:464-468(1965).
4. Stoewsand, G.S., H.A. Dymsza, S.M. Swift, M.A. Mehlman and D.G. Therriault. Effect of feeding polyhydric alcohols on tissue lipids and the resistance of rats to extreme cold. J. Nutrition 98:414-418(1966).
5. Dymsza, H.A., G.S. Stoewsand, M.A. Mehlman and D.G. Therriault. Fuel utilization in cold exposed rats. In: D.M. Zipf (Editor), Symposium on influence of cold on metabolic regulation. The Bulletin, New Jersey Academy of Science (March, 1969), p. 41.
6. Mehlman, M.A., R.B. Tobin and J.B. Johnston. Influence of dietary 1,3-butanediol on metabolites and enzymes involved in gluconeogenesis and lipogeneses in rats. J. Nutrition 100:1341(1970).
7. Mehlman, M.A., D.G. Therriault, W. Porter, G.S. Stoewsand and H.A. Dymsza. Distribution of lipids in rats fed 1,3-butanediol. J. Nutrition 88:215-218(1966).
8. Mehlman, M.A., R.B. Tobin and J.B. Johnston. Metabolic control of enzymes in normal, diabetic and diabetic insulin treated rats utilizing 1,3-butanediol. Metabolism 20:149(1971).
9. Tate, R.L., M.A. Mehlman and R.B. Tobin. Metabolic fate of 1,3-butanediol in the rat: conversion to β-hydroxybutyrate. J. Nutrition 101:1719-1726(1971).
10. Williamson, D.H. and J. Mellanby. D-(-)-β-hydroxybutyrate. In: H.U. Bergmeyer (Editor), Methods of enzymatic analysis, Academic Press, New York (1965), p. 459.
11. Mellanby, J. and D.H. Williamson. Acetoacetate. In: H.U. Bergmeyer (Editor), Methods of enzymatic analysis, Academic Press, New York (1965), p. 454.
12. Veech, R.L., L.V. Eggleston and H.A. Krebs. The redox state of free nicotinamide-adenine dinucleotide phosphate in the cytoplasm of rat liver. Biochem. J. 115:609-619(1969).
13. Hohorst, H.J. D-glucose-6-phosphate and D-fructose-6-phosphate-determination with glucose-6-phosphate dehydrogenase and phosphoglucose isomerase. In: H.U. Bergmeyer (Editor), Methods of enzymatic analysis, Verlag Chemie, Weinheim (1963), p. 134.
14. Bücher, Th., R. Czok, W. Lamprecht and E. Latzko. Pyruvate. In: H.U. Bergmeyer (Editor), Methods of enzymatic analysis, Verlag Chemie, Weinheim (1963), p. 253-259.
15. Bücher, Th. and H.J. Hohorst. Dihydroxyacetone phosphate, fructose-1,6-diphosphate and D-glyceraldehyde-3-phosphate; determination with glycerol-1-phosphate dehydrogenase, aldolase and triosephosphate isomerase. In: H.U. Bergmeyer (Editor), Methods of enzymatic analysis, Verlag Chemie, Weinheim (1963), p. 246.
16. Bergmeyer, H.U. and E. Bernt. α-Oxoglutarate. In: H.U. Bergmeyer (Editor), Methods of enzymatic analysis, Verlag Chemie, Weinheim (1963), p. 324.

17. Williamson, D.H., J. Mellanby and H.A. Krebs. Enzymic determination of D-(-)-β-hydroxybutyric acid and acetoacetic acid in blood. Biochem. J. 82:90-96(1962).
18. Czok, R. and L. Eckert. D-3-phosphoglycerate, D-3-phosphoglycerate, phosphoenolpyruvate. In: H.U. Bergmeyer (Editor), Methods of enzymatic analysis, Verlag Chemie, Weinheim (1963), p. 224.
19. Hohorst, H.J., F.H. Kreutz and Th. Bücher. Über Metabolitgehalte und Metabolit-konzentrationen in der Leber der Ratte. Biochem. Z. 332:18-46(1959).
20. Dagley, Stanley. Citrate; determination with citrase. In: H.U. Bergmeyer (Editor), Methods of enzymatic analysis, Verlag Chemie, Weinheim (1965), p. 313-317.
21. Lamprecht, W. and I. Trautschold. Determination with hexokinase and glucose-6-phosphate dehydrogenase. In: H.U. Bergmeyer (Editor), Methods of enzymatic analysis, Verlag Chemie, Weinheim (1963), p. 543.
22. Adam, H. Adenosine-5,-diphosphate and adenosine-5,-monophosphate. In: H.U. Bergmeyer (Editor), Methods of enzymatic analysis, Academic Press, New York (1965), p. 224.
23. Martin, J.B. and D.M. Doty. Determination of inorganic phosphate-modification of isobutyl alcohol procedure. Analyt. Chem. 21:965-967(1949).
24. Garland, P.B. Some kinetic properties of pig-heart oxoglutarate dehydrogenase that provide a basis for metabolic control of the enzyme activity and also a stoichiometric assay for coenzyme A in tissue extracts. Biochem. J. 92:10C-12C(1964).
25. Pearson, D.J. A source of error in the assay of acetyl-Coenzyme A. Biochem. J. 95:23C-24C(1965).
26. Folbergrova, J., J.V. Passoneau, D.H. Lowry and D.W. Schulz. Glycogen, ammonia and related metabolites in the brain during seizures evoked by methionine sulphoximine. J. Neurochem. 16:191-203(1969).
27. Veech, R.L. Dissertation, Oxford (1968).
28. Krebs, H.A. The equilibrium constants of the fumarase-aconitase system. Biochem. J. 54:78-86(1953).
29. Williamson, D.H., P. Lund and H.A. Krebs. The redox state of free nicotinamide-adenine dinucleotide in the cytoplasm and mitochondria of rat liver. Biochem. J. 103:514-527(1967).
30. Veech, R.L., L. Raijman and H.A. Krebs. Equilibrium relations between the cytoplasmic adenine nucleotide system and nicotinamide-adenine nucleotide system in rat liver. Biochem. J. 117:499-503(1970).

TABLE 1

COMPOSITION OF DIETS

	Basal		1,3-Butanediol	
	gm%	Cal%	gm%	Cal%
Protein	22.0	15.04	22.0	15.04
Corn oil	7.5	13.95	7.5	13.95
Lard	22.5	41.84	22.5	41.84
Dextrin	12.2	9.15	1.8	1.35
Dextrose	12.2	9.15	1.8	1.35
Sucrose	12.2	9.15	1.8	1.35
Minerals[a]	4.0	0	4.0	0
Vitamins[b]	2.5	1.72	2.5	1.72
1,3-Butanediol	0	0	20.0	23.4
Cellulose	5.1	0	16.1	0

[a]Mineral mix contained: (in g per kilogram of mix) $CaCO_3$, 292.9; KH_2PO_4, 343.1; NaCl, 250.6; $MgSO_4 \cdot 7H_2O$, 99.8; $CaHPO_4 \cdot 2H_2O$, 4.295; ferric citrate, 6.223; $CuSO_4$, 1.558; $MnSO_4 \cdot H_2O$, 1.209; $ZnCl_2$, 0.200; KI, 0.005; $(NH_4)_6 MoO_{24} \cdot 4H_2O$, 0.0025; and Na_2SeO_4, 0.015.

[b]Vitamin mix contained: (in IU per kilogram diet) Vitamin A, 5000; vitamin D, 500; dl-α-tocopheryl acetate, 100; and (in milligrams per kilogram) menadione, 5; thiamine·HCl, 10; riboflavin, 20; niacin, 50; ascorbic acid, 200; pyridoxine·HCl, 10; p-aminobenzoic acid, 100; biotin, 0.5; Ca-pantothenate, 50; folic acid, 2; inositol, 200; and vitamin B_{12}, 0.05.

TABLE 2

BODY WEIGHT AND FOOD CONSUMPTION OF ANIMALS FED 1,3-BUTANEDIOL-CONTAINING DIETS FOR 42 DAYS

Parameters Examined	Without 1,3-Butanediol (8)[b]	With 1,3-Butanediol (8)[a]
Initial body weight, g	122 ± 2.8	120 ± 4.3
Final body weight, g	307 ± 11.6	254 ± 14.8
Food consumed, g	469 ± 30	394 ± 64
Food efficiency[b]	39.4	35.0

All values reported as mean ± SEM.

[a]Number of animals in parentheses

[b]As g gain/g food intake x 100.

TABLE 3

URINARY EXCRETION OF 1,3-BUTANEDIOL AND KETONE BODIES

	Urinary BD Excreted	BD Metabolized	Ketones Excreted
	g	*percent*	$\mu moles$[a]
Without 1,3-butanediol (6)[b]	0	–	0.19 ± 0.03
With 1,3-butanediol (6)[b]	0.090 ± 0.042	95.33 ± 1.26	2.90 ± 0.20

All values reported as mean ± SEM

[a]Sum of acetoacetate and β-hydroxybutyrate

[b]Number of experimental animals per group

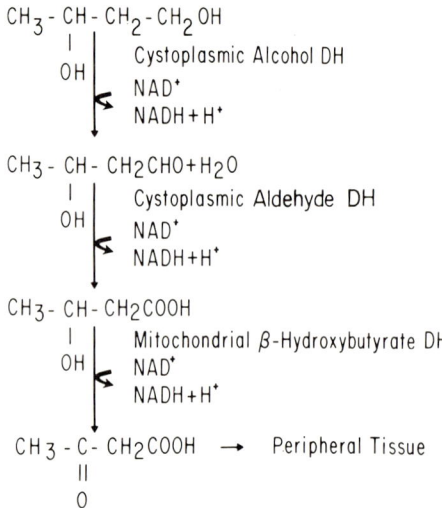

Fig. 1. *Proposed metabolic pathway for 1,3-butanediol utilization in the rat (8).*

METABOLIC REGULATION

	PURIFIED CONTROL DIET	P	20% 1,3 BUTANEDIOL	P	48 HOURS STARVED	WAYNE DIET	P	WAYNE DIET WITH 10 mM ETHANOL
NUMBER OF OBSERVATIONS	4		4		5	6		6
L - LACTATE	0.805 ± 0.052	$P < 0.001$	0.187 ± 0.017	$P < 0.05$	0.286 ± 0.026	0.586 ± 0.059	$P < 0.01$	1.28 ± 0.20
PYRUVATE	0.164 ± 0.010	$P < 0.001$	0.024 ± 0.001	$P < 0.01$	0.013 ± 0.002	0.107 ± 0.013	$P < 0.001$	0.031 ± 0.008
L - MALATE	0.285 ± 0.029		0.298 ± 0.029		0.337 ± 0.028	0.399 ± 0.037	$P < 0.001$	0.778 ± 0.062
PHOSPHOENOLPYRUVATE	0.099 ± 0.009		0.072 ± 0.015		0.078 ± 0.011	0.144 ± 0.014	$P < 0.001$	0.023 ± 0.007
3 - PHOSPHOGLYCERATE	0.296 ± 0.023	$P < 0.02$	0.166 ± 0.032		0.144 ± 0.022	0.314 ± 0.027	$P < 0.001$	0.050 ± 0.016
DIHYDROXYACETONE PHOSPHATE	0.035 ± 0.003	$P < 0.002$	0.014 ± 0.002		0.019 ± 0.001	0.024 ± 0.004		0.019 ± 0.001
α - GLYCEROPHOSPHATE	0.207 ± 0.017	$P < 0.05$	0.137 ± 0.015		0.170 ± 0.021	0.084 ± 0.005	$P < 0.001$	0.433 ± 0.078
GLUCOSE - 6 - PHOSPHATE	0.132 ± 0.004	$P < 0.01$	0.082 ± 0.009		0.057 ± 0.009	0.119 ± 0.004		0.117 ± 0.010
CITRATE	0.423 ± 0.043		0.299 ± 0.032		0.376 ± 0.031	0.483 ± 0.052		0.454 ± 0.063
ISOCITRATE	0.028 ± 0.003		0.020 ± 0.002		0.025 ± 0.002	0.032 ± 0.004		0.030 ± 0.004
α - OXOGLUTARATE	0.235 ± 0.037	$P < 0.02$	0.098 ± 0.016		0.089 ± 0.012	0.272 ± 0.002		0.251 ± 0.042
L - GLUTAMATE	3.49 ± 0.31		3.22 ± 0.35		2.77 ± 0.17	2.76 ± 0.16		2.72 ± 0.24
NH_4	0.437 ± 0.030		0.409 ± 0.040		0.32 ± 0.03			
ACETOACETATE	0.048 ± 0.004	$P < 0.001$	0.325 ± 0.028	$P < 0.002$	0.688 ± 0.056	0.094 ± 0.011	$P < 0.01$	0.055 ± 0.005
β - HYDROXYBUTYRATE	0.071 ± 0.010	$P < 0.001$	0.694 ± 0.083	$P < 0.001$	2.05 ± 0.16	0.129 ± 0.009		0.143 ± 0.021
ACETYL CoA	0.021 ± 0.005	$P < 0.05$	0.039 ± 0.003	$P < 0.01$	0.081 ± 0.009			
CoA	0.082 ± 0.006		0.084 ± 0.004					
ATP	1.80 ± 0.11		1.90 ± 0.09	$P < 0.002$	2.45 ± 0.06	2.83 ± 0.07		2.67 ± 0.04
ADP	0.976 ± 0.064		1.14 ± 0.007	$P < 0.01$	1.50 ± 0.05	1.06 ± 0.04		0.992 ± 0.056
AMP	0.239 ± 0.026		0.282 ± 0.034	N. S.	0.333 ± 0.020	0.212 ± 0.015	$P < 0.05$	0.272 ± 0.018
Pi	2.64 ± 0.12		3.09 ± 0.15	$P < 0.05$	3.84 ± 0.18	3.59 ± 0.13		3.31 ± 0.13

Values are given as mean ± S.E.M. P values refer to the significance of the difference between means judged by Student's T Test.
μmoles / g wet weight of liver

Fig. 2. *Metabolite levels in freeze-clamped rat liver from animals fed various diets.*

	PURIFIED CONTROL DIET	P	20% 1,3 BUTANEDIOL	P	48 HOURS STARVED	WAYNE DIET	P	WAYNE DIET WITH 10 mM ETHANOL
NUMBER OF OBSERVATIONS	4		4		5	6		6
NAD$^+$/NADH FROM LACTATE DEHYDROGENASE (CYTOPLASM)	1845 ± 93	P < 0.002	1164 ± 85	P < 0.001	426 ± 79	1620 ± 97	P < 0.001	212 ± 46
NADP$^+$/NADPH FROM ISOCITRATE DEHYDROGENASE (CYTOPLASM)	0.0082 ± 0.0008	P < 0.002	0.0049 ± 0.0005		0.0035 ± 0.0004	0.0087 ± 0.0009		0.0084 ± 0.0008
ATP/ADP × HPO$_4^{2-}$	1191 ± 122		931 ± 144		721 ± 55	1260 ± 94		1410 ± 125
NAD/NADH FROM β-HYDROXY-BUTYRATE DEHYDROGENASE (MITOCHONDRIA)	14.39 ± 1.74	P < 0.05	9.66 ± 0.57	P < 0.05	6.98 ± 0.77	14.83 ± 1.06	P < 0.01	8.51 ± 1.37
$\frac{ATP \times AMP}{(ADP)^2}$	0.449 ± 0.033		0.405 ± 0.019		0.365 ± 0.026	0.531 ± 0.020	P < 0.01	0.753 ± 0.053

For method of calculation, see text. Values are given as mean ± S.E.M. P values refer to the significance of the difference between the means as judged by the Student's T Test.

Fig. 3. *Redox and phosphorylation ratios in rat liver from animals fed various diets.*

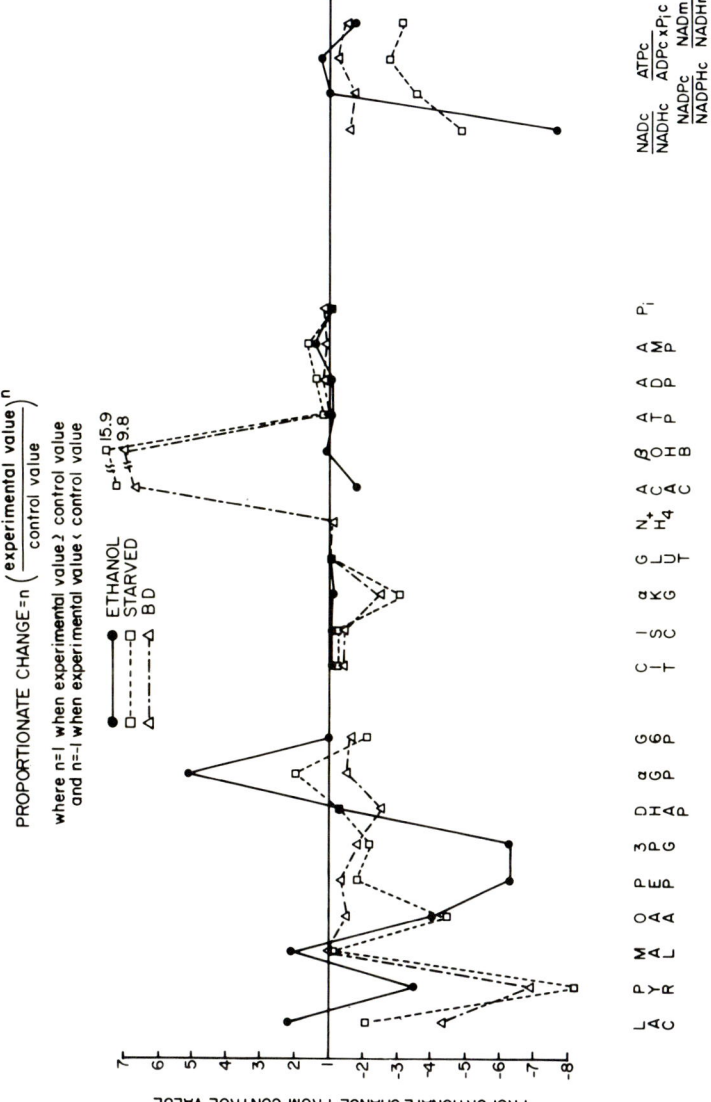

Fig. 4. *Changes in metabolite concentrations, and redox and phosphorylation ratios in liver from animals fed various diets.*

FEEDBACK CONTROL OF THE CITRIC ACID CYCLE

John R. Williamson, Colleen M. Smith, Kathryn F. LaNoue and Jadwiga Bryla

Introduction

Our approach to the investigation of the control properties of the citric acid cycle has been to measure flux through the various steps using intact rat heart mitochondria incubated under different experimental conditions which vary widely with respect to the intramitochondrial phosphate potential and pyridine nucleotide oxidation-reduction state (1). These results, which involve measurements of the rates of substrate uptake and accumulations of each cycle intermediate, have been correlated on the one hand with the contents of substrates and known regulators of specific enzyme steps of the citric acid cycle, and on the other hand with the kinetic properties of the purified enzymes.

The enzymes of the citric acid cycle are located in the inner membrane-matrix fraction of isolated rat liver mitochondria (2). A similar location may be presumed for citric acid cycle enzymes in heart mitochondria. The outer mitochondrial membrane appears to be freely permeable to sucrose, metabolic anions and cofactors (3, 4). The inner membrane of mammalian mitochondria has restricted permeability to most of the di- and tricarboxylic acids involved in the reactions of the citric acid cycle (5, 6); metabolic anions escape at different rates from the matrix during incubation of isolated mitochondrial suspensions until a steady state is reached when the rate of efflux for each species becomes equal to the rate of influx. Succinate, α-ketoglutarate, citrate and phosphate can be transported across the inner membrane in exchange with malate (7), and the concentration of malate in the medium determines the steady state concentration of the citric acid cycle anions in the extramitochondrial fluid with mitochondria under actively respiring conditions (8). During the interval prior to establishment of the steady state, flux through the individual steps of the citric acid cycle is unequal because the rate of formation of each intermediate is greater than its rate of utilization by the succeeding reaction step of the cycle. Consequently, in order to define the control properties of the citric acid cycle, it is necessary to measure the concentration of substrates and modulators accessible to the active sites of the enzymes. Clearly this is a formidable task not presently capable of completion with currently available techniques. However, methods for rapid separation of mitochondria from the incubation medium have recently been developed [see review by Klingenberg (9)] which, when suitably applied, permit measurements of the content of citric acid cycle intermediates and coenzymes in the inner membrane-matrix fractions. As a first approximation, the assumption has to be made that the intermediates are uniformly distributed in the sucrose-impenetrable space of the mitochondria, herein defined as the matrix volume. This assumption that all the [^{3}H]-exchangeable matrix water is available as solvent at the high protein concentration in the matrix is probably incorrect, and intramitochondrial concentrations so calculated will tend to be underestimated. Furthermore, all intermediates will be present in equilibrium between free and bound forms (enzyme bound or otherwise). When the matrix content of an intermediate is low compared with the concentrations of the enzymes to which it may be bound, as is the case with oxalacetate, the calculated and measured concentrations differ by several orders of magnitude. In fact, the apparent discrepancy between the measured oxalacetate content of mitochondria and the calculated values based on the near-equilibrium of malate dehydrogenase and glutamate-oxalacetate transaminase have led us to suggest a direct coupling between malate dehydrogenase and citrate synthase.

Data obtained with intact rat heart mitochondria (10) have suggested the possibility of an important regulatory role for succinyl-CoA in coordinating flux through the separate spans from oxalacetate plus acetyl-CoA to α-ketoglutarate and from α-ketoglutarate to malate in the citric acid cycle. The proposed mechanism which is developed further in this paper relates to control of the succinyl-CoA concentration by the guanine and adenine nucleotide phosphate potentials, with negative feedback effects exerted both at α-ketoglutarate dehydrogenase and at citrate synthase. The feasibility of this postulate has been confirmed by measurements of the succinyl-CoA content of mitochondria and by kinetic studies with the purified isolated enzymes. It is suggested that this mechanism supplements control of citrate synthase activity by oxalacetate availability, particularly at highly oxidized states of the mitochondrial pyridine nucleotides. In this paper the results of the kinetic studies with purified citrate synthase will be presented first, followed by data obtained with isolated rat heart mitochondria.

Methods

Preparation of rat heart mitochondria and the general procedures used for mitochondrial incubations and analytical methods have been described elsewhere (1, 8). Except when noted in the legends, the incubation medium contained 130 mM KCl, 20 mM tris-chloride, 5 mM $MgCl_2$ and 30 mM glucose, adjusted to pH 7.2. Metabolic fluxes were determined at 28° using 1 to 3 mg mitochondrial protein per ml in a total volume of 6 ml with 1 ml aliquots removed into 0.3 ml of 1.5 N perchloric acid after 0, 1, 2, 4 and 6 minutes of incubation. The different respiratory conditions were obtained by the following additions to the buffer: 1) *state 4:* no addition; 2) *oligomycin-inhibited*: hexokinase (1 mg/ml), ADP (100 μM) and oligomycin (up to 1 μg/mg protein); 3) *state 3*: hexokinase (1 mg/ml) and ADP (100 μM); 4) *uncoupled*: FCCP, p-trifluoromethoxyphenylhydrazone of carbonylcyanide (∼0.2 μM); and 5) *uncoupled plus oligomycin*: FCCP (∼0.2 μM) and oligomycin (1 μg/mg protein).

For separation of mitochondria from the medium, suspensions containing 4 to 6 mg protein/ml were used, and 0.5 ml aliquots were rapidly centrifuged through silicone oil (specific gravity of 1.05) into 0.1 ml of 1.5 N perchloric acid. The buffer included 8% (w/v) dextran which aided in obtaining a clean separation of the mitochondria. After centrifugation the top layer was immediately added to perchloric acid (0.6 N final concentration) to prevent metabolic changes due to any residual enzyme activity in the supernatant fluid. Pooled aliquots containing extracts from a series of runs from the equivalent of about 20 mg mitochondrial protein were used for metabolic analyses. In some experiments, a second filtration procedure employing Millipore filtration was used (1). Analytical procedures for the fluorometric assay of intermediates of the citric acid cycle have been described by Williamson and Corkey (11). Succinyl-CoA was assayed by a new procedure involving the coupling of β-hydroxybutyryl-CoA dehydrogenase with beef heart α-ketoacyl-CoA transferase prepared in our laboratory (12).

Enzyme kinetics of crystalline beef heart and purified rat liver citrate synthase and purified pig heart α-ketoglutarate dehydrogenase (13) were followed using NADH fluorescence in a modified Eppendorf fluorometer at a sensitivity of 0.5 μM NADH full scale recorder deflection. For the citrate synthase velocity measurements a concentration ratio of malate to NAD^+ of approximately 1:1 was used; an excess of malate dehydrogenase ensured a stoichiometry of 2 between the rate of NADH change and the rate of citrate synthesis (14). Initial reaction rates were obtained using about 20% of the recorder chart. CoA, acetyl-CoA, succinyl-CoA and propionyl-CoA were purchased from P-L Biochemicals, Inc. The succinyl-CoA was 90-92% pure as determined from the ratio of its concentration measured enzymatically to the calculated concentration measured by its absorbance at 260 nm. Acetyl-CoA from P-L Biochemicals was kinetically identical to that prepared from acetic anhydride and CoA purified by the DEAE cellulose and Biogel P-2 chromatography system of C. Fung and M.F. Utter (personal communication).

Results and Discussion

Kinetics of purified citrate synthase.

Over the range of concentrations of 1 μM to 10 μM oxalacetate and 2 μM to 100 μM acetyl-CoA, double reciprocal plots of initial velocity of beef heart citrate synthase against one substrate at fixed concentrations of the cosubstrate intersected on the abscissa, consistent with a mechanism of random order addition (15). The K_m values for oxalacetate and acetyl-CoA were found to be 1.6 μM and 8 μM, respectively. These values are similar to those obtained by Shepherd and Garland (14) for the rat liver enzyme. Fig. 1 demonstrates inhibition by ATP, which was noncompetitive with respect to oxalacetate. Straight lines were obtained using a Dixon plot of the reciprocal of initial velocity against the ATP concentration, even with oxalacetate concentrations down to 1 μM (Fig. 1A). Fig. 1B shows that double-reciprocal plots of 1/v against acetyl-CoA at a series of fixed concentrations of ATP were also linear and intersected to the left of the ordinate, indicating an incomplete reversal of ATP inhibition by acetyl-CoA. The exact position of the intersection relative to the ordinate may depend upon the ionic strength, since the inhibition appeared to be competitive when ionic strengths slightly higher than 0.1 were used. No evidence of cooperative effects of ATP inhibition were obtained [*cf.* (14), however]. In the absence of Mg^{++}, the K_i for ATP was 950 μM. Addition of stoichiometric amounts of Mg^{++} greatly decreased the inhibitory effects of ATP, as reported by Kosicki and Lee (16).

Fig. 2A shows that succinyl-CoA is an inhibitor of citrate synthase, competitive with respect to acetyl-CoA, with a K_i of about 130 μM. CoA itself is also an inhibitor, exhibiting mixed inhibition against acetyl-CoA (Fig. 2B). The inhibition of succinyl-CoA with respect to oxalacetate was noncompetitive. Preliminary studies have indicated that the inhibitory effect of succinyl-CoA is not reversed by the presence of 1 mM Mg^{++}. Further work is in progress to elucidate the effects of ions on the interaction of CoA and succinyl-CoA with citrate synthase using a non-coupled assay system. At present, the significance of the CoA inhibition is not clear. The finding of mixed inhibition with respect to acetyl-CoA suggests the possibility that an impurity in the CoA may be involved. The inhibitory effects of succinyl-CoA, on the other hand, are clear-cut and reproducible with purified succinyl-CoA. Citrate was also found to be an effective inhibitor, competitive with respect to oxalacetate and noncompetitive with respect to acetyl-CoA (Fig. 3). The K_i for citrate inhibition was 1.6 mM.

Preliminary data with purified rat liver citrate synthase indicate a kinetic behavior very similar to the beef heart enzyme. A summary of the kinetic constants and types of inhibition with respect to oxalacetate and acetyl-CoA using beef heart and rat liver enzymes are given in Table 1. It may be seen from Table 1 that propionyl-CoA is also a strong inhibitor of citrate synthase, competitive with respect to acetyl-CoA and noncompetitive with respect to oxalacetate. Probably most short chain acyl-CoA derivatives affect the enzyme similarly. This finding may be of particular significance with respect to abnormalities of metabolism induced by vitamin B_{12} deficiency or in patients with an inborn deficiency of methylmalonyl-CoA mutase (17, 18). Both pathological conditions are associated with methylmalonic aciduria and may result in elevated tissue levels of propionyl-CoA and methylmalonyl-CoA. Ketone body production is often high in these patients, and diminished activity of the citric acid cycle has been advanced as an explanation for some of the overall metabolic disturbances, which include hypoglycemia and neurological changes (19). Inhibition of citrate synthase by propionyl-CoA or methylmalonyl-CoA could provide a rationale for the complications induced by an enzyme defect in the apparently nonessential pathway of propionate to succinate metabolism.

Regulation of citric acid cycle flux by oxalacetate and succinyl-CoA.

Fig. 4 shows a schematic diagram of the citric acid cycle and malate-aspartate H-transport shuttle as applied to heart tissue which incorporates the new finding of succinyl-CoA inhibition of citrate synthase. Carbohydrate in the form of glucose and lactate, free fatty acids and ketone bodies serve as the major respiratory fuels of cardiac muscle both *in vitro* and *in vivo* (20, 21, 22). During the metabolism of lactate or glucose to pyruvate, equivalent amounts of NADH are generated in the cytosol, and these reducing equivalents are transported into the mitochondria for oxidation by the respiratory chain. Control of the malate-aspartate H-transport shuttle in rat heart mitochondria has been discussed previously (1). During metabolic transitions involving a switch of respiratory substrates but little change of energy utilization and oxygen consumption by the myocardium, an increasing contribution of NADH via the malate-aspartate shuttle to the oxygen consumption must be associated with an appropriate decrease of flux in the citric acid cycle between citrate and malate. This regulation involves control at the citrate synthase step. During operation of the malate-aspartate shuttle, citrate synthase and intramitochondrial glutamate-oxalacetate transaminase compete for oxalacetate during the formation of citrate and aspartate. α-Ketoglutarate is formed both from isocitrate dehydrogenase and as a transamination product from glutamate. A number of factors have already been described which regulate the metabolism of α-ketoglutarate to succinate via the citric acid cycle and its rate of efflux from the mitochondria (1, 23). Our purpose in this section is to present evidence obtained with isolated mitochondria incubated under a variety of metabolic states which illustrates the potential importance of succinyl-CoA as a feedback inhibitor of citric acid cycle flux by regulation of the activity of α-ketoglutarate dehydrogenase and citrate synthase.

Control of succinyl-CoA levels.

Succinyl-CoA is a substrate for succinate thiokinase, which in mammalian systems is a GDP-linked reaction with GDP and GTP being competitive with each other in the forward and back reactions (24). It is to be expected, therefore, that the succinyl-CoA level will be determined in part by the intramitochondrial GTP/GDP ratio. Although this ratio is not readily measurable, it will be a reflection of the ATP/ADP ratio if the activity of nucleoside diphosphate kinase in the inner membrane matrix fraction of heart mitochondria is great enough to maintain the adenine and guanine nucleotides in equilibrium (3, 25). An alternative mechanism for transphosphorylation via GTP-AMP phosphotransferase requires a source of intramitochondrial AMP, which could be supplied by the ATP-linked fatty acyl-CoA synthase.

In order to describe the factors controlling the succinyl-CoA content of isolated mitochondria and delineate the role of oxalacetate, ATP, CoA, acetyl-CoA and succinyl-CoA in the regulation of flux through citrate synthase, rat heart mitochondria were incubated with substrate in various metabolic states of differing intramitochondrial phosphate potentials, as described under **Methods**. Fig. 5 shows the changes of ATP and succinyl-CoA contents of heart mitochondria during a 6-minute incubation period with 5 mM acetylcarnitine and 1 mM malate present as substrates. Since the total content of CoA and derivatives in rat heart mitochondria, as determined by alkaline hydrolysis, is about 2 μmoles/mg protein, the data show that up to 70% of the total intramitochondrial CoA can be in the form of succinyl-CoA. As the steady state is approached, it may be seen that succinyl-CoA levels are high when the ATP level is high, *e.g.* state 4 and the uncoupled plus oligomycin state, but that there is a poor kinetic correlation between changes of ATP and succinyl-CoA. This suggests that although the guanine nucleotide phosphate potential does not equilibrate rapidly with the adenine nucleotide phosphate potential, it is strongly influenced by it.

Control of α-ketoglutarate dehydrogenase activity.

Succinyl-CoA has been reported to inhibit the activity of α-ketoglutarate dehydrogenase, the extent of the inhibition being greater in the presence of NADH (26). We have confirmed this effect using purified pig heart enzyme in the assay system of Hirashima *et al.* (27) at saturating levels of α-ketoglutarate and NAD^+, and have demonstrated that even in the absence of added NADH, succinyl-CoA is inhibitory, competitive with respect to CoA, with a K_i of approximately 10 μM. The K_m of pig heart α-ketoglutarate dehydrogenase for CoA under the same conditions was 3.6 μM, rather than below 0.1 μM as reported by Massey (28). Using a value of 1 μl/mg protein for the volume of the matrix space, the data in Fig. 5 show that in isolated mitochondria the succinyl-CoA concentration rises well above both the K_m for succinate thiokinase [10 to 60 μM (24)] and the K_i for α-ketoglutarate dehydrogenase, if activity coefficients of unity are assumed. Therefore, changes of succinyl-CoA will have a weak influence on the velocity of succinate thiokinase but a strong inhibitory effect on α-ketoglutarate dehydrogenase. Furthermore, since α-ketoglutarate dehydrogenase has a large free energy change in the forward direction, competitive inhibition of succinyl-CoA with CoA will prevent acylation of all the free CoA to succinyl-CoA, with consequent loss of availability of CoA to other CoA-requiring enzymes which have a K_m for CoA higher than that of α-ketoglutarate dehydrogenase.

α-Ketoglutarate produced within the mitochondria may either be converted to succinate or be released into the extramitochondrial space, so that the activity of α-ketoglutarate dehydrogenase relative to its rate of formation is reflected by the accumulation of α-ketoglutarate in the medium. As the extramitochondrial α-ketoglutarate concentration increases, its rate of efflux will be offset by an increasing rate of influx, until a steady state is reached when the two fluxes become equal. Studies with [5-^{14}C]–α-ketoglutarate added to mitochondria incubated with pyruvate, 1mM malate and steady state levels of α-ketoglutarate, showed that in state 3 the rate of α-ketoglutarate:α-ketoglutarate exchange across the mitochondrial membrane was 165 nmoles/min/mg protein, or about 60% faster than the rate of α-ketoglutarate formation. It appears probable, therefore, that during the course of mitochondrial incubations, the α-ketoglutarate concentration in the matrix increases as the extramitochondrial α-ketoglutarate concentration increases, and thereby increases flux through α-ketoglutarate dehydrogenase at a given inhibitory level of succinyl-CoA. However, a strict correlation is not to be expected since the concentration gradient of α-ketoglutarate across the inner mitochondrial membrane changes with different energy states. Detailed measurements of the intramitochondrial α-ketoglutarate concentration under all the different incubation conditions have not yet been made, but the range appears to be from about 30 μM in the oligomycin-inhibited state to about 0.4 mM in state 3. Unfortunately, there is some ambiguity about the K_m of α-ketoglutarate dehydrogenase for α-ketoglutarate, with values ranging from 13 μM (28) to 110 μM (27). The latter value would appear a more reasonable one.

In order to arrive at a better evaluation of the cause and effect relationships between the intramitochondrial concentration of possible enzyme modulators and flux through a particular enzyme step, it is necessary to compare changes with time within a given respiratory state in addition to changes between different metabolic states. Fig. 6 shows a comparison of the accumulation of α-ketoglutarate in the five respiratory states with either 1 mM pyruvate plus 1 mM malate or 5 mM acetylcarnitine plus 1 mM malate as substrates. In fact, it was this set of experiments which first suggested the possibility of a feedback interaction between succinyl-CoA and citrate synthase. Certain aspects of this data have been discussed previously, notably the relative activation of α-ketoglutarate dehydrogenase in the oligomycin-inhibited state compared with state 4 (1). It is now possible to document our earlier suggestion that this was caused by a fall of succinyl-CoA due to the lowered ATP/ADP ratio by direct analytical measurements of intramitochondrial succinyl-CoA, as illustrated in Fig. 5.

The total accumulation of α-ketoglutarate, but not its rate of accumulation, is increased in the uncoupled state as compared with state 3. This is probably due to a diminished gradient of α-ketoglutarate between the matrix and medium spaces in the uncoupled state, so that a higher medium concentration is needed before the intramitochondrial α-ketoglutarate concentration can build up sufficiently to overcome the succinyl-CoA inhibition of α-ketoglutarate dehydrogenase and allow input flux through this enzyme step to equal output flux. The accumulation of α-ketoglutarate in the uncoupled plus oligomycin state is much greater with pyruvate and malate than with acetylcarnitine and malate as substrates (Fig. 6). Measurements of pyruvate uptake and carnitine release from acetylcarnitine showed that flux through citrate synthase was greatly inhibited in the uncoupled plus oligomycin state with acetylcarnitine as substrate. Thus, the lower α-ketoglutarate accumulation under these conditions is probably a reflection of the diminished input flux of α-ketoglutarate. Data obtained so far indicate that the changes of intramitochondrial ATP and succinyl-CoA in the different metabolic states with mitochondria oxidizing pyruvate and malate are similar to those shown in Fig. 5 which were obtained with mitochondria oxidizing acetylcarnitine and malate. However, acetyl-CoA levels were about twice as high with pyruvate than with acetylcarnitine as substrate (10), so that a similar concentration of succinyl-CoA would provide a more effective inhibition of citrate synthase at the lower acetyl-CoA concentration because of the competitive nature of the kinetics.

Fluxes through citrate synthase and α-ketoglutarate dehydrogenase in mitochondria incubated with pyruvate plus malate and acetylcarnitine plus malate over the different time intervals of the experiments are presented in Table 2 for reference. Using these data, the change of α-ketoglutarate dehydrogenase activity with time can be normalized for different rates of α-ketoglutarate formation by expressing the rate of loss of α-ketoglutarate to the medium as a function of the rate of α-ketoglutarate formation from isocitrate dehydrogenase. The rate of α-ketoglutarate formation was calculated from the citrate synthase flux by subtracting the small rate of citrate accumulation. The loss of α-ketoglutarate to the medium is initially high during early incubation times because of the rapid rate of formation of succinyl-CoA and diminishes during the incubation as the rate of succinyl-CoA formation decreases, and presumably as the intramitochondrial α-ketoglutarate concentration increases. This relationship is shown in Fig. 7 for the acetylcarnitine plus malate experiments using mitochondria incubated in state 3, the uncoupled and uncoupled plus oligomycin states. Under these conditions, the pyridine nucleotides are highly oxidized, so that there is relatively little complication due to changes of the $NAD^+/NADH$ ratio as a potential modulator of α-ketoglutarate dehydrogenase activity.

Control of citrate synthase activity.

Flux through citrate synthase varies over a 10-fold range when mitochondria are incubated under different metabolic states with approximately constant concentrations of substrate. Since ketone body production by heart mitochondria is very low, citric acid cycle activity must be closely linked to the rate of respiration. It appears that various feedback influences from the electron transport chain combine to regulate the input flux to the citric acid cycle, and no simple pattern of control is to be expected. Regulation of citrate synthase may be mediated through availability of either of its substrates or by effects of inhibitors on the apparent Michaelis constants for the substrates.

In Table 3, flux through citrate synthase between 2 and 4 minutes of incubation in rat heart mitochondria incubated in different metabolic states with 5 mM acetylcarnitine plus 1 mM malate as substrate is compared with the intramitochondrial content of various regulators measured after 2 minutes of incubation. However, it may be noted that some of these parameters change with time (*cf.* Fig. 5). The concentration of oxalacetate available to citrate synthase is probably the most important regulator since the intramitochondrial free oxalacetate concentration is below the K_m value. At least three factors contribute to control the oxalacetate availability. These are the intramitochondrial malate concentration, the pH, and the NADH/NAD$^+$ ratio. On the basis of previous work with rat heart mitochondria oxidizing pyruvate or pyruvate plus malate as substrates (8), we concluded that the state 4 to state 3 transition was associated with an increase of intramitochondrial oxalacetate due to the fall of the NADH/NAD$^+$ ratio. However, as shown in Table 3, the NADH/NAD$^+$ ratio, as measured by direct tissue analyses of the reduced and oxidized pyridine nucleotides, decreased only 3-fold while flux through citrate synthase increased 7-fold. The changes of CoA, succinyl-CoA and acetyl-CoA are all in the direction to decrease citrate synthase activity during the state 4 to state 3 transition. The decrease of ATP would cause an activation to the extent that ATP is present in the free form. However, as described later in this paper, most of the ATP is probably present as the Mg^{++}-ATP complex, even in state 4. The remaining factors to consider, excluding pH changes, are the intramitochondrial concentrations of citrate and malate. These parameters have not been measured in mitochondria incubated under all the different metabolic states with acetylcarnitine and malate as substrates. In state 4, the matrix concentrations of citrate and malate were 1.26 mM and 0.25 mM, respectively, under conditions similar to those of Table 3. It must be concluded, therefore, that during the state 4 to state 3 transition, either the malate concentration must rise or the citrate concentration must fall, in addition to the fall of the NADH/NAD$^+$ ratio, in order to account for the observed flux increase. No change of the intramitochondrial malate concentration was observed between state 4 and state 3 respiration with mitochondria incubated for 2 or 4 minutes with pyruvate and 0.1 mM malate (Table 4); in experiments with pyruvate plus 1 mM malate, the intramitochondrial citrate content fell from 2.6 to 0.9 μmoles/mg protein after 2 minutes of incubation. Since citrate is a competitive inhibitor with oxalacetate, this change would result in an effective increase of oxalacetate availability to citrate synthase.

In state 4 and in the oligomycin-inhibited state, where citrate synthase fluxes are the same but ATP levels are 10-fold different, the rise of the NADH/NAD$^+$ ratio in the oligomycin-inhibited state is balanced by an increase of intramitochondrial malate (Table 4). Oxalacetate probably remains rate-limiting, so that the fall of succinyl-CoA has no influence on flux because the acetyl-CoA concentration remains far above the apparent K_m value.

A comparison of the state 3, uncoupled and uncoupled plus oligomycin respiratory states (Table 3) more clearly reveals the conditions under which regulation of citrate synthase activity by succinyl-CoA becomes important. At low NADH/NAD$^+$ ratios, oxalacetate availability will be expected to be relatively high except to the extent that malate is lost from the matrix, so that factors which affect the apparent K_m for acetyl-CoA assume an added significance, especially at low substrate levels for acetyl-CoA. Thus, the marked fall of flux through citrate synthase between the uncoupled and uncoupled plus oligomycin states is associated with no change of acetyl-CoA but a rise of succinyl-CoA and ATP. The possible influence of ATP concentration as a control parameter is discounted both by the kinetics of the ATP changes and by the fact that Mg^{++}-ATP is a much weaker inhibitor than free ATP.

As a further assessment of the relative roles of ATP and succinyl-CoA in the control of citrate synthase in intact rat heart mitochondria, data from experiments with a variety of substrates, in which mitochondria were incubated under conditions of state 3, uncoupled and uncoupled plus oligomycin respiration, were used to correlate flux with effector concentration at different acetyl-CoA levels. Under these conditions, the intramitochondrial oxalacetate concentration may be expected to be relatively constant. A Dixon plot of the reciprocal of flux through citrate synthase as a function of succinyl-CoA concentration is shown in Fig. 8. Three different ranges of acetyl-CoA levels were used for the plot: 0.15 to 0.3 mM, 0.3 to 0.6 mM and greater than 0.6 mM. Considering the nature of the variables involved, straight lines could reasonably be drawn which intersected to the left of the ordinate. Since no other variables are taken into account in this plot, it appears that inhibition of citrate synthase by CoA, as observed with the isolated enzyme, may not be of physiological significance in the intact mitochondria. Marked deviation was only obtained with the low range of acetyl-CoA at very high succinyl-CoA concentrations. An operational K_i in the intact mitochondria of about 270 μM was obtained for the correlation of citrate synthase inhibiton by succinyl-CoA. This value is in reasonable agreement with the value of 130 μM obtained with isolated citrate synthase for the K_i of succinyl-CoA, and suggests that kinetic constants for enzymes in the intact mitochondria may not be dissimilar to the values obtained with isolated enzymes.

A similar type of plot of the reciprocal of flux through citrate synthase as a fucntion of ATP concentration is shown in Fig. 9. This figure shows that flux through citrate synthase was apparently independent of ATP concentrations up to 5 mM, again suggesting that ATP is not a physiological regulator of this enzyme step.

Equilibria and disequilibria of intramitochondrial enzyme systems.

A fairly complete set of measurements have been made of the intramitochondrial concentrations of intermediates and cofactors for rat heart mitochondria incubated for 6 and 8 minutes in state 4 with 10 mM acetylcarnitine and 0.4 mM malate (initial concentrations) present as substrates (Table 5). These data permit certain calculations which relate directly to the above discussion of the control of the citric acid cycle.

Comparison between calculated and measured oxalacetate concentrations.

The mitochondrial oxalacetate concentration may be calculated by assuming equilibrium at either the malate dehydrogenase step or at glutamate-oxalacetate transaminase. In order to calculate oxalacetate from the equation

$$K_{MDH} = \frac{[NADH][H^+]}{[NAD^+]} \times \frac{[OAA]}{[Mal]}$$

it is necessary to define the intramitochondrial pH in addition to the ratio of free $NADH/NAD^+$. For simplicity, we will assume a pH of 7.4, which is within the range of measured values according to Addanki *et al.* (29). If equilibrum of β-hydroxybutyrate, dehydrogenase is assumed, the ratio of free $NADH/NAD^+$ in the mitochondria can be calculated from the measured values of β-hydroxybutyrate, acetoacetate and the equilibrium constant. However, inspection of the data in Table 5 shows that a problem is immediately apparent because the ratio of β-hydroxybutyrate/acetoacetate in the medium and matrix (0.39 and 0.054, respectively) differ by an order of magnitude due to the relatively low concentration of β-hydroxybutyrate in the matrix space. This finding raises questions concerning the possibility of different pools of NAD^+ within the mitochondria and the validity of the calculation method. If the medium β-hydroxybutyrate/acetoacetate ratio is taken as the more reasonable value, the ratio of free $NADH/NAD^+$ is 0.05, using an equilibrium constant of 0.12 for β-hydroxybutyrate dehydrogenase at pH 7.4 (30). The measured value for the ratio of total $NADH/NAD^+$ was 0.28 under similar experimental conditions. By substituting the value of 0.05 for the $NADH/NAD^+$ ratio in the malate dehydrogenase equilibrium reaction, and using 7×10^{-5} for the equilibrium constant at pH 7.4, a matrix oxalacetate concentration of 0.35 μM is calculated with a matrix malate concentration of 0.25 mM. This value is two orders of magnitude lower than the measured concentration of 36 μM. The oxalacetate concentration may also be calculated assuming equilibrium of glutamate-oxalacetate transaminase:

$$\frac{[\alpha Kg]}{[OAA]} \frac{[Asp]}{[Glut]} = 6.6$$

By substituting measured values for α-ketoglutarate (αKg), aspartate (Asp) and glutamate (Glut) in this equation, an oxalacetate concentration of 10 μM is obtained, which is only 3.6-fold lower than the measured value.

These calculations raise the possibility that different pools of oxalacetate are available to malate dehydrogenase and glutamate dehydrogenase, or that gross disequilibrium of mitochondrial malate dehydrogenase occurs even for the low flux conditions of state 4 respiration. Alternatively, malate dehydrogenase and citrate synthase may interact in such a way that oxalacetate remains enzyme-bound, and therefore may be considered in a separate pool from oxalacetate available to glutamate-oxalacetate transaminase. This conclusion is reinforced by a comparison between the equilibrium constant (31) and the measured mass action ratio (K_{app}) for the combined malate dehydrogenase-citrate synthase reaction, as shown below:

$$K_{eq} = \frac{[Cit][CoA][NADH][H^+]^2}{[Mal][Acetyl\text{-}CoA][NAD^+][H_2O]} = 6.86 \times 10^{-16}$$

At pH 7.4, K'_{eq} = 0.43 L·Mol^{-1}. By substituting values from Table 5, using the value of 0.05 for the free NADH/NAD$^+$ ratio, and 55.5 M for the molarity of water, K_{app} = 4.6 × 10^{-3} L·Mol^{-1}. This value differs from the K'_{eq} value by two orders of magnitude. If equilibrium is assumed, the unreasonable value of 5.8 is calculated for the matrix pH.

Calculation of intramitochondrial free Mg^{++}.

England *et al.* (32) have reported that Mg^{++} affects the equilibrium citrate/isocitrate concentration ratio for rat heart aconitase, being 7.8 at zero Mg^{++} and maximally 33 at 2 mM Mg^{++} at pH 7.3. From the data in Table 5, a mean value of 8.7 is obtained for the citrate/isocitrate ratio in the matrix, while a value of 33 is obtained for the ratio in the medium, which contained 5 mM Mg^{++}. The equilibrium citrate/isocitrate ratio of 8.7 corresponds to a matrix free Mg^{++} concentration of about 0.2 mM for state 4 conditions.

Guanosine nucleotide phosphate potential.

From the equilibrium of succinate thiokinase at pH 7.4, 20° (24)

$$K_{eq} = 0.27 = \frac{[GDP][P_i][Succinyl\text{-}CoA]}{[GTP][Succinate][CoA]}$$

and substituting mean values from Table 5, a value of 68,800 L·Mol^{-1}, is obtained for the ratio [GTP]/[GDP] [P$_i$]. This value is more than an order of magnitude greater than values reported for the intramitochondrial adenine nucleotide phosphate potential (33), but the different experimental conditions make comparisons difficult.

Summary

The role of succinyl-CoA and other metabolites in the regulation of flux through the citric acid cycle has been investigated. Studies with purified pig heart α-ketoglutarate dehydrogenase at saturating concentrations of NAD^+ and α-ketoglutarate showed that succinyl-CoA inhibited competitively with respect to CoA (K_m for CoA of 3.6 μM; K_i for succinyl-CoA of 10.5 μM). Studies of the kinetic properties of purified beef heart and rat liver citrate synthase indicated a random order of addition of substrates, and demonstrated a number of new inhibitors. Inhibitions by succinyl-CoA (K_i of 130 μM) and propionyl-CoA (K_i of 50 μM) were competitive with respect to acetyl-CoA and noncompetitive with respect to oxalacetate. Inhibition by citrate (K_i of 1.6 mM) was noncompetitive with respect to acetyl-CoA and competitive with respect to oxalacetate.

Experiments with isolated rat heart mitochondria incubated with pyruvate or acetylcarnitine, in the presence of malate, under a variety of metabolic states which differed with respect to energy state, pyridine nucleotide redox state and rates of respiration, are described which demonstrate the dual role of succinyl-CoA as a negative feedback regulator of flux through α-ketoglutarate dehydrogenase and citrate synthase. It is concluded that flux through citrate synthase in the intact mitochondria is controlled by the intramitochondrial concentration of either oxalacetate or acetyl-CoA, depending on substrate and phosphate acceptor availability, and that regulation is achieved by feedback inhibition of metabolites which differentially increase the apparent Michaelis constant for the substrates.

Presented by John R. Williamson. The experimental work reported in this paper was supported by the United States Public Health Service Grants GM-12202 and AM-15120, the American Heart Association, and the American Medical Association Education and Research Foundation. John R. Williamson and Kathryn F. LaNoue are Established Investigators of the American Heart Association (K.F.L., William D. Stroud Established Investigator).

References

1. LaNoue, K.F. and J.R. Williamson. Interrelationships between malate-aspartate shuttle and citric acid cycle in rat heart mitochondria. Metabolism 20:119-140(1971).

2. Greville, G.D. Intracellular compartmentation and the citric acid cycle. In: J.M. Lowenstein (Editor), Citric acid cycle control and compartmentation, Marcel Dekker, New York (1969), pp. 1-136.
3. Klingenberg, M. and E. Pfaff. Structural and functional compartmentation in mitochondria. In: J.M. Tager, S. Papa, E. Quagliariello, and E.C. Slater, (Editors), Regulation of metabolic processes in mitochondria; BBA library volume 7, Elsevier, Amsterdam (1966), pp. 180-201.
4. Haddock, B.A., D.W. Yates and P.B. Garland. The localization of some coenzyme A-dependent enzymes in rat liver mitochondria. Biochem. J. 119:565-573(1970).
5. Chappell, J.B. Systems used for the transport of substrates into mitochondria. Brit. Med. Bull. 24:150-157(1968).
6. Klingenberg, M. Mitochondria metabolite transport. FEBS Letters 6:145-154(1970).
7. Papa, S., N.E. Lofrumento, E. Quagliariello, A.J. Meijer and J.M. Tager. Coupling mechanisms in anionic substrate transport across the inner membrane of rat-liver mitochondria. Bioenergetics 1:287-307(1970).
8. LaNoue, K., W.J. Nicklas and J.R. Williamson. Control of citric acid cycle activity in rat heart mitochondria. J. Biol. Chem. 245:102-111(1970).
9. Klingenberg, M. Metabolite transport in mitochondira: an example for intracellular membrane function. In: P.N. Campbell and F. Dickens (Editors), Essays in biochemistry, Vol. VI, Academic Press, New York (1970), pp. 119-159.
10. LaNoue, K.F., J. Bryla and J.R. Williamson. Feedback interactions in the control of citric acid cycle activity in rat heart mitochondria. J. Biol. Chem., 247:667-679 (1972).
11. Williamson, J.R. and B.E. Corkey. Assays of intermediates of the citric acid cycle and related compounds by fluorometric enzyme methods. In: J.M. Lowenstein (Editor), Methods in enzymology, Vol. XIII, Academic Press, New York (1969), pp. 434-513.
12. Smith, C.M., K.F. LaNoue and J.R. Williamson. In preparation.
13. Sanadi, D.R., J.W. Littlefield and R.M. Bock. Studies on α-ketoglutaric oxidase II. Purification and properties. J. Biol. Chem. 197:851-862(1952).
14. Shepherd, D. and P.B. Garland. The kinetic properties of citrate synthase from rat liver mitochondria. Biochem. J. 114:597-610(1969).
15. Smith, C.M. and J.R. Williamson. Inhibition of citrate synthase by succinyl-CoA and other metabolites. FEBS Letters 18, 35(1971).
16. Kosicki, G.W. and L.P.K. Lee. Effect of divalent metal ions on nucleotide inhibition of pig heart citrate synthase. J. Biol. Chem. 241:3571-3574(1966).
17. White, A.M. and E.V. Cox. Methylmalonic acid excretion and vitamin B_{12} deficiency in the human. Ann. N.Y. Acad. Sci. 112:915-921(1964).
18. Rosenberg, L.E., A. Lilljequist and Y.E. Hsia. Methylmalonic aciduria. An inborn error leading to metabolic acidosis, long-chain ketonuria and intermittent hyperglycinemia. New Eng. J. Med. 278:1319-1322(1968).
19. Oberholzer, V.G., B. Levin, E.A. Burgess and W.F. Young. Methylmalonic aciduria: an inborn error of metabolism leading to chronic metabolic acidosis. Arch. Dis. Childh. 42:492-504(1967).
20. Williamson, J.R. and H.A. Krebs. Acetoacetate as fuel for respiration in the perfused rat heart. Biochem. J. 80:540-547(1961).
21. Bing, R.J. Cardiac metabolism. Physiol. Rev. 45:171-213(1965).
22. Opie, L. Metabolism of the heart in health and disease. Part I. Amer. Heart J. 76:685-698(1968).

23. LaNoue, K.F. and J.R. Williamson. Control of the malate-aspartate shuttle in rat heart mitochondria. In: Energy transduction in respiration and photosynthesis, Adriatica Editrice, Bari, in press.
24. Cha, S. and R.E. Parks, Jr. Succinic thiokinase I. Purification of the enzyme from pig heart. J. Biol. Chem. 239:1961-1977(1964).
25. Tager, J.M., E.J. de Haan, and E.C. Slater. The metabolism of α-ketoglutarate. In: J.M. Lowerstein (Editor), Citric acid cycle control and compartmentation, Marcel Dekker, New York (1969), pp. 213-247.
26. Garland, P.B. Some kinetic properties of pig-heart oxoglutarate dehydrogenase that provide a basis for metabolic control of the enzyme activity and also a stoichiometric assay for coenzyme A in tissue extracts. Biochem. J. 92:10C-12C(1964).
27. Hirashima, M., T. Hayakawa and M. Koike. Mammalian α-keto acid dehydrogenase complexes II. An improved procedure for the preparation of 2-oxoglutarate dehydrogenase complex from pig heart muscle. J. Biol. Chem. 242:902-907(1967).
28. Massey, V. The composition of the ketoglutarate dehydrogenase complex. Biochim. Biophys. Acta 38:447-460(1960).
29. Addanki, S., F. Cahill and J.F. Sotos. Determination of intramitochondrial pH and intramitochondrial-extramitochondrial pH gradient of isolated heart mitochondria by the use of 5,5-dimethyl-2,4-oxazolidinedione I. Changes during respiration and adenosine triphosphate-dependent transport of Ca^{++}, Mg^{++}, and Zn^{++}. J. Biol. Chem. 243:2337-2348(1968).
30. Krebs, H.A. and R.L. Veech. Pyridine nucleotide interrelations. In: S. Papa, J.M. Tager, E. Quagliariello and E.C. Slater (Editors), The energy level and metabolic control in mitochondria, Adriatica Editrice, Bari, pp. 329-382(1969).
31. Stern, J.R., S. Ochoa and F. Lynen. Enzymatic synthesis of citric acid V. Reaction of acetyl coenzyme A. J. Biol. Chem. 198:313-321(1952).
32. England, P.J., R.M. Denton and P.J. Randle. The influence of magnesium ions and other bivalent metal ions on the aconitase equilibrium and its bearing on the binding ou magnesium ions by citrate in rat heart. Biochem. J. 105:32C-33C(1967).
33. Duée, E.D. and P.V. Vignais. Kinetics of phosphorylation of intramitochondrial and extramitochondrial adenine nucleotides as related to nucleotide translocation. J. Biol. Chem. 244:3932-3940(1969).
34. Srere, P. Citrate synthase. In: J.M. Lowenstein (Editor), Methods in enzymology, Vol. XIII, Academic Press, New York (1969), pp. 3-11.

TABLE 1

Beef heart citrate synthase (specific activity = 160) was isolated and crystallized according to Srere (34). The rat liver enzyme was purified according to Shepherd and Garland (14). Initial reaction velocities were measured by following NADH fluorescence increase in the coupled assay system using NAD^+, malate and malate dehydrogenase as discussed by Shepherd and Garland (14). The Michaelis constants for acetyl-CoA with the beef liver and rat liver enzymes were 8 μM and 10 μM, respectively. The K_m for oxalacetate with the beef heart enzyme was 1.6 μM. Each parameter was independent of concentration of cosubstrate consistent with a random order of addition.

Enzyme source	Inhibitor	K_i	Inhibition type	
			With respect to acetyl-CoA	With respect to oxalacetate
		μM		
Beef heart	ATP	950	competitive or mixed	noncompetitive
Beef heart	CoA	67	mixed	noncompetitive
Beef heart	succinyl-CoA	130	competitive	noncompetitive
Beef heart	propionyl-CoA	50	competitive	noncompetitive
Beef heart	citrate	1600	noncompetitive	competitive
Rat liver	succinyl-CoA	140	competitive	noncompetitive

TABLE 2

FLUX THROUGH CITRATE SYNTHASE (CS) AND α-KETOGLUTARATE DEHYDROGENASE (αKgDH) IN RAT HEART MITOCHONDRIA INCUBATED IN DIFFERENT METABOLIC STATES

A. 5mM Acetylcarnitine plus 1 mM malate as substrate

Time interval	State 4		Oligo-ADP		Time interval	State 3		Uncoupled		Uncoupled + oglio	
	CS	αKgDH	CS	αKgDH		CS	αKgDH	CS	αKgDH	CS	αKgDH
min	nmoles/min/mg protein				min	nmoles/min/mg protein					
0-2	14.4	6.8	29.1	19.5	0-1	86	48	75	40	48	11
2-4	14.0	6.6	15.4	19.2	1-2	116	90	88	51	54	36
4-6	10.0	7.1	18.1	16.8	2-4	94	88	76	59	33	29
6-8	13.9	8.9	15.7	15.3	4-6	106	105	75	68	38	36

B. 1mM Pyruvate plus 1 mM malate as substrate

Time interval	State 3		Uncoupled		Uncoupled + oligo	
	CS	αKgDH	CS	αKgDH	CS	αKgDH
min	nmoles/min/mg protein					
0-1	146	97	144	97	141	89
1-2	98	86	121	84	72	29
2-4	87	86	86	72	66	31
4-6	99	97	98	84	64	30

TABLE 3

REGULATION OF CITRATE SYNTHASE IN RAT HEART MITOCHONDRIA

Mitochondria were incubated with 5 mM acetylcarnitine plus 1 mM malate under conditions as described in **Methods**. The mitochondria were rapidly separated from the medium after 2 minutes of incubation by filtration through Millipore filters (0.65 μ porosity) and the filter was immediately extracted with 0.6 N perchloric acid.

Metabolic state	Citrate synthase flux (2 to 4 min)	Intramitochondrial values after 2 minutes						
		NADH	CoA	Acetyl-CoA	Succinyl-CoA	ATP	$\frac{ATP}{ADP}$	$\frac{NADH}{NAD^+}$
	nmoles/min/mg	nmoles/mg protein						
State 4	14	1.36	0.47	0.90	1.17	5.8	6.8	0.28
Oligo-ADP	15	2.71	0.62	0.77	0.39	0.6	0.1	0.78
State 3	94	0.56	1.18	0.19	0.62	3.5	0.7	0.10
Uncoupled	76	0.16	1.21	0.20	0.58	2.9	0.8	0.03
Uncoupled + oligo	33	0.11	0.77	0.19	1.39	5.0	3.6	0.02

TABLE 4

INTRAMITOCHONDRIAL MALATE CONCENTRATIONS AS A FUNCTION OF THE METABOLIC STATE

Mitochondria (5 to 7 mg per ml) were incubated in different respiratory states as described under **Methods**. The incubation medium included 8% (w/v) dextran in the buffer. Aliquots (0.2 ml) were removed after 2 min and 4 min for centrifugation of the mitochondria through a silicone oil layer (1.05 specific gravity) into 0.04 ml of 1.5 N perchloric acid. The upper and lower layers from triplicate centrifugations were pooled for metabolite assays. The matrix water volume was measured by addition of $[^3H]-H_2O$ (2 μCi per ml) and ^{14}C-sucrose (0.2 μCi/ml) to the reaction medium.

Condition	Substrate					
	2 mM Pyruvate			2 mM Pyruvate + 0.1 mM Malate		
	Matrix malate	Matrix H$_2$O	Malate in / Malate out	Matrix malate	Matrix H$_2$O	Malate in / Malate out
	mM	μl/mg		mM	μl/mg	
State 4	0.22 ± 0.03	0.81 ± 0.11	46 ± 10	0.38 ± 0.07	0.97 ± 0.11	4.2 ± 0.7
Oligomycin-ADP	0.37 ± 0.04	0.93 ± 0.07	39 ± 4	0.76 ± 0.05	1.07 ± 0.18	7.9 ± 0.4
State 3	0.28 ± 0.05	0.79 ± 0.05	84 ± 15	0.30 ± 0.03	1.18 ± 0.08	7.0 ± 1.0
Uncoupled	0.09 ± 0.01	0.91 ± 0.06	51 ± 6	0.12 ± 0.02	1.28 ± 0.12	1.8 ± 0.5
Uncoupled + oligomycin	0.05 ± 0.01	1.03	36 ± 4	0.06	1.27 ± 0.12	1.0

TABLE 5

INTRAMITOCHONDRIAL CONCENTRATIONS OF METABOLITES IN RAT HEART MITOCHONDRIA

Mitochondria (3 to 4 mg/ml) were incubated in the presence of 10 mM acetylcarnitine and 0.4 mM malate in state 4, and separated by centrifugation through silicone oil into perchloric acid. Extracts from multiple runs were pooled for metabolic analyses. [1-^{3}H]-D-Mannitol was used to measure the volume of extra-matrix water transferred through the silicone oil layer. A matrix water volume of 1 µl/mg protein was used to calculate intramitochondrial concentrations.

Metabolite	Metabolite concentrations			
	After 6 min incubations		After 8 min incubation	
	In matrix	In medium	In matrix	In medium
	µM	µM	µM	µM
Malate	250	239	120	212
Oxalacetate	37	0.36	35	0.40
Citrate	1160	56	1350	57
Isocitrate	170	1.5	120	1.8
α-Ketoglutarate	110	32	140	44
Succinate	100	8	70	7
Glutamate	860	48	1080	58
Aspartate	430	38	580	38
Pyruvate	90	1.2	112	1.2
CoA	270	—	430	—
Acetyl-CoA	660	—	420	—
Succinyl-CoA	550	—	680	—
β-Hydroxybutyrate	30	1.9	10	1.7
Acetoacetate	400	4.5	300	4.9

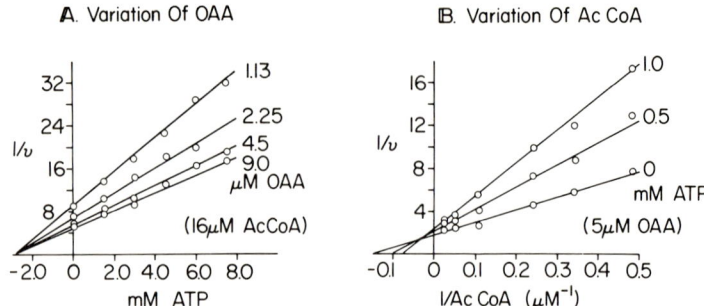

Fig. 1. *Inhibition of purified beef heart citrate synthase by ATP.* The reactions were performed at 21° in 0.1 M Tris-Cl at pH 7.4. Citrate synthase concentration is 1 to 10×10^{-5} mg/ml. Oxalacetate levels are fixed by the equilibrium of the malate dehydrogenase reaction and adjusted by varying malate and NAD^+.

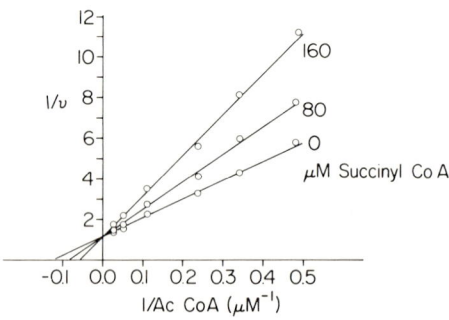

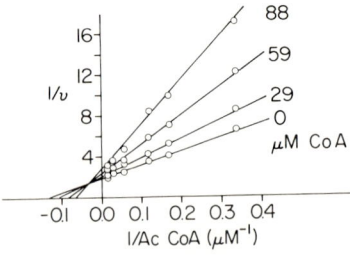

Fig. 2. *Inhibition of purified beef heart citrate synthase by succinyl-CoA and CoA.* The reaction conditions were the same as those of Fig. 1.

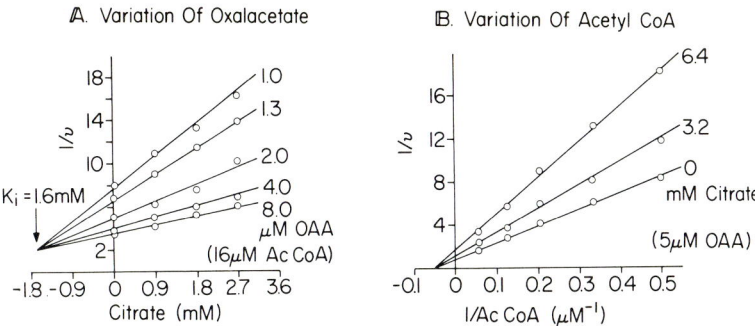

Fig. 3. *Inhibition of purified beef heart citrate synthase by citrate.* The reaction conditions were the same as those of Fig. 1. In this experiment the K_m for acetyl-CoA was 20 μM.

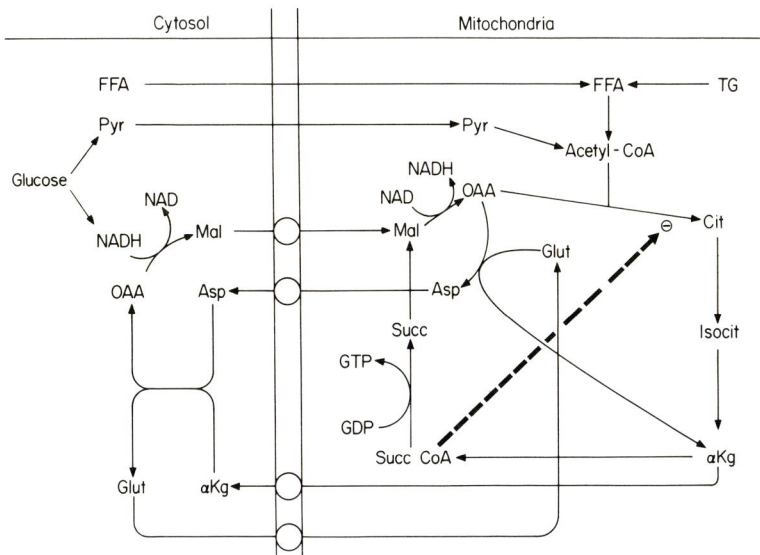

Fig. 4. *Scheme of the interrelationship of the malate-aspartate H-transport shuttle with the citric acid cycle in heart.*

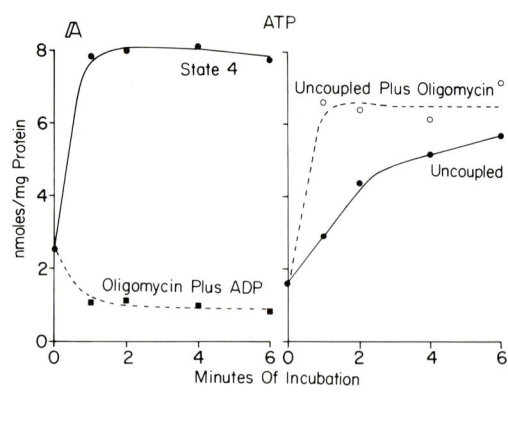
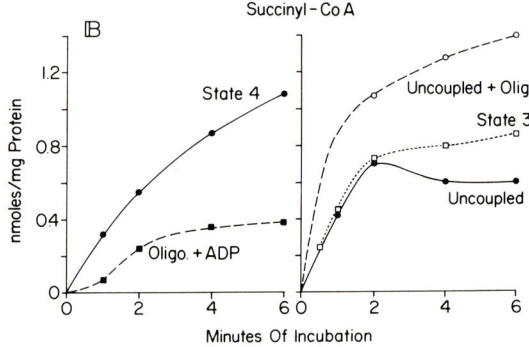

Fig. 5. *Changes of ATP and succinyl-CoA in isolated rat heart mitochondria incubated in different metabolic states as described in* **Methods.**

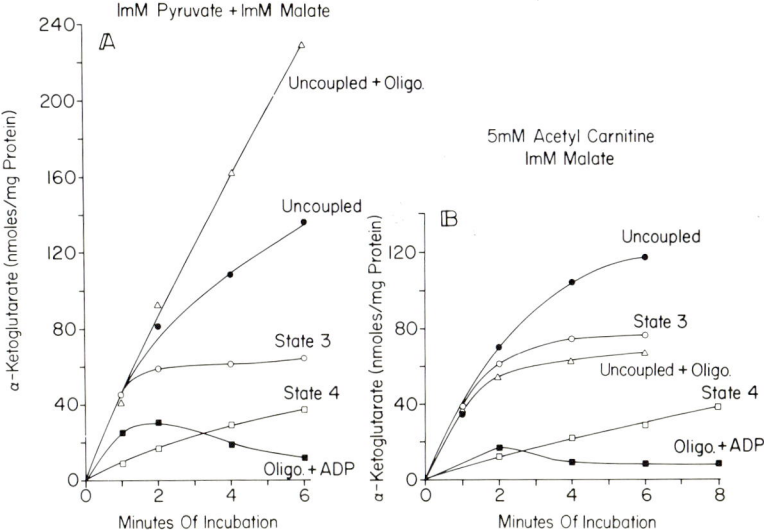

Fig. 6. *Accumulation of α-ketoglutarate by rat heart mitochondria incubated with* A, *1 mM pyruvate plus 1 mM malate and* B, *5 mM acetylcarnitine plus 1 mM malate, in different metabolic states as described in* **Methods.**

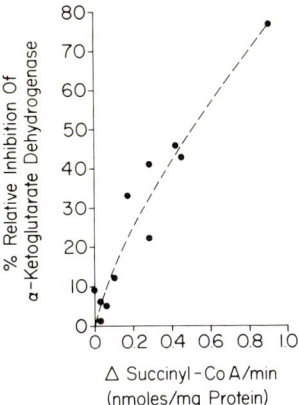

Fig. 7. *Relationship between the relative change of flux through α-ketoglutarate dehydrogenase with the rate of change of succinyl-CoA in isolated rat heart mitochondria.* Values shown were obtained with mitochondria incubated for 1, 2, 4 and 6 minutes with 5 mM acetylcarnitine and 1 mM malate under conditions of state 3, uncoupled and uncoupled plus oligomycin respiration. The percentage relative inhibition of α-ketoglutarate dehydrogenase was calculated from the ratio of the rate of α-ketoglutarate accumulation in the medium to the rate of α-ketoglutarate formation via the citric acid cycle.

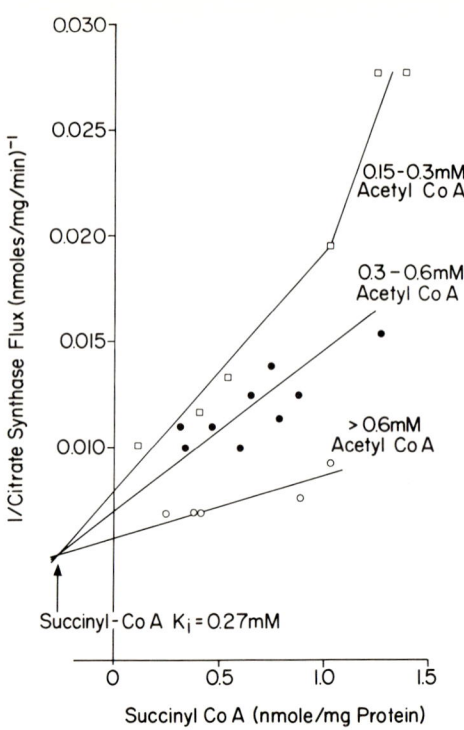

Fig. 8. Correlation plots of the relationship of succinyl-CoA content to flux through citrate synthase at various ranges of acetyl-CoA concentrations in intact rat heart mitochondria.

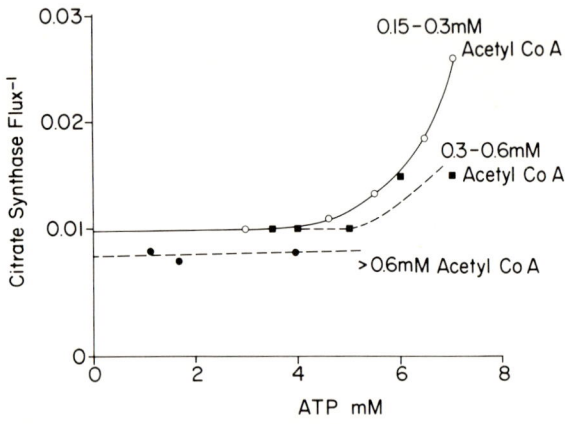

Fig. 9. Correlation plot of the relationship of ATP concentration to flux through citrate synthase at various ranges of acetyl-CoA concentrations in intact rat heart mitochondria.

SOURCES AND DISPOSITION OF AEROBICALLY-GENERATED INTERMEDIATES IN HEART MUSCLE

E. Jack Davis, Reneé C. Lin and David Li-Shan Chao

Introduction

Muscle metabolism is restricted in the sense that it carries out no glucose synthesis and little or no fatty acid synthesis [*cf.* (1)]. This follows, since the key enzymes required for these pathways are almost, if not totally, missing in muscle (1-5). This fact was indicated very early when it was found that the respiration of muscle minces oxidizing succinate (6), endogenous substrates (7) or pyruvate (8) was almost completely inhibited by malonate. Indeed, this singular nature of the aerobic metabolism of muscle was very important in structuring the concept of, and eventually proving the existence of, the citric acid cycle (9). More recently, detailed studies on the stoichiometry of the oxidation of pyruvate and glutamate by isolated heart muscle mitochondria have revealed that reactions capable of net feed-in or feed-out of citric acid cycle intermediates from these substrates are minimal (10-13).

It is thoroughly documented, however, that changes in the dietary or other physiological state of an animal will produce marked fluctuations in the steady-state level of citric acid cycle intermediates in muscle (14, 15). These alterations can be mimicked in isolated perfused hearts (16-20). This is illustrated in Table 1. These data are compiled from reports in the literature on the effect of various substrates on the steady-state levels of citrate, malate and α-oxoglutarate found in perfused rat hearts oxidizing glucose. The level of all three acids is elevated due to the presence of pyruvate, fatty acid or β-hydroxybutyrate. In addition, hearts from alloxan diabetic animals had elevated levels of intermediates. It should be pointed out that these values represent a net increase in citric acid cycle intermediates, and not merely a re-shuffling of those already present, since citrate and malate together account for a large fraction of the total intermediates present at any time.

We would like to raise the questions, (1) What are the sources of these new intermediates? (2) How is the feed-in of intermediates compensated by feed-out? These two questions will be considered in turn.

Methods

Heart mitochondria were prepared as described previously (12) and liver mitochondria as described by Johnson and Lardy (21).

Mitochondria were subjected to sonic disruption at 0° using a Bronson sonifier operating at 20 KHz and 100 watts output. Sonication was carried out for 2 min at 30 sec intervals to prevent heating. Malic enzyme activities were followed spectrophotometrically at 340 nm as a measure of formation of NADH or NADPH.

Guinea pig hearts were perfused as described by Davis and Quastel (22). Perfusion was terminated by clamping the hearts in aluminum tongs which had been precooled in liquid nitrogen. The frozen hearts were pulverized in a percussion mortar and extracted three times by suspending in a glass-teflon homogenizer with 1 M $HClO_4$. The perfusion medium was concentrated by freeze-drying and extracted with $HClO_4$. Pyruvate, malate, α-oxoglutarate, citrate, aspartate and glutamate were measured as reported previously (23). Alanine was determined spectrophotometrically with lactate dehydrogenase and alanine aminotransferase in the presence of a large excess of α-oxoglutarate and NH_4Cl (20 mM of each). Pyruvate was first removed with lactate dehydrogenase, followed by addition of alanine aminotransferase. It was necessary to use a large amount of the latter enzyme (2.5 enzyme units) in order to obtain quantitative measurement of alanine. Recovery of metabolites by the $HClO_4$ extractions of hearts and perfusates was in most cases more than 95%.

Protein determinations were by the biuret method (24) after first dispersing with deoxycholate. Specific activities of the more purified enzyme fractions were based on absorbance at 280 nm, using purified egg albumin as a standard.

Coenzymes and enzymes for substrate assays were obtained from Sigma Chemical Company, St. Louis, Missouri.

Sources of citric acid cycle intermediates in muscle

Bowman (17) observed that a rise in level of intermediates in perfused rat hearts coincided with a fall in the level of free aspartate and a corresponding increase in the level of glutamate. He proposed that the oxaloacetate required for extra citrate synthesis is derived at least partly from aspartate by transamination. Randle et al. (20) seem to concur in this interpretation. It should be emphasized, however, that a single transamination between aspartate and α-oxoglutarate does not of itself affect the level of total intermediates. This is illustrated in reactions 1-4 below:

1) aspartate + α-oxoglutarate ⇌ oxaloacetate + glutamate

2) acetyl-CoA + oxaloacetate → citrate + CoASH

3) citrate + 1/2 O_2 → α-oxoglutarate + CO_2 + H_2O

---

4) Sum: acetyl-CoA + aspartate + 1/2 O_2 → glutamate + CO_2 + H_2O + CoASH

However, if the aspartate aminotransferase reaction is coupled to a second transamination with pyruvate, one can observe a net change in citric acid cycle intermediates (reactions 1, 5 and 6).

1) aspartate + α-oxoglutarate ⇌ oxaloacetate + glutamate

5) glutamate + pyruvate ⇌ α-oxoglutarate + alanine

---

6) Sum: aspartate + pyruvate ⇌ oxaloacetate + alanine

Net synthesis of oxaloacetate by the above reactions would require a ready source and perhaps an elevated concentration of pyruvate, and an obligatory coupled accumulation or excretion of alanine. Convincing evidence that the above reactions do indeed occur in muscle has been provided by Felig and co-workers. They have found that skeletal muscle of fasted man excretes large amounts of alanine, and have emphasized that alanine excretion by muscle coincides with preferential extraction of alanine by the liver which is used for new glucose synthesis (25, 26). Carlsten *et al.* (27) had observed earlier that the concentration of plasma alanine increased during passage through the coronary circulation of the heart in man. Felig *et al.* (26) have drawn attention to the role of alanine as a carrier both of nitrogen and carbon to the liver from muscle for their conversion to urea and glucose.

We have extended this concept to accommodate also the fate of the α-oxoacids derived from the catabolism of amino acids in muscle. This is summarized below (reactions 5, 7-10). In this simplified scheme we have indicated that many amino acids transaminate with α-oxoglutarate to yield glutamate. Not shown is the fact that, in addition to this pathway, the carbon skeletons of several amino acids are also incorporated into glutamate without undergoing a transamination reaction. Reaction 8 represents the fate of a number of amino acids, including most of those which are glucogenic. The facility with which muscle is able to catalyze these reactions (reaction 8) appears not to be well established. However, recent reports of Bremer and co-workers (28-30) give evidence that heart as well as liver and kidney mitochondria have branched chain α-oxoacid dehydrogenases and all of the enzymes of branched chain amino acid pathways. Connelly et al. (31) have studied the branched chain α-oxoacid dehydrogenases of a number of tissues, including heart muscle.

7) amino acid + α-oxoglutarate ⇌ α-oxoacid + glutamate

5) glutamate + pyruvate ⇌ α-oxoglutarate + alanine

A second pathway capable of net synthesis of citric acid cycle intermediates in muscle has recently been described by Lowenstein and Tornheim (32). As a result of this pathway, nitrogen from amino acids is released in the form of ammonia. These workers reported that extracts of rat skeletal muscle produce ammonia from aspartate in a cyclical reaction sequence. The cycle utilized the reactions catalyzed by adenylic acid deaminase, adenylosuccinate synthetase and adenylosuccinase. The direct effect of this cycle, as depicted in reactions 11-14, is the formation of fumarate and ammonia from aspartate. More indirectly, other amino acids are the sources of both fumarate and ammonia (reactions 7, 1 and 8). As pointed out (32), this sequence would be energetically favorable and, in addition, may well play a role in regulation of the ATP level of the muscle cells.

7) amino acid + α-oxoglutarate ⇌ α-oxoacid + glutamate

1) glutamate + oxaloacetate ⇌ α-oxoglutarate + aspartate

8) α-oxoacid → oxaloacetate

11) $AMP + H_2O \rightarrow IMP + NH_3$

12) $IMP + aspartate + GTP \rightarrow adenylosuccinate + GDP + P_i$

13) Adenylosuccinate → AMP + fumarate

14) Sum: Amino acid + H_2O + GTP → fumarate + NH_3 + GDP + P_i

A third source of intermediates to muscle is probably the diet. Nearly all vegetables and fruit that we eat contain large stores of citrate or malate, or both (33). Presumably some of these would find their way to cells of muscle.

What then are the sources of amino acids which are catabolized by muscle? Undoubtedly, some of these can come from the diet. During starvation or other conditions of negative nitrogen balance a large portion of the amino acids are released from muscle protein. Cahill and co-workers (25, 26) have repeatedly emphasized that muscle protein is the major carbon source for new glucose during starvation.

Pathways for removal of citric acid cycle intermediates by mitochondria

Malate oxidation by isolated mitochondria

Since it is established that there is, at least under certain circumstances, net input of carbon compounds into the citric acid cycle in muscle, we have raised the question of what are the main mechanisms by which these "extra" intermediates are removed? All types of isolated mitochondria are capable of the oxidation of added L-malate, but to varying degrees. Many types of plant mitochondria are capable of rapid oxidation of malate (34-38), whereas mitochondria isolated from most mammalian tissues oxidize malate much more slowly than other substrates. Tables 2 and 3 illustrate the stoichiometry of oxidation of added L-malate by guinea pig heart and liver mitochondria, respectively. Arsenite was present to prevent removal of pyruvate formed. In the case of heart mitochondria, pyruvate production was essentially stoichiometric (1:1) with oxygen consumption. Respiration, even at this low rate, was limited by the rate of coupled electron transfer as evidenced by the fact that addition of ADP or an uncoupler stimulated respiration and pyruvate production. With liver mitochondria, both phosphoenolpyruvate and pyruvate were formed, and their *sum* was equal to the amount of oxygen consumed (Table 3). Added phosphate acceptor (ADP or ATP) stimulated respiration and accumulation of both pyruvate and phosphoenolpyruvate. Formation of the latter was favored under these conditions. However, when respiration was stimulated by dinitrophenol, most of the respiration was accounted for by the production of pyruvate.

At the time this work was begun about two years ago, we were convinced that heart mitochondria must contain an oxaloacetic decarboxylase which could compensate for input of extra intermediates. For such an enzyme to have physiological significance (for the expressed purpose), it must be able to act on oxaloacetate at concentrations perhaps as low as 1 µM, since its concentration inside the mitochondria is always very low.

Oxaloacetic decarboxylase activities

Corwin (39) described an enzyme in liver mitochondria capable of decarboxylation of oxaloacetate, but its activity was quite low, and its apparent K_m for oxaloacetate was about 1 mM. The enzyme isolated more recently from codfish muscle (40) also had a high K_m for oxaloacetate ($\sim$ 1 mM).

The assay we chose to use was one which was coupled to malic dehydrogenase, and in which the concentrations of both L-malate and oxaloacetate were known. This is illustrated in Fig. 1. In the presence of L-malate and NAD^+, oxaloacetate was produced on addition of malic dehydrogenase. The equilibrium concentration of oxaloacetate is given by the change in absorbance at 340 nm. Then on addition of an enzyme from heart mitochondria, more NADH was produced. This assay does not, however, readily discriminate which of the following reactions is taking place (reactions 15-17):

15) malate + NAD^+ ⇌ pyruvate + CO_2 + NADH + H^+

16) malate + $NADP^+$ ⇌ pyruvate + CO_2 + NADPH + H^+

17) (a) malate + NAD^+ ⇌ oxaloacetate + NADH + H^+

    (b) oxaloacetate → pyruvate + CO_2

---

15) Sum of (a) and (b): malate + NAD^+ → pyruvate + CO_2 + NADH + H^+

Reaction 15 would be catalyzed by an NAD-specific "malic enzyme," and reaction 16 by an NADP-specific "malic enzyme" which could react to some degree with NAD^+. Reactions 17a and 17b would give the same products as reaction 15, but would be catalyzed by two enzymes, malic dehydrogenase [E.C. 1.1.1.37 (reaction 17a)], and an oxaloacetic decarboxylase (reaction 17b). Ochoa (41) mentioned that an NADP-dependent "malic enzyme" is present in heart mitochondria, but otherwise these early observations have never been published. L-malic-NADP enzyme occurs also in the mitochondria of rat liver and kidney [(42); but contrast (43)]. More recently, Estabrook and co-workers have described this enzyme in the mitochondria from adrenal cortex and beef heart (44-46). The mitochondrial enzyme appears to be distinct from the one found in the cytosol.

Malic enzyme in heart mitochondria

Figs. 2 and 3 show the effects of disruption of heart mitochondria and of added metal ion on the reduction of added NAD^+ and $NADP^+$, respectively. Fig. 2 shows that, other than the NADH produced by malic dehydrogenase, there is very little further reduction of added NAD^+, whether in the presence or absence of added Mg^{++}. When the mitochondria are broken by sonic oscillations, added NAD^+ is reduced in the presence, but not absence, of added metal ion. Thus, the activity observed is not found on the surface of the mitochondria, since it is unable to react with added NAD^+. Fig. 3 shows a similar result for the L-malate-dependent reduction of added $NADP^+$.

Table 4 shows the early steps in the separation and purification of the two enzyme activities from rabbit heart mitochondria. As seen from this table, almost all of both activities was found in the fraction made soluble by sonic disruption of the mitochondria. On further fractionation of the soluble protein by salt precipitation, nearly all of the NAD-dependent activity was found in the fraction precipitating between 33 and 55% saturation of $(NH_4)_2SO_4$, whereas the NADP-requiring activity was found mainly in the fraction precipitating between 55 and 75% saturation. Clearly then, there are two (or more) enzymes (other than malic dehydrogenase) responsible for the reduction of NAD^+ and $NADP^+$ by malate.

The P_2 fraction, containing mostly NAD-specific activity, was purified further on a DEAE-cellulose column, using a pH and salt gradient for elution. A typical separation is shown in Fig. 4. As seen in this figure, all of the activity was found in one peak which did not coincide with any of the major protein peaks. Obviously, it could be purified further, but this has not yet been done due to the low yield of protein from the present source. The active peak was free of NADP-specific activity and, as will be seen later, contained only a very weak contamination with malic dehydrogenase.

Table 5 shows the results of an attempt to demonstrate reversibility of the two activities (*i.e.,* malate production) using the protein fraction which was solubilized by sonic disruption of guinea pig heart mitochondria. In the presence of a high concentration of CO_2, but not in its absence, malate was produced from pyruvate in the presence of NADPH. In no case was malate produced if NADH was the coenzyme. There was a slow disappearance of NADH in both buffer systems which was probably due to lactate production in the presence of a contamination of the extract with lactate dehydrogenase, since oxidation of NADH was dependent on the presence of pyruvate. More recent experiments with more purified enzyme from rabbit heart mitochondria have shown that this is the correct explanation.

Table 6 shows that pyruvate is the major product of oxidation of malate in the presence of NAD^+ or $NADP^+$, and that both reactions require the presence of added divalent metal ion. The small amount of NADH and pyruvate produced in the absence of Mg^{++} is due to equilibrium of the malic dehydrogenase reaction followed by a slow spontaneous decarboxylation of oxaloacetate.

Table 7 summarizes the metal ion requirements of the malate-dependent reduction of NAD^+ and $NADP^+$. The kinetics were carried out for reduction of NAD^+ using both the P_2 fraction (see Table 4) and the center of the active peak from the DEAE-cellulose column (corresponding to tubes 18-22 of Fig. 4) with essentially identical results. The P_3 fraction (Table 4) was used in the studies with the NADP-specific "malic enzyme" for purposes of comparison. Mn^{++} was much the most effective metal ion tested for both activities. The apparent K_m of Mg^{++}, Mn^{++} and Ni^{++} was several fold higher for the NAD-specific oxidation of malate than that for the activity with $NADP^+$. The V_{max} for the former with Mn^{++} was three-fold higher than with Mg^{++}.

The following experiments were designed to show unambiguously whether the NAD-specific activity is a NAD-specific "malic enzyme" (reaction 15), or an oxaloacetic decarboxylase (reactions 17a and 17b). The most desirable way to do this would be to remove all of the contaminating malic dehydrogenase from the enzyme preparation. If this were accomplished, and the enzyme were an oxaloacetic decarboxylase, production of the NADH and pyruvate (reaction 15) would be blocked owing to the absence of substrate for the reaction (reactions 17a and 17b).

The stoichiometry of production of NADH and pyruvate by the purified NAD-enzyme in the absence and presence of added malic dehydrogenase is compared in Fig. 5 and Table 8. L-malate and NAD^+ were present at the beginning of the reaction. In curve 1, malic dehydrogenase was added at the beginning, and no other additions were made. NADH produced was due to equilibration of the malic dehydrogenase reaction. For curves 2 and 3, the reaction was initiated after 14 min by the NAD-specific enzyme in the presence (curve 2) and absence (curve 3) of malic dehydrogenase. Reactions were terminated when the total amount of NADH produced in reactions 2 and 3 was the same. As seen from these curves, both reactions were linear, but the rate of reaction 3 was somewhat higher than that for reaction 2. This difference in rate was due to the slow equilibration of malic dehydrogenase present as a contaminant in the NAD-malic enzyme. A comparison of pyruvate and NADH produced in these reactions is recorded in Table 8. The rate of pyruvate production was unaffected by the presence of excess malic dehydrogenase.

All attempts to detect oxaloacetic decarboxylase activity, using freshly prepared solutions from crystalline oxaloacetic acid, were negative. However, it was of some concern that relatively high concentrations of oxaloacetate must be used for this assay which might inhibit the reaction, and also that crystalline oxaloacetate is in the *enol* form. The rate of its tautomerization to the *oxo* form is not conveniently controlled. We therefore chose to generate oxaloacetate enzymatically (presumably the *oxo* form) using bacterial citrate lyase. The results of such an experiment are shown in Table 9. The purified NAD-specific enzyme from rabbit heart did not catalyze production of pyruvate from oxaloacetate.

The redox potential of the 3-acetyl pyridine adenine dinucleotide (3-acetyl PAD$^+$)-reduced 3-acetyl pyridine adenine dinucleotide (3-acetyl-PADH) couple is about 70 mV more positive than is the NAD$^+$-NADH couple (47). 3-Acetyl-PAD$^+$ was substituted for NAD$^+$ in the coupled reaction with malic dehydrogenase and the NAD-specific enzyme from rabbit heart mitochondria in order to generate higher levels of oxaloacetate than was convenient with NAD$^+$. It was found that the NAD-specific mitochondrial enzyme did not cause further reduction of 3-acetyl PAD$^+$. Fig. 6 shows the results of such an experiment carried out comparing the activities of this enzyme, using NAD$^+$ and 3-acetyl-PAD$^+$ as coenzyme. The relative concentrations of NAD$^+$ and 3-acetyl-PAD$^+$ were selected to give equal steady-state concentrations of oxaloacetate when the malic dehydrogenase reaction was brought to equilibrium. When 30 μg of purified NAD-enzyme was added, there was rapid reduction of added NAD$^+$, whereas the rate of reduction of 3-acetyl-PAD$^+$ was not significantly greater than background. Thus, since the NAD-enzyme catalyzes the production of NADH and pyruvate from NAD$^+$ and malate, and since it does not form products from 3-acetyl-PAD$^+$ and malate, it must be concluded that the reaction studied is an NAD-specific oxidative decarboxylation of malate to pyruvate.

Finally, an experiment was carried out at three different levels of NAD$^+$, and concentrations of L-malate from 0-40 mM at each of the concentrations of NAD$^+$. After equilibration with malic dehydrogenase, the initial rate of reduction of NAD$^+$ by the mitochondrial NAD-specific enzyme was then plotted against the concentrations of malate and of oxaloacetate (Fig. 7). The results show that the rate of NADH production follows simple saturation kinetics with respect to malate concentration, and is independent of the concentration of NAD$^+$ between 1 and 5 mM. The average half-maximal rate, taken from this and four other similar experiments at pH 7.4, was 3.9 mM

L-malate. The rate of reaction was indeed very poorly correlated with oxaloacetate concentration.

The apparent Michaelis constant of the NAD-specific "malic enzyme" (NAD-ME) for NAD^+ at pH 7.4 was estimated in the presence of 30 mM L-malate. Excess α-oxoglutarate, ammonia and glutamic dehydrogenase were also present in order to keep NAD^+ completely in the oxidized form. The results of this experiment are shown in Fig. 8. The apparent K_m for NAD^+ is about 80 μM.

Finally, a pH activity curve (Fig. 9) carried out in the presence of excess malate and NAD^+ shows that, under the conditions of the experiment, the NAD-ME from rabbit heart mitochondria exhibits a fairly sharp pH optimum between pH 6.5 and 7.0.

Ochoa and co-workers (48, 49) reported that the NADP-dependent "malic enzymes" (NADP-ME) isolated from pigeon liver and *Lactobacillus arabinosus* were capable of rapid decarboxylation of oxaloacetate at low pH. This activity was stimulated several fold by added $NADP^+$. Rutter and Lardy (50) repeated and extended these observations with a more purified preparation of the enzyme from pigeon liver. Saz and Hubbard (51) described a malic enzyme from the muscle of *Ascaris lumbricoides*. NAD^+ was a better co-substrate than was $NADP^+$. The enzyme did not decarboxylate oxaloacetate. We therefore examined the ability of the two malic enzymes from rabbit heart mitochondria to decarboxylate oxaloacetate at pH 4.5. Although the NADP-ME was capable of appreciable stimulation of CO_2 production from oxaloacetate, all attempts to date using the NAD-ME have yielded negative results.

Estimation of the maximal rates of input of net citric acid cycle intermediates and their removal by the NAD- and NADP-specific malic enzymes of heart muscle

Heart muscle is capable of the oxidation of propionate to CO_2 at a very high rate. In fact, its rate of oxidation per carbon is about the same as that for acetate (22). We have used propionate as an example of a compound the oxidation of which will result in a stoichiometric net feed-in of intermediates into the citric acid cycle. Therefore, when propionate is oxidized there must be an accumulation of intermediates equal to the propionate consumed, or alternatively there must be a mechanism for rapid removal of the extra intermediates. This carbon balance is summarized in reactions 18-22, in which propionate is oxidized in the presence of a source of acetyl-CoA:

18) propionate + CO_2 → succinate

19) succinate + H_2O + O_2 → oxaloacetate + 2 H_2O

20) acetate + oxaloacetate → citrate

21) citrate + H_2O + 2 O_2 → oxaloacetate + 4 H_2O + 2 CO_2

---

22) Sum: propionate + acetate + 3 O_2 → oxaloacetate + CO_2 + 4 H_2O

Without removal by some mechanism, there is net production of a mole of oxaloacetate (or a precursor) for every mole of propionate consumed.

Table 10 shows the effect of propionate on the levels of pyruvate, citric acid cycle intermediates and free amino acids in guinea pig hearts perfused with glucose. Propionate caused a significant accumulation of pyruvate, citrate, malate and α-oxoglutarate, and shifted the steady-state position of the aspartate aminotransferase reaction toward aspartate production and glutamate removal. The main point that I would like to emphasize is that for the period of these experiments, these hearts have oxidized approximately 120 μmoles of propionate/g dry weight (22). Since no more than 5 μmoles of total intermediates accumulated during this period due to propionate oxidation, more than 100 μmoles of citric acid cycle intermediates (per g dry weight of heart) were removed during this 45 min perfusion period.

Table 11 summarizes approximations of the rate of propionate oxidation by perfused hearts, and these rates have been compared with the capacities of the "malic enzyme" activities found in heart mitochondria. According to this approximation, each of these two enzymes, the NAD- and NADP-specific malic enzymes, is capable of catalyzing the formation of pyruvate from malate at a rate equal to the observed rate of propionate oxidation by the intact beating heart.

In conclusion, we have evaluated the possible pathways available to muscle for anaplerotic maintenance of citric acid cycle intermediates required for its normal function and, conversely, for pathways which will provide a mechanism for complete combustion of intermediates produced in excess of the catalytic amounts required for functioning of the aerobic cycle. It is proposed that amino acid catabolism, with production and catabolism of α-oxoacids coupled with nitrogen excretion in the form of alanine may be a major pathway in muscle under certain conditions. In addition to the above pathway, the purine nucleotide cycle (32) probably contributes to the supply of intermediates by a different mechanism.

Two enzymes capable of net removal of intermediates have been isolated from heart mitochondria. In addition to the NADP-linked mitochondrial "malic enzyme" recently described (44, 46), a separate NAD-dependent activity, which is apparently an NAD-specific "malic enzyme," has been purified 100-200 fold from rabbit heart mitochondria and has been partially characterized. The maximal activity of each of these enzymes has been estimated to be adequate for removal of malate produced during the oxidation of propionate by the intact heart.

Presented by E. Jack Davis. The assistance of Miss Suzanne Baugh is gratefully acknowledged. The experimental work reported in this paper was supported by U.S.P.H.S. Grants AM13939, HE06308 and HE04219 and a grant from the Indiana Heart Association.

References

1. Krebs, H.A. Gluconeogenesis. The Croonian Lecture, 1963. Proc. Roy. Soc. of London, Series B 159:545-564(1964).
2. Evans, E.A. and L. Slotin, Jr. Carbon dioxide utilization by pigeon liver. J. Biol. Chem. 141:439-450(1941).
3. Lowenstein, J.M. The tricarboxylic acid cycle. In: D.M. Greenberg (Editor), Metabolic pathways, Vol. I, Third Ed., Academic Press, New York (1967), pp. 147-270.
4. Wit-Peeters, E.M. Synthesis of long-chain fatty acids in mitochondria. Biochim. Biophys. Acta 176:453-462(1969).
5. Wit-Peeters, E.M., H.R. Scholte and H.L. Elenbaas. Fatty acid synthesis in heart. Biochim. Biophys. Acta 210:360-370(1970).
6. Quastel, J.H. and A.H.M. Wheatley. Biological oxidations in the succinic acid series. Biochem. J. 25:117-128(1931).
7. Szent-Györgyi, A. Mechanism of respiration. Nature 135:305(1935).
8. Krebs, H.A. and L.V. Eggleston. The oxidation of pyruvate in pigeon breast muscle. Biochem. J. 34:442-459(1940).
9. Krebs, H.A. and W.A. Johnson. The role of citric acid in intermediate metabolism in animal tissues. Enzymologia 4:148-156(1937).
10. Davis, E.J. On the oxidation of acetate and pyruvate by guinea-pig heart sarcosomes. Biochim. Biophys. Acta 96:217-230(1965).
11. Slater, E.C., C. Tamblyn-Hague and W. Davis-van Thienen. The oxidation of pyruvate by isolated heart sarcosomes. Biochim. Biophys. Acta 96:206-216(1965).
12. Davis, E.J. The effect of pyruvate on cyclic oxidations by heart sarcosomes. Biochim. Biophys. Acta 143:26-36(1967).
13. Davis, E.J. On the nature of malonate-insensitive oxidation of pyruvate and glutamate by heart sarcosomes. Biochim. Biophys. Acta 162:1-10(1968).
14. Parmeggiani, A. and R.H. Bowman. Regulation of phosphofructokinase activity by citrate in normal and diabetic muscle. Biochem. Biophys. Res. Commun. 12:268-273(1963).
15. Kraupp, O., L. Adler-Kastner, H. Niessner and B. Plank. The effects of starvation and of acute and chronic alloxan diabetes on myocardial substrate levels and on liver glycogen in the rat *in vivo*. Europ. J. Biochem. 2:197-214(1967).

16. Williamson, J.R. and E.A. Jones. Inhibition of glycolysis by pyruvate in relation to the accumulation of citric acid cycle intermediates in the perfused rat heart. Nature 203:1171-1174(1964).
17. Bowman, R.H. Effect of diabetes, fatty acids, and ketone bodies on tricarboxylic acid cycle metabolism in the perfused rat heart. J. Biol. Chem. 241:3041-3048(1966).
18. Williamson, J.R. Glycolytic control mechanisms. I. Inhibition of glycolysis by acetate and pyruvate in the isolated perfused rat heart. J. Biol. Chem. 240:2308-2321(1965).
19. Williamson, J.R. Glycolytic control mechanisms. III. Effects of iodoacetamide and fluoroacetate on glucose metabolism in the perfused rat heart. J. Biol. Chem. 242:4476-4485(1967).
20. Randle, P.J., P.J. England and R.M. Denton. Control of the tricarboxylate cycle and its interactions with glycolysis during acetate utilization in rat heart. Biochem. J. 117:677-695(1970).
21. Johnson, D. and H. Lardy. Isolation of liver or kidney mitochondria. In: R.W. Estabrook and M.E. Pullman (Editors), Methods in enzymology, Vol. X, Academic Press, New York (1967), pp. 94-96.
22. Davis, E.J. and J.H. Quastel. The effect of short-chain fatty acid and starvation on the metabolism of glucose and lactate by the perfused guinea pig heart. Canad. J. Biochem. 42:1605-1621(1964).
23. Davis, E.J. and D.M. Gibson. Regulation of the metabolism of rabbit liver mitochondria by long chain fatty acids and other uncouplers of oxidative phosphorylation. J. Biol. Chem. 244:161-170(1969).
24. Gornall, A.G., C.J. Bardawill and M.M. David. Determination of serum proteins by means of the biuret reaction. J. Biol. Chem. 177:751-766(1949).
25. Felig, P., O.E. Owen, J. Wahren and G.F. Cahill, Jr. Amino acid metabolism during prolonged starvation. J. Clin. Invest. 48:584-594(1969).
26. Felig, P., T. Pozefsky, E. Marliss and G.F. Cahill, Jr. Alanine: key role in gluconeogenesis. Science 167:1003-1004(1970).
27. Carlsten, A., B. Hallgren, R. Jagenburg, A. Svanborg and L. Werkö. Amino acids and free fatty acids in plasma in diabetes. II. The myocardial arterio-venous differences before and after insulin. Acta Med. Scand. 179:631-639(1966).
28. Bøhmer, T. and J. Bremer. Propionylcarnitine physiological variations *in vivo*. Biochim. Biophys. Acta 152:559-567(1968).
29. Bremer, J. Pyruvate dehydrogenase, substrate specificity and product inhibition. Europ. J. Biochem. 8:535-540(1969).
30. Solberg, H.E. and J. Bremer. Formation of branched chain acylcarnitines in mitochondria. Biochim. Biophys. Acta 222:372-380(1970).
31. Connelly, J.L., D.J. Danner and J.A. Bowden. Branched chain α-keto acid metabolism. I. Isolation, purification, and partial characterization of bovine liver α-ketoisocaproic:α-keto-β-methylvaleric acid dehydrogenase. J. Biol. Chem. 243:1198-1203(1968).
32. Lowenstein, J. and K. Tornheim. Ammonia production in muscle: the purine nucleotide cycle. Science 171:397-400(1971).
33. Chemical composition of foodstuffs. Documenta Geigy, 5th ed. (English), S. Karger, New York (1959), pp. 230-243.
34. Wiskich, J.T. and W.D. Bonner, Jr. Preparation and properties of sweet potato mitochondria. Plant Physiol. 38:594-604(1963).
35. Ikuma, H. and W.D. Bonner, Jr. Properties of higher plant mitochondria. II. Effects of DNP, M-CL-CCP and oligomycin on respiration of mung bean mitochondria. Plant Physiol. 42:67-75(1967).

36. Lance, C., G.E. Hobson, R.E. Young and J.B. Baile. Metabolic processes in cytoplasmic particles of avocado fruit. IX. Oxidation of pyruvate and malate during climacteric cycle. Plant Physiol. 42:471-478(1967).
37. Sarkissian, I.V. and H.K. Srivastava. On methods of isolation of active tightly coupled mitochondria of wheat seedlings. Plant Physiol. 43:1406-1410(1968).
38. Muecke, P.S. and J.T. Wiskich. Respiratory activity of mitochondria from legume root nodules. Nature 221:674-675(1969).
39. Corwin, L.M. Oxaloacetic decarboxylase from rat liver mitochondria. J. Biol. Chem. 234:1338-1341(1959).
40. Schmitt, A., I. Bottke and G. Siebert. Eigenschaften einer Oxalacetat-Decarboxylase aus Dorschmuskulatur. Hoppe-Seyler's Z. Physiol. Chem. 347:18-34(1966).
41. Ochoa, S. Enzymic synthesis of citric acid and other reactions of the tricarboxylic acid cycle. 2nd Congr. Intern. Biochim., Symposium cycle tricarboxylique, Paris (July 21-27, 1952), pp. 73-88.
42. Stern, J.R. Personal communication (1970).
43. Utter, M.F. The role of CO_2 fixation in carbohydrate utilization and synthesis. Ann. N.Y. Acad. Sci. 72:451-461(1959).
44. Simpson, E.R., W. Cammer and R.W. Estabrook. The role of malic enzyme in bovine adrenal cortex mitochondria. Biochem. Biophys. Res. Commun. 31:113-118(1968).
45. Simpson, E.R. and R.W. Estabrook. Mitochondrial malic enzyme: the source of reduced nicotinamide adenine dinucleotide phosphate for steroid hydroxylation in bovine adrenal cortex mitochondria. Arch. Biochem. Biophys. 129:384-395(1969).
46. Frenkel, R. Separation and characterization of two malic enzymes from beef heart. Fed. Proc. 29:878(1970).
47. Kaplan, N.O., M.M. Ciotti and F.E. Stolzenback. Reaction of pyridine nucleotide analogues with dehydrogenases. J. Biol. Chem. 221:833-844(1956).
48. Veiga Salles, J.B. and S.C. Ochoa. Biosynthesis of dicarboxylic acids by carbon dioxide fixation. II. Further study of the properties of the "malic" enzyme of pigeon liver. J. Biol. Chem. 187:849-861(1950).
49. Korkes, S., A. del Campillo and S. Ochoa. Biosynthesis of dicarboxylic acids by carbon dioxide fixation IV. Isolation and properties of an adaptive "malic" enzyme from *Lactobacillus arabinosus*. J. Biol. Chem. 187:891-905(1950).
50. Rutter, W.J. and H.A. Lardy. Purification and properties of pigeon liver malic enzyme. J. Biol. Chem. 233:374-382(1958).
51. Saz, H.J. and J.A. Hubbard. The oxidative decarboxylation of malate by *Ascaris lumbricoides*. J. Biol. Chem. 225:921-933(1957).

TABLE 1

THE EFFECT OF VARIOUS CONDITIONS ON THE LEVELS OF CITRIC ACID CYCLE INTERMEDIATES IN RAT HEARTS PERFUSED WITH GLUCOSE

Conditions/Additions	Percentage of Control			Perfusion time	Reference
	Citrate	Malate	α-Oxo-glutarate		
				min	
Pyruvate	28.5	13.9	17.0	15	(18)
Acetate	10.1	1.5	2.1	15	(18)
Octanoate	7.0	4.7	0.8	20	(17)
Acetate	7.5	3.0	2.1	20	(17)
β-Hydroxybutyrate	2.2	1.7	1.9	20	(17)
Alloxan diabetes	6.4	5.0	2.6	8	(17)
Acetate	3.7	2.0	2.1	12	(20)
Pyruvate	13.4	15.0	10.5	10	(19)
Pyruvate + Fluorocitrate	98.5	2.7	3.1	10	(19)

TABLE 2

STOICHIOMETRY OF OXIDATION OF MALATE BY GUINEA PIG HEART MITOCHONDRIA IN THE PRESENCE OF ARSENITE

Reactions were carried out in a Gilson 'Oxygraph' at 25°. Incubation media contained 225 mM sucrose, 10 mM KCl, 10 mM potassium phosphate (pH 7.4) and 1 mM sodium arsenite. Mitochondria (3.25 mg protein) were added and, after 5 min pre-incubation, potassium L-malate (10 mM) was added to initiate respiration. After 10 min the reactions were terminated with $HClO_4$ and the contents assayed for pyruvate. Other additions, when present were: ADP, 5 μmoles; ATP, 5 μmoles; 2,4-dinitrophenol (DNP), 50 μM final concentration.

Additions	Δ Pyruvate	Δ Oxygen	Respiration Rate
	nmoles	*natoms*	*natoms O min^{-1} mg $prot^{-1}$*
None	143	167	5.0
ADP	348	369	11.4
ATP	139	142	5.0
DNP	294	365	11.2
DNP, ATP	291	362	11.1
DNP, ADP	278	362	11.1

TABLE 3

STOICHIOMETRY OF MALATE OXIDATION BY GUINEA PIG LIVER MITOCHONDRIA IN PRESENCE OF ARSENITE

Conditions were as described in Table 2, except that the incubation media initially contained 50 mM KCl, 25 mM Tris·HCl (pH 7.4), 5 mM $MgCl_2$, 5 mM potassium phosphate (pH 7.4), and 1 mM sodium arsenite. Reactions in presence of L-malate were for 6 min with 9 mg mitochondrial protein.

Additions	ΔO	Δ Pyruvate	ΔPEP	Sum	Rate of respiration
	natoms	nmoles			natoms $mg\ prot^{-1}\ min^{-1}$
None	170	48	98	146	3.1
ADP	419	192	225	417	7.7
DNP	350	258	74	332	6.3
ATP	350	116	240	356	6.3
DNP, ATP	290	116	160	276	5.2

TABLE 4

FRACTIONATION OF THE NAD- AND NADP-SPECIFIC 'MALIC ENZYME' ACTIVITIES FROM RABBIT HEART MITOCHONDRIA

Assays were carried out at 25° in 50 mM Tris·HCl (pH 7.4) containing 30 mM potassium malate, 1 mM NAD^+ or 1 mM $NADP^+$ and 1 mM $MnCl_2$. For the assays with NAD^+ present, malic dehydrogenase was added initially to prevent interference by this reaction. Production of NADH or NADPH was recorded at 1 min intervals for 10 min.

Fraction	Protein	Specific activity x 10^3		Total enzyme units	
		with NAD^+	with $NADP^+$	with NAD^+	with $NADP^+$
	mg	units/mg protein			
Broken Mitochondria	302	9.2	8.3	2.78	2.51
Soluble (from sonication)	42	66	50	2.77	2.10
Insoluble (from sonication)	220	0.8	1.4	0.17	0.33
Subfractions of soluble protein					
P_1 [0-33% $(NH_4)_2SO_4$]	10.9	4.0	2.5	0.04	0.03
P_2 [33-55% $(NH_4)_2SO_4$]	15.8	203	21	3.20	0.33
P_3 [55-75% $(NH_4)_2SO_4$]	9.2	12	152	0.11	1.40

TABLE 5

REVERSIBILITY OF THE NAD- AND NADP-SPECIFIC ACTIVITIES OF 'MALIC ENZYME' FROM GUINEA PIG HEART MITOCHONDRIA

Reactions were carried out as indicated in either 50 mM Tris·HCl (pH 7.4), or NaHCO$_3$ (200 mM) equilibrated with 50% CO$_2$ -50% N$_2$ (pH 7.2-7.4), each containing 5 mM MgCl$_2$ and 0.5 mg protein solubilized by sonication of guinea pig heart mitochondria. Where indicated, the following substrates were present: NADH or NADPH, 0.2 mM; NAD$^+$ or NADP$^+$, 5 mM; potassium malate, 30 mM; and potassium pyruvate, 50 mM. Changes in the content of NADH or NADPH were followed for 10 min, after which assays were performed, where appropriate, for malate.

				Δ Malate	
Buffer	Substrate	ΔNADH	ΔNADPH	with NADH	with NADPH
		nmoles/min/mg protein			
Tris·HCl	Malate	+95	+82	–	–
HCO$_3^-$-CO$_2$	Pyruvate	−27	−19	0	18
Tris-HCl	Pyruvate	−25	− 1	0	0

TABLE 6

STOICHIOMETRY OF THE NAD- AND NADP-SPECIFIC 'MALIC ENZYMES' FROM GUINEA PIG HEARTS: PRODUCTION OF NAD(P)H AND PYRUVATE

Incubations contained 50 mM Tris HCl, 10 mM potassium L-malate and 1 mM of NAD$^+$ or NADP$^+$. An aliquot from guinea pig heart mitochondria which was made soluble by sonic disruption was added to initiate the reactions. Change in absorbance at 340 nm was followed for 15 min. Reactions were terminated with HClO$_4$ and assayed for pyruvate. NADH produced is the *sum* of that formed by equilibration of the malic dehydrogenase reaction, spontaneous decarboxylation of oxaloacetate, and activity of the mitochondrial extract. Where present, MgCl$_2$ was 5 mM.

Additions	NADH	NADPH	Pyruvate
	nmoles produced		
None (± NAD$^+$ or NADP$^+$)	0	0	0
Malate + NAD$^+$	95	–	55
Malate + NAD$^+$ + MgCl$_2$	510	–	475
Malate + NADP$^+$	0	−10	0
Malate + NADP$^+$ + MgCl$_2$	–	399	418

TABLE 7

METAL ION REQUIREMENTS FOR THE NAD- AND NADP-SPECIFIC 'MALIC ENZYMES' FROM RABBIT HEART MITOCHONDRIA

Incubations were carried out at 25° in Tris·HCl (pH 7.4) in the presence of 30 mM potassium malate and 1 mM NAD$^+$ or NADP$^+$ and excess (2 E.U.) of malic dehydrogenase. Reactions were initiated with purified NAD-ME (from DEAE cellulose column) for the NAD-specific activity, and the P_3 ammonium sulfate fraction (Table 4) for the NADP-specific activity. Initial rates of reactions were followed for 10 min at 340 nm.

	K_m		V_{max}	
Ion	NAD-Enzyme	NADP-Enzyme	NAD-Enzyme	NADP-Enzyme
	mM		*nmoles/min*	
Mg^{2+}	4.1	0.23	100	100
Mn^{2+}	0.2	0.01	310	98
Ni^{2+}	14.7	0.32	35	–
Cd^{2+}	3.7	–	18	–
Ca^{2+}	–	–	0	–

TABLE 8

STOICHIOMETRY OF PRODUCTION OF NADH AND PYRUVATE BY PURIFIED NAD-SPECIFIC 'MALIC ENZYME' IN THE PRESENCE AND ABSENCE OF ADDED MALIC DEHYDROGENASE

		Products formed			
Reaction	Additions	Total NADH	NADH due to malic dehydrogenase	NADH due to malic enzyme	Pyruvate
		nmoles			
1	Malic dehydrogenase	215	215	0	32
2	Malic dehydrogenase + NAD-specific 'malic enzyme'	732	215	517	566
3	NAD-specific 'malic enzyme'	724	(215)	509	560

TABLE 9

DEMONSTRATION OF THE FAILURE OF NAD-SPECIFIC 'MALIC ENZYME' TO DECARBOXYLATE ENZYMATICALLY-GENERATED OXALOACETATE AT pH 7.4

Reaction mixtures contained 25 μmoles Tris-HCl (pH 7.4), 1 μmole $MnCl_2$, 0.250 μmoles potassium citrate and, when present, 0.25 enzyme units of NAD-specific 'malic enzyme' and 1 μmole of NAD^+ in a volume of 1.0 ml. Reactions were started with 10 enzyme units of citrate lyase. Incubation time, 5 min at 25°C.

			Products formed	
Sample	NAD-Malic Enzyme	NAD^+	Oxaloacetate	Pyruvate
			nmoles	
1	−	−	236	18
2	−	−	241	19
3	+	−	247	16
4	+	+	243	16

TABLE 10

THE EFFECT OF PROPIONATE ON THE LEVEL OF CITRIC ACID CYCLE INTERMEDIATES AND AMINO ACIDS IN PERFUSED GUINEA PIG HEARTS AND IN PERFUSING MEDIUM

Hearts were perfused as previously described (22) in the presence of 5 mM glucose and, when present, 5 mM potassium propionate. Perfusion time, 45 min. The data are the mean values of 4 to 7 determinations ± S.E.M.

		Perfused with	
Intermediate	Compartment	Glucose	Glucose + Propionate
Pyruvate	tissue (μmoles/gm dry wt)	1.2 ± 0.2	1.8 ± 0.3
	perfusate (total μmoles)	12.4 ± 2.7	29.0 ± 1.5
Citrate	tissue	1.3 ± 0.1	2.8 ± 0.3
	perfusate	1.1 ± 0.2	0.4 ± 0.3
Malate	tissue	0.7 ± 0.2	4.2 ± 0.7
	perfusate	0.7 ± 0.1	1.3 ± 0.2
α-Ketoglutarate	tissue	0.14 ± 0.04	0.38 ± 0.06
	perfusate	0.66 ± 0.12	0.66 ± 0.14
Glutamate	tissue	20.4 ± 2.2	9.0 ± 1.0
	perfusate	3.4 ± 0.3	2.1 ± 0.6
Aspartate	tissue	2.3 ± 0.2	10.4 ± 1.9
	perfusate	2.3 ± 0.1	1.4 ± 0.4
Alanine	tissue	24.4 ± 1.6	30.0 ± 1.9
	perfusate	5.2 ± 0.7	3.2 ± 0.5

TABLE 11

CALCULATION OF MAXIMAL RATES OF OXIDATION OF ^{14}C-PROPIONATE BY NON-WORKING ISOLATED HEARTS AND ESTIMATED ACTIVITIES OF NAD- AND NADP-SPECIFIC 'MALIC ENZYMES' OF HEART MITOCHONDRIA

Preparation	Temperature	Rate of Propionate Oxidation
Perfused heart	37°C	$\sim$ 2.3 μmoles/g total protein/min (22)
		Enzyme Activitiy
NAD-specific mitochondrial 'malic enzyme'	25°C	$\sim$ 9 μmoles/g mitochondrial protein/min
NADP-specific mitochondrial 'malic enzyme'	25°C	$\sim$ 8 μmoles/g mitochondrial protein/min
Estimated sum of the NAD-and NADP-specific 'malic enzymes' of mitochondria,	37°C	$\sim$ 35 μmoles/g mitochondrial protein/min, or $\sim$ 9 μmoles/g total protein/min

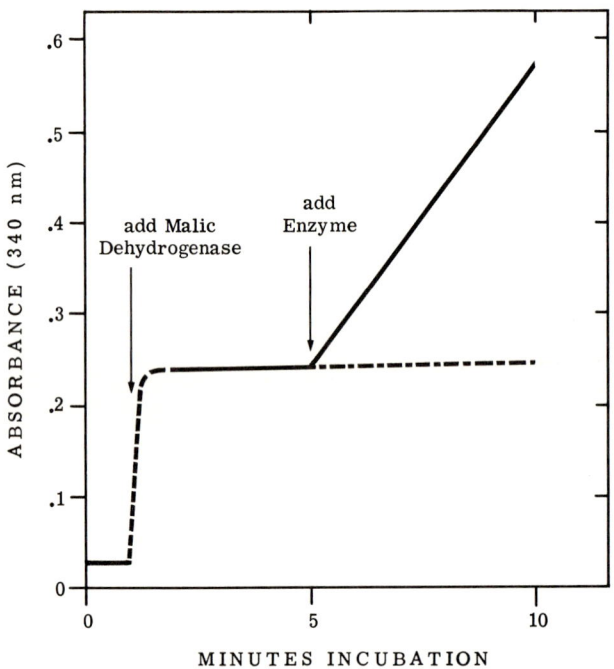

Fig. 1. *Conditions for routine assay of the crude NAD-specific "malic enzyme" from heart mitochondria.* Incubations were carried out in 50 mM Tris·HCl (pH 7.4), 30 mM potassium L-malate, 1 mM NAD$^+$ and 1 mM MnCl$_2$ in a volume of 3.0 ml. Malic dehydrogenase was added initially to bring oxaloacetate and NADH in equilibrium with the malate and NAD$^+$ present. The preparation containing NAD-specific "malic enzyme" activity was then added (solid line) to initiate the further reduction of NAD$^+$. The broken line represents the slow spontaneous decarboxylation of oxaloacetate without the second enzyme.

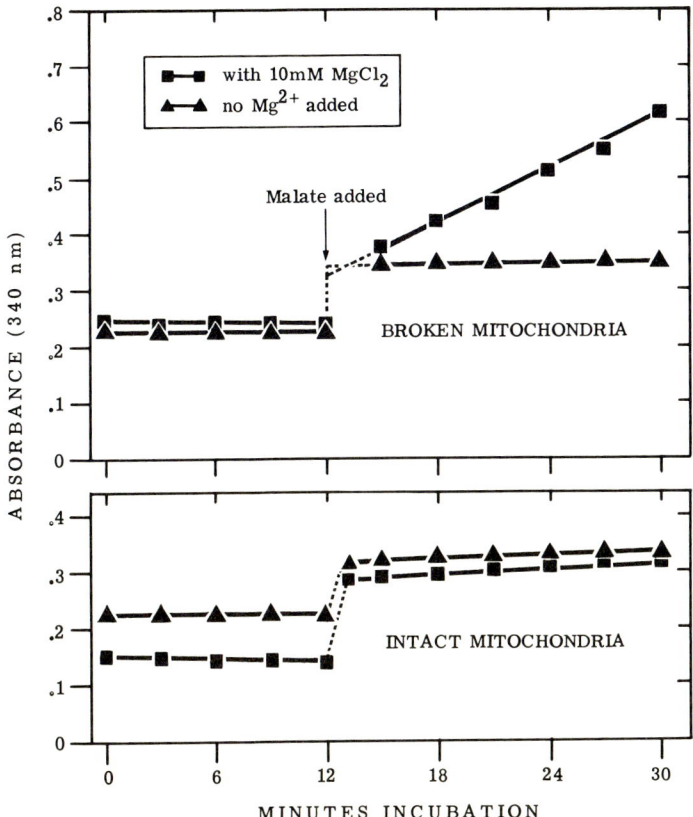

Fig. 2. *The malate-dependent reduction of added NAD^+ by intact and broken guinea pig heart mitochondria in the presence and absence of added Mg^{++}.* Cuvettes contained 250 mM sucrose, 20 mM Tris·HCl (pH 7.4), 5 mM NAD^+, 2 μg of rotenone, dialyzed malic dehydrogenase (2 E.U.), and 0.43 mg mitochondrial protein (whole mitochondria) or 0.33 mg protein (broken mitochondria). When present (5), $MgCl_2$ was 10 mM. The reference cuvette contained all components except malate. Mitochondria were disrupted by sonication. Reaction volume, 3.0 ml.

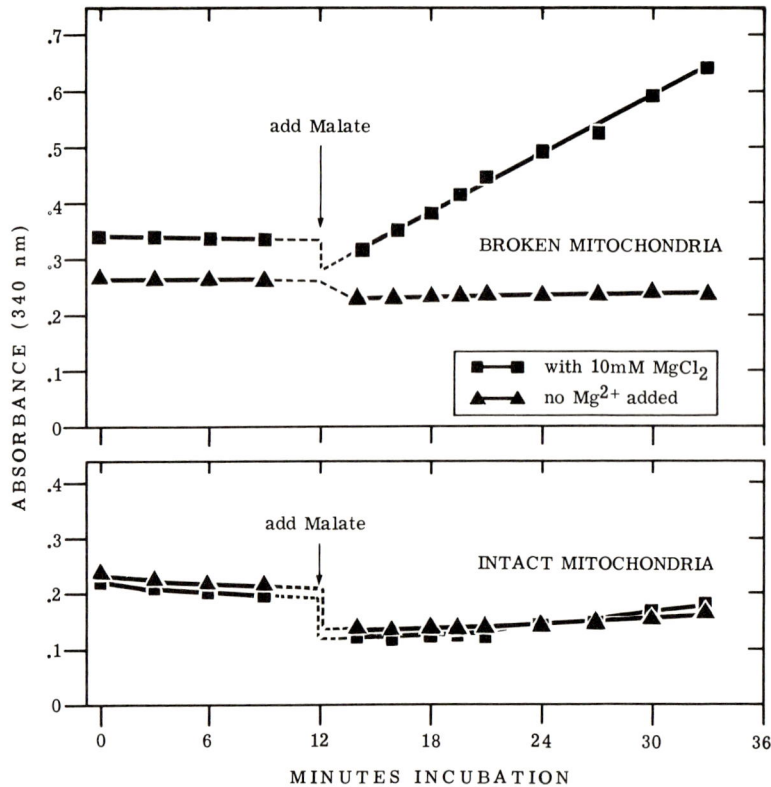

Fig. 3. *The malate-dependent reduction of added $NADP^+$ by intact and broken guinea pig heart mitochondria in the presence and absence of added $MgCl_2$.* Conditions were the same as in Fig. 2, except $NADP^+$ (5 mM) was substituted for NAD^+.

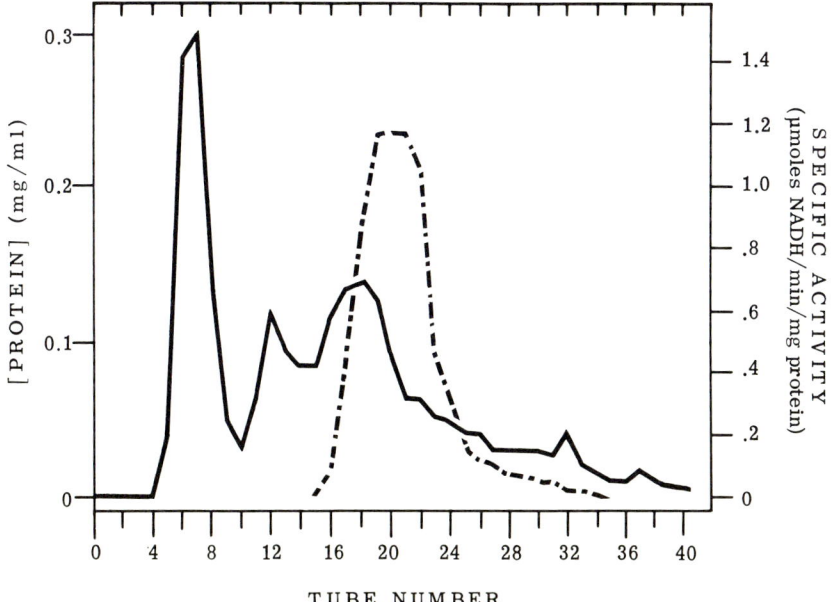

Fig. 4. *DEAE-Cellulose column separation of the NAD^+-specific "malic enzyme" from rabbit heart mitochondria.* Mitochondria were isolated from eight New Zealand white rabbits weighing approximately 2 kg. After sonic disruption, the soluble protein (42 mg) was precipitated into P_1, P_2 and P_3 fractions using $(NH_4)_2SO_4$ (see Table 4). The P_2 fraction (16 mg), containing most of the NAD-specific activity, was desalted by passing it through a Sephadex-G-25 column and placed on a DEAE-cellulose column (13 × 1 cm) which had been equilibrated with 10 mM K_2HPO_4-1 mM EDTA (pH 8.0). A pH and salt gradient was contrived which contained 100 ml of 10 mM $KHPO_4$-1 mM EDTA (pH 8.0) in the vessel adjacent to the column and 100 ml of 10 mM KH_2PO_4, 1 mM EDTA and 1.0 M NaCl in the eluting vessel. Samples (2.5 ml) were collected from the columns at a rate of about 0.3 ml/min, using an LKB fraction collector and automatic recorder (280 nm). Enzyme assays were performed on aliquots from each sample.

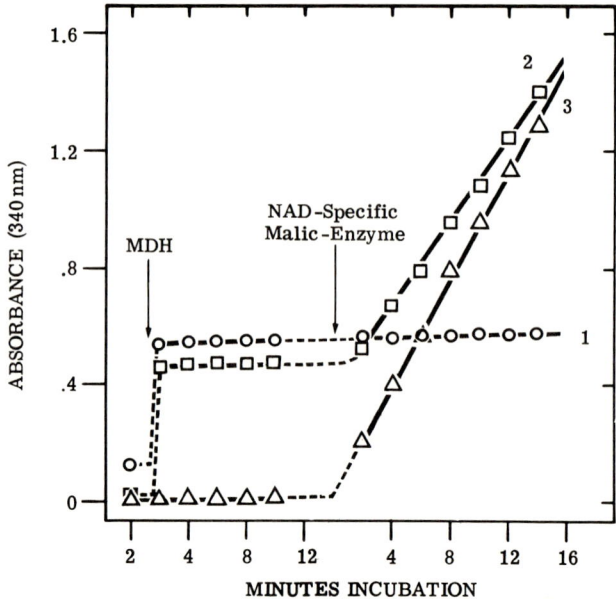

Fig. 5. *Stoichiometry of NADH production by NAD-specific "malic enzyme" in presence and absence of added malic dehydrogenase.* Incubations contained 100 mM Tris·HCl (pH 7.4), 5 mM NAD$^+$, 1 mM MnCl$_2$ and 30 mM potassium L-malate in a volume of 3.0 ml. Malic dehydrogenase (2 E.U.) was added to reactions 1 and 2. After 14 min incubation, 25 µg purified NAD-specific "malic enzyme" was added to reactions 2 and 3. Reactions were terminated after an additional 16 min, and the contents were assayed for pyruvate. These data are summarized in Table 8.

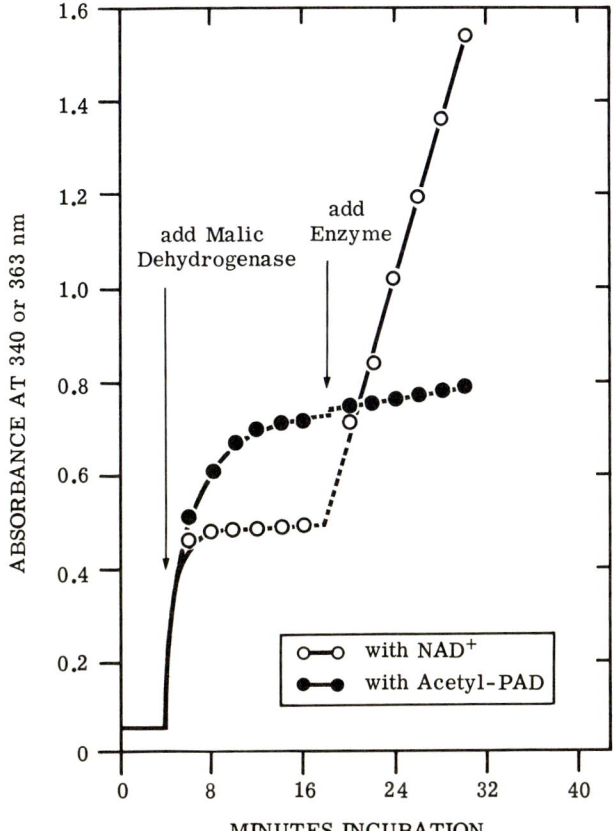

Fig. 6. *Comparison of malic enzyme activities with NAD^+ and 3-acetyl-PAD^+.* Incubations contained 100 mM Tris·HCl (pH 7.4), 30 mM potassium L-malate, 1 mM $MnCl_2$ and 5 mM NAD^+ (o) or 150 µM 3-acetyl-PAD^+ (•). Malic dehydrogenase (0.5 E.U.) was added after 4 min. NAD-specific malic enzyme (38 µg protein) was added after 18 min. Production of NADH and 3-acetyl-PADH was followed at 340 nm and 363 nm, respectively.

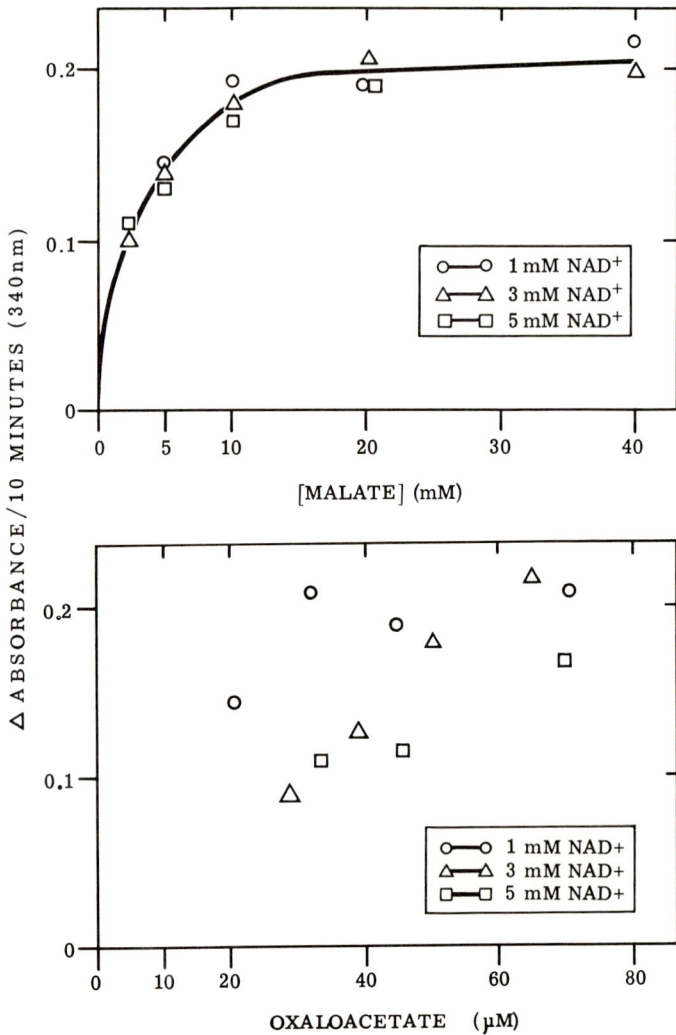

Fig. 7. *The rates of malate-dependent NADH production plotted as a function of the concentrations of malate and oxaloacetate.* Incubations contained 100 mM Tris·HCl (pH 7.4), 5 mM $MnCl_2$, malate concentrations as indicated (0-40 mM) and NAD^+ to final concentrations of 1, 3, and 5 mM. Malic dehydrogenase (1 E.U.) was added initially in order to observe the equilibrium concentration of oxaloacetate and to avoid interference by this reaction. NAD-specific malic enzyme was then added, and the reactions were followed for 15 min.

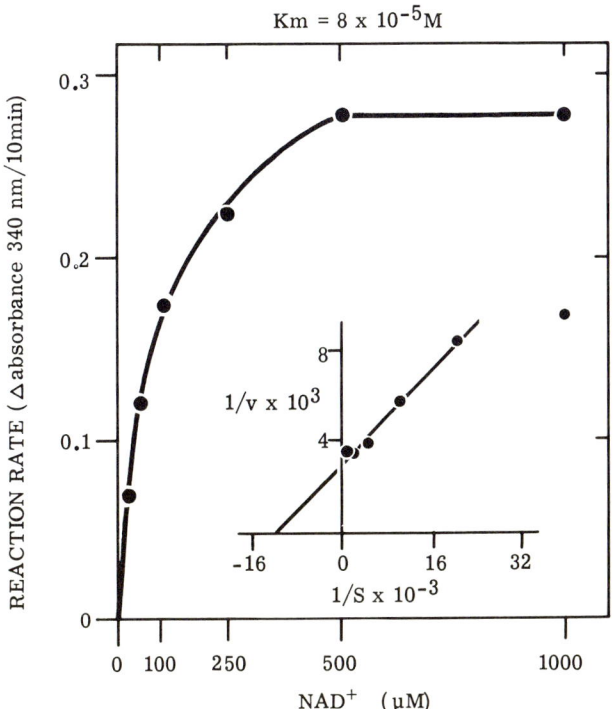

Fig. 8. *The apparent Michaelis constant of the NAD-specific "malic enzyme" from rabbit heart mitochondria for NAD^+ at pH 7.4.* Reaction mixtures contained 10 mM Tris·HCl, 30 mM potassium L-malate, 1 mM $MnCl_2$, 10 mM α-oxoglutarate, 10 mM NH_4Cl, 4 E.U. of highly purified glutamate dehydrogenase, and NAD^+ at the concentrations indicated to a final pH of 7.4. Reactions were initiated with 25 μg purified NAD-ME which was devoid of any activity with $NADP^+$, and essentially free of malic dehydrogenase. Duplicate incubations were carried out at 25° for periods up to 20 min. Pyruvate was determined in the samples after the reactions were terminated.

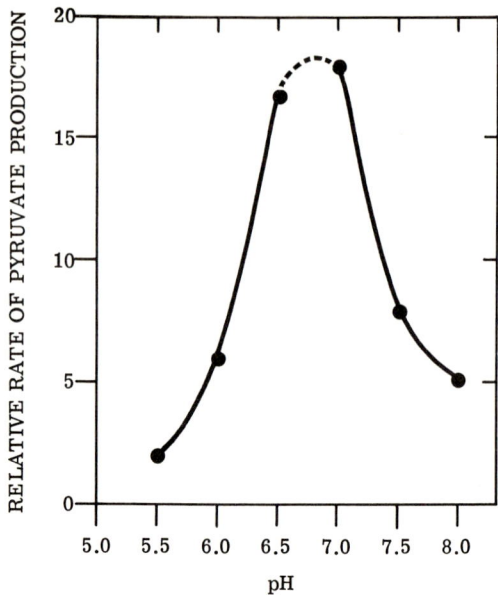

Fig. 9. *pH Activity curve of the NAD-specific "malic enzyme" from rabbit heart mitochondria.* Incubations contained 200 mM K_2HPO_4-KH_2PO_4 buffer at the appropriate pH, 30 mM potassium malate, 1 mM $MnCl_2$, 10 mM α-oxoglutarate, 10 mM NH_4Cl, 4 E.U. of highly purified glutamic dehydrogenase and 1 mM NAD^+. Reactions were initiated with 30 µg purified NAD-ME. Duplicate incubations were carried out at 25° for periods up to 20 min. Pyruvate was determined after the reactions were terminated.

HORMONAL REGULATION OF PYRUVATE METABOLISM IN RAT LIVER MITOCHONDRIA

Robert C. Haynes, Jr.

Introduction and methods

During the past ten years, evidence has been accumulated indicating that hormonal control of hepatic gluconeogenesis is exerted at some point or points in the initial stages of synthesis of glucose. Two types of experiments have contributed to this conclusion. The first, typified by the work of Uete and Ashmore (1), demonstrates that glucose precursors fed into the synthetic pathway at the triose level support a more rapid glucose synthesis than do lactate or pyruvate as substrates, but no hormonal stimulation is observed. Second, analysis of the concentrations of various intermediate compounds involved in gluconeogenesis indicates that hormonal stimulation results in a "crossover" between pyruvate and the four carbon dicarboxylic acids. That is, the level of pyruvate is depressed, and the level of the dicarboxylic acids is raised, suggesting that there is a facilitation of the pathway at this locus (2, 3).

On the basis of this and other evidence, Peter Adam and I decided to look at what is generally considered to be the first reaction of gluconeogenesis, the pyruvate carboxylase reaction in which CO_2 is fixed to the pyruvate molecule to form oxaloacetate. Attempts to find stable changes in the pyruvate carboxylase enzyme (PC) after hormone treatment were not successful, although hormonally-induced increases have been reported by others (4). Since we could not find any alteration in the enzyme after treatment of rats with glucagon, epinephrine, or cortisol, neither in assays employing saturating amounts nor in assays employing less than saturating amounts of substrates and cofactors, we began to study the reaction in a slightly more physiological system, the isolated hepatic mitochondria.

It was found immediately that treatment of rats with epinephrine or glucagon for 15-25 minutes nearly doubles the rate of CO_2 fixation in mitochondria isolated from these animals in comparison with mitochondria from control rats (5).

The assay system is quite simple. Basically, mitochondria are incubated with pyruvate and bicarbonate-^{14}C. The reaction is stopped by adding trichloroacetic acid which also liberates unreacted $^{14}CO_2$. The radioactivity in aliquots of the TCA-soluble mixture is determined by scintillation counting. We believe this technique is a satisfactory assay of PC activity in the mitochondria. Two other enzymes that could contribute to CO_2 fixation, acetyl-CoA carboxylase and the malic enzyme, are both located in the cytosol of the liver cell. In addition, the omission of pyruvate from the assays decreases the fixation of CO_2 by more than 90%.

The amount of radioactive oxaloacetate accumulated during these assays is too small to be detected. Lardy *et al.* (6) and Haynes (7) showed in incubations with liver mitochondria that ^{14}C from labelled CO_2 appears primarily as citrate, malate and fumarate. When the levels of citrate and malate were measured fluorometrically after incubations of mitochondria from hormone-treated rats, it was found that the production of citrate and malate was increased by the prior treatment.

In addition, the rate of disappearance of pyruvate during incubation of the mitochondria was found to be enhanced by prior treatment with hormones. The amount of pyruvate used during the incubations was more than twice that accounted for by fixation of CO_2. To learn more about the overall fate of pyruvate in the mitochondria, we estimated pyruvate dehydrogenase (PD) activity by measuring $^{14}CO_2$ liberated from pyruvate-1-^{14}C, and the level of activity of the tricarboxylic acid cycle by the amount of $^{14}CO_2$ released from pyruvate-2-^{14}C.

Treatment of rats with epinephrine or glucagon did not significantly increase oxidation of pyruvate via the tricarboxylic cycle, but it stimulated the rate of decarboxylation of pyruvate about two fold, or to about the same extent as the stimulation of the pyruvate carboxylase reaction. At this stage of the investigation, we had made the following observations: treatment of rats with epinephrine, glucagon, or cortisol produced no demonstrable alteration in the pyruvate carboxylase enzyme, but treatment with hormones did stimulate the pyruvate carboxylase reaction when measured in intact mitochondria. Treatment with these hormones also increased the rate of utilization of pyruvate by mitochondria and increased the decarboxylation of pyruvate.

Two alternative hypotheses of hormonal action are consistent with these observations. The first hypothesis is that the hormones act primarily to increase the activity of PD; a consequence of this is a rise in the intramitochondrial acetyl-coenzyme A level which secondarily activates PC.

A second hypothesis is that there is a restriction to the passage of pyruvate into mitochondria so that under control conditions neither PC nor PD is saturated in respect to pyruvate. With hormonal stimulation there is an increase in the rate of entry of pyruvate into the mitochondria resulting in an acceleration of both reactions as more stubstrate becomes available.

These hypotheses were tested in two experiments. The first hypothesis that implicates a stimulation of PD as the primary event implies that hormonal stimulation leads to an increased concentration of acetyl-CoA in the mitochondria. When the content of acetyl-CoA was measured in mitochondria from rats previously treated with epinephrine, it was found that there was a 30% increase in level, an increase probably inadequate to stimulate the carboxylase reaction to the extent usually found, although this is not certain. The activation of PC by acetyl-CoA follows cooperative kinetics (8), and if the concentration of acetyl-CoA in the mitochondria happens to lie at a point at which the activation curve is exceptionally steep, a 30% increase could conceivably activate PC 70-100%, the range of stimulation found after hormonal treatment. When cortisol was given, no change in acetyl-CoA levels could be detected in the mitochondria in spite of elevated rates of carboxylation and decarboxylation of pyruvate.

A second test of these alternative hypotheses was made in the following way. It was reasoned that if hormones act primarily to increase the rate of movement of pyruvate into mitochondria, then facilitating the penetration of pyruvate by another technique should mimic the action of the hormones. Incubation of mitochondria in hypotonic media appeared to be one way in which permeability of the mitochondria might be increased, since hypotonicity leads to some swelling of the organelles with a consequent stretching of the membranes. Decreasing the osmolality of the incubation mixture does lead to the changes in all metabolic parameters of the system that occur when hormonal treatment is given. That is, hypotonicity increases the fixation of CO_2, increases the decarboxylation of pyruvate and increases the rate of utilization of pyruvate by the mitochondria. In addition the effects of hypotonicity and hormone treatment are not additive, so that the hormonal effect tends to decrease or disappear as tonicity is lowered.

As a consequence of these two experiments, the measurement of acetyl-CoA in the mitochondria and the stimulation of pyruvate metabolism by hypotonicity, it was concluded that the second hypothesis, which proposes that hormonal control is exerted on the entrance of pyruvate into mitochondria, fits the experimental observations better than the first hypothesis, which proposes that the primary event is a stimulation of the pyruvic dehydrogenase reaction. Now that Wieland and Siess (9) have reported an activation of PD by cyclic-AMP, it may be worthwhile to reconsider the first hypothesis. At the present time, we are preparing to look at PD obtained from mitochondria of glucagon-treated rats. If there is a stable change in the enzyme brought about by cyclic AMP, it should be possible to detect it after hormonal treatment.

Results and Discussion

As part of our study of the hormonal control of pyruvate metabolism in mitochondria some of the biochemical characteristics of the mitochondrial system have been investigated. The results that follow addition of various metabolites to the incubation mixture are presented in Table 1. It is evident that only succinate and α-ketoglutarate stimulate fixation of CO_2 above the level found with pyruvate alone. As noted above, in the absence of added pyruvate, fixation of CO_2 decreases to less than 10% of control levels. Succinate added in the absence of pyruvate does not affect this low basal level. α-Ketoglutarate increases it slightly, but not sufficiently to account for the stimulatory effect seen in the presence of pyruvate. When succinate is added, mitochondria sustain fixation of CO_2 for a longer period of time. These observations, together with the inhibition of the succinate stimulation by added malonate, indicate that the metabolism of succinate is required for its stimulatory effect. It is possible that mitochondria are incapable of generating enough ATP from pyruvate alone to continue fixing CO_2 at a rapid rate. Metabolism of succinate may provide a second source of energy to sustain high ratios of ATP/ADP.

It can also be seen that the additions of acetate, butyrate, and octanoate were inhibitory at the concentrations tested. If these compounds had enhanced the fixation of CO_2, this would have provided support for the concept of regulation of carboxylation by the level of acetyl-CoA. The negative results are obviously much more difficult to interpret. These observations made with added substrates agree closely with similar findings of Mehlman *et al.* (10).

Some interesting relationships between added magnesium ion and ATP have been found. In Table 2 it can be seen that the addition of $MgCl_2$ has only a small effect on CO_2 fixation. ATP added in the absence of Mg^{++} is stimulatory, but in the presence of 12.5 mM Mg^{++} it acts as an inhibitor of the reaction. ADP can be seen to inhibit at all levels tested with or without added Mg^{++}. The mechanism of the effects of added ATP is not obvious. One possibility is that in the presence of added Mg^{++} an ATPase is activated that converts some of the added ATP to the inhibitory ADP. In the absence of added Mg^{++} the ATP may increase the ATP/ADP ratio and activate PC. The stimulating effect of ATP is mimicked by added succinate, again suggesting that the level of ATP itself may be influencing the activity of the carboxylase enzyme. An alternative explanation might be that in the presence of Mg^{++}, enzyme. An alternative explanation might be that in the presence of Mg^{++}, ATP contracts the mitochondria making them less permeable to pyruvate.

The fixation of CO_2 is remarkably sensitive to inhibition by calcium ions. Table 3 shows the results of an experiment in which various concentrations of calcium chloride were added to the incubation mixture. It is apparent that even though 0.68 mM EDTA was present, calcium added at concentrations as low as 10^{-5} M was inhibitory. This is in agreement with Kimmwich and Rasmussen (11), who reported an inhibition of pyruvate carboxylase by 2.5 μM calcium.

The utilization of pyruvate in the pyruvate carboxylase reaction as measured in intact, isolated mitochondria follows saturation kinetics with a maximal velocity far below that of the enzyme extracted from disrupted mitochondria. For example, in one experiment the V_{max} for the CO_2 fixation reaction was calculated by the Wilkinson method (12) to be 0.64 ± 0.007 μmoles/g/min. The pyruvate carboxylase freed from the same preparation of mitochondria by the detergent Lubrol had a V_{max} of 4.5 ± 0.12 μmoles/g/min. It is obvious that the enzyme operates under great restraint in the isolated mitochondria. Whether this restraint is the result of less than saturating levels of pyruvate within the mitochondria as we have proposed or is the result of some other control mechanism remains to be determined.

We have estimated K_m values for pyruvate in the CO_2 fixation reaction in the intact mitochondria. These values obviously are of somewhat ambiguous meaning, as they are derived from a complex system, but they were of interest as a possible indicator of the nature of the changes brought about by hormonal treatment. Substrate-velocity determinations have been made and the results analyzed by the method of Wilkinson (12). The apparent K_m values for pyruvate have ranged from .044 to .078 mM in several experiments. Table 4 shows the data from two experiments indicating that changes in osmolality and prior treatment with glucagon do not significantly affect the apparent K_m's for pyruvate.

As just noted, the K_m values found for pyruvate vary between 0.044 and 0.078 mM. These values are somewhat lower than the K_m values reported for PC itself. In the experiment described above in which the V_{max} of the mitochondrial reaction was compared with that of the solublized enzyme, the K_m for pyruvate in the intact mitochondrial system was 0.078 ± 0.007 mM; for the solublized enzyme, it was 0.22 ± 0.026 mM. This difference of nearly 3-fold is consistent with the hypothesis that pyruvate does not have free access to the mitochondrial enzyme in the intact mitochondria. Recently Papa et al. (13) have presented evidence that the penetration of pyruvate into mitochondria follows saturation kinetics and that the rate of penetration increases with decreasing pH. This finding also supports our hypothesis of a barrier to the free entrance of pyruvate into hepatic mitochondria.

The alteration of hepatic mitochondria by treatment with glucagon and epinephrine is presumably mediated via cyclic adenylic acid. To test this idea directly, the dibutyryl derivative of cyclic AMP (4 mg) was given to rats intravenously. Ten minutes later the animals were killed and liver mitochondria prepared. As can be seen in Table 5, this treatment stimulated the CO_2 fixation reaction just as treatment with glucagon or epinephrine does. It is also apparent that the addition of succinate to the assay medium stimulates the basal level of CO_2 fixation, but does not significantly alter the absolute increase resulting from the treatment with dibutyryl cyclic AMP.

In our experiments, glucagon and epinephrine have routinely been administered 15 to 30 minutes before the rats are killed. Cortisol has been given for two hours. The research groups using perfused rat livers as test objects report that glucagon increases the rate of gluconeogenesis within a few minutes after it is added, and Rinard et al. (3) could detect no lag period in stimulation of gluconeogenesis when epinephrine was added to liver slices. Therefore, it was felt most important to see if the mitochondrial changes could be seen at early time periods after hormone treatment. If the mitochondrial changes occur after the onset of increased gluconeogenesis, then the hypothesis that these changes control gluconeogenesis would have to be discarded.

As yet, only glucagon has been studied in relationship to the time required for development of the mitochondrial changes. Rats were given glucagon in phosphate buffer intravenously; control rats received buffer alone. Animals were killed and mitochondria prepared at various intervals following treatment. Fig. 1 presents the results of one experiment of this type. Each point represents the mitochondrial assay from a single rat. A slight tendency for a downward drift in values can be noted with control rats. It can be seen that by 4 minutes after treatment there is complete activation of the mitochondrial reaction. If one ignores the value of one treated rat which is aberrant, it is evident that treatment produces no effect until 3 minutes, when partial activation is present.

At this point in our studies we felt it might be possible to complete the link between hormonal stimulation and final effect on mitochondrial activity. At least one hormone acts very rapidly; cyclic adenylic acid is almost certain to mediate the action of the hormones, and the mitochondrial response is reasonably well defined. It seemed that one might be able to add cyclic AMP directly to isolated mitochondria and by demonstrating an activation, complete the chain of events that connects hormone to response.

A large number of experiments and a great amount of time have been devoted to this goal which has, unfortunately, as yet proved elusive. Conditions have not been found under which cyclic AMP stimulated CO_2 fixation when added directly to mitochondria.

The activity of mitochondria can be altered, but no effect on these changes has been demonstrated.

In Table 6 data are presented from an experiment in which mitochondria were incubated with various fractions of a liver homogenate prepared in 0.135 M KCl, 0.01 M potassium phosphate, pH 7.4, and dithiothreitol, 0.5 mg/ml. The mitochondria were themselves prepared in a 0.31 M sucrose– 5 mM TES mixture, pH 7.4. Equal volumes of the mitochondrial preparation and fractions of the liver homogenate were incubated together with added ATP (2 mM) and succinate (4 mM); aliquots were removed and assayed in the presence of ATP and succinate for the rate of fixation of CO_2. It can be seen that mitochondria mixed with KCl-phosphate buffer have an initially high rate of CO_2 fixation that declines rapidly. On the other hand, mitochondria incubated with the microsomal fraction exhibit a low level of activity initially which increases somewhat during the incubation. The soluble fraction obtained at the time the microsomal fraction was prepared behaves in a fashion that is intermediate between the simple buffer and the microsomal fraction. The nature of the inhibitory material in the microsomal fraction is not known. Commercial glycogen was tested and did not inhibit the reaction.

A similar rise in activity can also be seen when mitochondria are incubated with the cell fractions obtained as supernatant when a 1:1 liver homogenate is centrifuged at 9,000 × g. Such an experiment is presented in Fig. 2. In this study, cyclic AMP was added and had no effect on the rate of increase of activity in the mitochondria.

It was considered possible that an activation-inactivation cycle exists for the mitochondria and that cyclic AMP acts on the inactivation limb. To explore this possibility, mitochondria were incubated with a 9,000 × g supernatant of a liver homogenate without added ATP or Mg^{++}. Under these conditions the activity of the mitochondria drifts downward as can be seen in Fig. 3. The addition of 2.5 mM cyclic AMP sustained the activity of the mitochondria at the original level. However, 2.5 mM 5'-AMP was nearly as active. When the concentration of these nucleotides was dropped to 0.25 mM, neither had a detectable effect on maintenance of mitochondrial activity.

One other general approach has been tried without success. The experiment described that explored the time required for the effect of glucagon to be manifested indicated that 3 minutes after treatment with glucagon the liver is in a transition state, with the activity of the mitochondria rapidly increasing toward the final elevated level. It was thought that extracts prepared from livers at precisely this stage in their response to glucagon might be able to bring about an increase in activity in mitochondria from untreated rats.

Extracts [9,000 × g supernatants of 1:1 (w/v) homogenates] were prepared from animals killed 3 minutes after treatment with glucagon or phosphate buffer. Mitochondria were prepared from a control rat and incubated with the fortified tissue extracts. Table 7 shows that in this small experiment there was no hint of any difference between control extracts and those from glucagon-treated rats in their ability to activate mitochondria.

The conclusion must be that, as yet, it has not been possible to complete the linkage between hormone action and mitochondrial stimulation. Since the mediator of the hormones is presumably cyclic AMP, one suspects that a great deal of information about the control of mitochondria would be obtained if one could discover the necessary conditions that would permit cyclic AMP to act on the mitochondria *in vitro* as it presumably does *in vivo*. Elucidation of the action of epinephrine and glucagon via cyclic AMP would still leave the problem of cortisol action unsolved, however.

Mr. Rudnick, working in my laboratory, has discovered that high concentrations of glucagon stimulate isolated mitochondria, both pyruvate carboxylation and decarboxylation (19). This effect is specific for glucagon in that a number of other proteins tested are inactive. The biological significance of this finding is uncertain, but it provides an interesting model that may be helpful in understanding the control of mitochondrial metabolism.

The experimental work presented in this paper was supported by Grants A-1256 and A-14347 from the National Institute of Arthritis and Metabolic Disease, U.S.P.H.S.

References

1. Uete, T. and J. Ashmore. Effects of triamcinolone on carbohydrate synthesis by rat liver slices. J. Biol. Chem., 238:2906-2911(1963).
2. Exton, J.H. and C.R. Park. Control of gluconeogenesis in liver. III. Effects of L-lactate, pyruvate, fructose, glucagon, epinephrine and adenosine 3',5'-monophosphate on gluconeogenic intermediates in the perfused rat liver. J. Biol. Chem., 244:1424-1433(1969).
3. Rinard, G.A., G. Okuno and R.C. Haynes, Jr. Stimulation of gluconeogenesis in rat liver slices by epinephrine and glucocorticoids. Endocrinol., 84:622-631(1969).
4. Henning, H.V., I. Seiffert and W. Seubert. Cortisol induzierter Anstieg der Pyruvatcarboxylaseaktivität in der Rattenleber. Biochim. Biophys. Acta, 77:345-348(1963).
5. Adam, P.A.J. and R.C. Haynes, Jr. Control of hepatic mitochondrial CO_2 fixation by glucagon, epinephrine, and cortisol. J. Biol. Chem., 422:6444-6450(1969).
6. Lardy, H.A., V. Paetkau and P. Walter. Paths of carbon in gluconeogenesis and lipogenesis: the role of mitochondria in supplying precursors of phosphoenolpyruvate. Proc. Natl. Acad. Sci., U.S., 53:1410-1414(1965).

7. Haynes, R.C., Jr. The fixation of carbon dioxide by rat liver mitochondria and its relation to gluconeogenesis. J. Biol. Chem., 240:4103-4106(1965).
8. Scrutton, M.C. and M.F. Utter. Pyruvate carboxylase. IX. Some properties of the activation by certain acyl derivatives of Coenzyme A. J. Biol. Chem., 242:1723-1735(1967).
9. Wieland, O. and E. Siess. Interconversion of phospho- and dephospho- forms of pig heart pyruvate dehydrogenase. Proc. Natl. Acad. Sci., U.S., 65:947-954(1970).
10. Mehlman, M.A., P. Walter and H.A. Lardy. Paths of carbon in gluconeogenesis and lipogenesis. VII. The synthesis of precursors for gluconeogenesis from pyruvate and bicarbonate by rat liver mitochondria. J. Biol. Chem., 242:4594-4602(1967).
11. Kimmich, G.A. and H. Rasmussen. Regulation of pyruvate carboxylase activity by calcium in intact rat liver mitochondria. J. Biol. Chem., 244:190-199(1969).
12. Wilkinson, G.N. Statistical estimations in enzyme kinetics. Biochem. J., 80:324-332(1961).
13. Papa, S., A. Francavilla, G. Paradies and B. Meduri. The transport of pyruvate in rat liver mitochondria. FEBS Letters, 12:285-288(1971).
14. Rudnick, S.A. and R.C. Haynes, Jr. Stimulation of mitochondrial CO_2 fixation by glucagon *in vitro*. Fed. Proc., 30:315(1971).

TABLE 1

EFFECT OF ADDITION OF VARIOUS SUBSTRATES ON FIXATION OF CO_2 BY MITOCHONDRIA

Fixation of CO_2 by intact mitochondria recovered from 9.1 mg of liver was assayed at 37° in 0.21M sucrose, 3.4mM N-Tris (hydroxymethyl) methyl-2-aminoethane sulfonic acid (TES) 0.68 mM EDTA, 5 mM sodium pyruvate, 12 mM $KH^{14}CO_3$, 12.5 mM $MgCl_2$, 2.5 mM K_2SO_4, and 8 mM potassium phosphate. Volume was 0.5 ml, pH 7.6, and the incubation period was for 5 minutes. Each value represents the mean of two incubations. These assay conditions are those used in all experiments reported unless otherwise noted.

Addition	Experiment 1 Substrates 1 mM	Experiment 2 Substrates 5 mM
	μmoles CO_2 fixed/g/min	
None	.64	.72
Malate	.57	.44
α-Ketoglutarate	.76	1.06
Fumarate	.68	.69
Acetate	.63	.73
Butyrate	.56	.53
Citrate	.75	.53
Glutamate	.69	.72
Octanoate	.39	.16
Malonate	.58	.53
Succinate	1.11	.98
Succinate + malonate	.70	.70

TABLE 2

THE EFFECTS OF MAGNESIUM AND ATP ON CO_2 FIXATION

Conditions of assay were the same as described in Table 1 except for magnesium concentrations.

Additions		No Mg++	12.5mM Mg++
		μmoles CO_2 fixed/g/min	
Experiment 1			
None		1.07	1.17
ATP	.5mM	1.11	.92
	1mM	1.25	.83
	2mM	1.42	.80
ADP	.5mM	.47	.22
	1mM	.33	.14
	2mM	.25	.12
Experiment 2			
None		1.03	1.13
ATP 2mM		1.44	.73
Succinate 4mM		1.41	1.44
ATP 2mM + Succinate 4mM		1.67	1.43

TABLE 3

EFFECT OF CALCIUM IONS ON CO_2 FIXATION BY MITOCHONDRIA

Addition	
	μmoles of CO_2 fixed/g/min
None	.50
$CaCl_2$ 1μm	.54
$CaCl_2$ 5μm	.54
$CaCl_2$ 10μm	.43
$CaCl_2$ 50μm	.26
$CaCl_2$ 100μm	.13
$CaCl_2$ 500μm	.009

TABLE 4

EFFECT OF OSMOLALITY AND HORMONE TREATMENT ON KINETIC PARAMETERS

For experiment 1 mitochondria from a single rat were assayed in mixtures containing different amounts of sucrose.

For experiment 2 groups of 4 rats were given 0.1ml 0.1M sodium phosphate pH 7.4 i.v. or the same amount of buffer containing 20 µg glucagon. After 5 minutes, the rats were killed and mitochondria prepared for each rat. For kinetic measurements equal aliquots from members of each group were pooled and the two pools were assayed.

	K_m pyruvate	V_{max}
	mM	*µmoles CO_2/g/min*
Experiment 1		
Osmolality = 378m osm	0.061 ± 0.025	0.114 ± 0.006
Osmolality = 252m osm	0.051 ± 0.001	0.503 ± 0.023
Experiment 2		
Control	0.044 ± 0.007	0.27 ± 0.008
Glucagon treated	0.061 ± 0.005	0.62 ± 0.009

TABLE 5

EFFECT OF DIBUTYRYL CYCLIC ADENYLIC ACID (dbc AMP) ON FIXATION OF CO_2

Four rats were used in each group. Four mg of dbc AMP was given i.v. 10 minutes before the rats were killed.

	µmoles CO_2 fixed/g/min	
Control	0.55 ± 0.04	$p = < 0.025$
dbc AMP	0.99 ± 0.14	
Assayed with succinate added		
Control	0.98 ± 0.08	$p = < 0.025$
dbc AMP	1.59 ± 0.16	

TABLE 6

EFFECT OF BUFFER AND VARIOUS CELLULAR FRACTIONS ON CO_2 FIXATION

Mitochondria were prepared in 0.31M sucrose, 5mM TES pH 7.4, 1 gram liver to 10 ml sucrose-TES. A homogenate of liver was prepared in 0.135M KCl, 0.01M potassium phosphase, pH 7.4, 0.5 mg/ml dithiothreitol, one gram of liver to 1 ml of buffer. This homogenate was centrifuged 9000 x g for 20 minutes and the precipitate was discarded. The supernatant was centrifuged 100,000 x g for 45 minutes. The supernatant was used directly; the precipitate, the microsomal fraction, was suspended in KCl-phosphate buffer. For incubation, an equal volume of mitochondrial suspension and the test solution were mixed and fortified with 2mM ATP, 2mM sodium succinate and 12mM magnesium ion. Aliquots were removed for assay at 0, 6, and 12 minutes and assayed with 2mM ATP and 4mM sodium succinate added to the usual assay system.

Additions	Minutes Incubation		
	0	6	12
	μmoles CO_2 fixed/g/min		
KCl-P	1.61	1.08	1.0
Microsomes	.91	1.01	1.14
Soluble	1.35	1.18	1.21

TABLE 7

LACK OF A DISTINCTIVE EFFECT OF LIVER EXTRACTS FROM GLUCAGON-TREATED RATS

The preparation and incubation of mitochondria and liver homogenate fractions was as in figure 2. The incubations labelled glucagon were made with homogenate fractions prepared from rats killed 3 minutes after treatment with glucagon.

	Minutes Incubation			
	0	4	8	12
	μmoles CO_2 fixed/g/min			
Control	1.31	1.17	1.59	1.66
Glucagon	1.32	1.30	1.59	1.43
Glucagon	1.59	1.28	1.58	1.76
Control	1.30	1.26	1.55	1.63

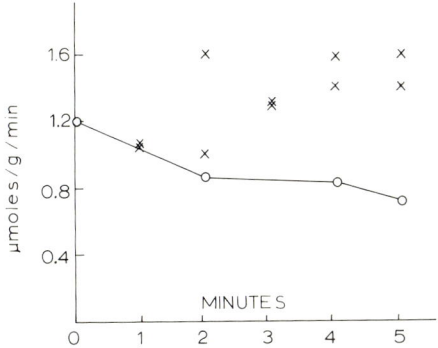

Fig. 1. *Time of response to glucagon.* Anesthetized rats serving as controls were given 0.1 ml of 0.1 M phosphate buffer pH 7.4 in the tail vein at time zero. Treated rats received the same buffer containing 20 µg of glucagon. At various intervals the animals were killed, hepatic mitochondria prepared and assayed for fixation of CO_2. Values of individual control animals are presented as open circles. Values of individual glucagon-treated animals are indicated by **x** symbols.

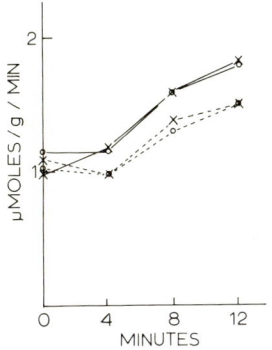

Fig. 2. *Increase in activity of mitochondria during incubation with a crude microsomal, soluble fraction.* Mitochondria were prepared as in Table 6, and were incubated in the supernatant of a 1:1 liver homogenate centrifuged 9,000 **x** g for 20 minutes. Conditions of incubation were those of Table 6 except 2.5 mM creatine phosphate and 50 µg of creatine phosphokinase were present. Values connected by solid lines were obtained with 7 mM caffeine present. Open circles represent controls; **x** symbols represent values obtained with 0.1 mM cyclic AMP added.

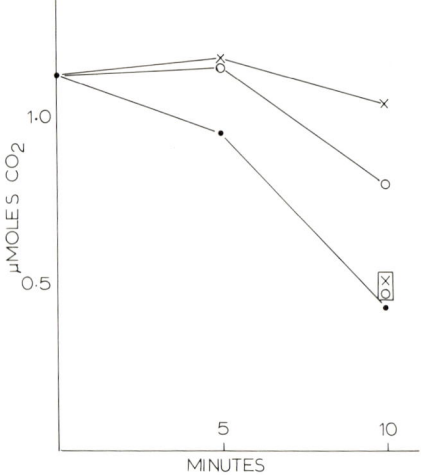

Fig. 3. *Effect of cyclic AMP and 5'-AMP on the decline in mitochondrial activity.* Mitochondria were prepared and incubated with a homogenate fraction as described in Fig. 2. Neither ATP nor Mg^{++} was added to the incubation mixture. Succinate (2 mM) and theophylline (2.5 mM) were present. Solid circles represent control values. Open circles represent values obtained when 5'-AMP was present; x symbols represent values when cyclic AMP was present. The upper two curves were obtained when the nucleotides were added at 2.5 mM. The two values in a box represent 10 minute values when the nucleotides were added at 0.25 mM. Five minute values were identical to control levels and were omitted from the graph for clarity.

PYRUVATE DEHYDROGENASE COMPLEX: STRUCTURE, FUNCTION, AND REGULATION

Lester J. Reed, Tracy C. Linn, Flora H. Pettit, Robert M. Oliver, Ferdinand Hucho, John W. Pelley, Douglas D. Randall, and Thomas E. Roche

Introduction

Pyruvate dehydrogenase systems have been isolated from *Escherichia coli* (1) avian (2) and mammalian tissues (3, 4, 5) and *Neurospora crassa* (6) as functional units with molecular weights in the millions. The bacterial, avian, and mammalian pyruvate dehydrogenase complexes have been separated into three enzymes: pyruvate dehydrogenase, dihydrolipoyl transactylase, and dihydrolipoyl dehydrogenase (a flavoprotein), and functional units resembling the native complexes have been reassembled from the isolated enzymes (7, 8, 9). These three enzymes act in a coordinated manner as indicated in Fig. 1 (10, 11).

Composition and structure of pyruvate dehydrogenase complexes

Major features of the structures of the *E. coli* and mammalian pyruvate dehydrogenase complexes have been elucidated by biochemical and electron microscopic studies (12). Both complexes contain a core consisting of the dihydrolipoyl transacetylase, to which the pyruvate dehydrogenase and the flavoprotein are joined (Figs. 2 and 3). Thus the core enzyme plays both a catalytic and a structural role. The transacetylase core of the *E. coli* pyruvate dehydrogenase complex is made up of 24 identical polypeptide chains of molecular weight about 60,000 [L. Hamilton, P. Munk, M.H. Eley, and L.J. Reed, unpublished observations; (13)]. Its design appears to be based on octahedral [(432)] symmetry, and it has the appearance of a cube (Fig. 2D through 2G). It apparently consists of 8 morphological units, which correspond to 8 groups of three polypeptide chains clustered about the three-fold axes of symmetry. It appears that 24 pyruvate dehydrogenase units of molecular weight about 90,000 [L. Hamilton, D.J. Cox, G. Namihira, and L.J. Reed, unpublished observations; (14)] and 24 flavoprotein units of molecular weight about 55,000 are distributed in a regular manner around the transacetylase cube, one unit of each of the former components being attached to each transacetylase chain (Fig. 2B and 2C).

The mammalian dihydrolipoyl transacetylase is made up of 60 similar, if not identical, polypeptide chains of molecular weight about 51,000 (L. Hamilton, P. Munk, G. Namihira, and L.J. Reed, unpublished observations). Its design appears to be based on icosahedral [(532)] symmetry, and it has the appearance of a pentagonal dodecahedron (Fig. 3A through 3C). It apparently consists of 20 morphological units, which correspond to 20 groups of three polypeptide chains clustered about the three-fold axes of shmmetry.

Recent investigations (5, 15) have revealed that the mammalian pyruvate dehydrogenase complex also contains two regulatory enzymes, a kinase and a phosphatase. There is some evidence that the kinase is attached to the transacetylase, but the site of attachment of the phosphatase is not yet clear. The two regulatory enzymes, and the dihydrolipoyl dehydrogenase as well, tend to dissociate from the complex during extraction and purification of the latter from mitochondria. For this reason, it has been difficult to determine the stoichiometry of the mammalian pyruvate dehydrogenase complex. The transacetylase core of the complex consists of 60 subunits. Assuming that there are 60 units of each of the other four enzymes (pyruvate dehydrogenase, flavoprotein, kinase, and phosphatase) and that all five enzymes comprise a functional unit, the resulting "native" molecule of the pyruvate dehydrogenase complex would have a molecular weight of about 16 million. We have estimated that a bovine kidney mitochondrion contains approximately 15 molecules of the pyruvate dehydrogenase complex. It appears that these molecules are localized in the matrix space (16, 17, 18).

Regulation of pyruvate dehydrogenase complexes

Pyruvate occupies a central position in metabolism (Fig. 4). Of particular importance are its oxidation to acetyl-CoA and its carboxylation to oxalacetate. Oxidation of acetyl-CoA via the tricarboxylic acid cycle leads to the generation of ATP. The converison of pyruvate to oxalacetate is a major mechanism for the net synthesis of compounds of the tricarboxylic acid cycle which serve as primary biosynthetic intermediates and is, as well, the first step toward gluconeogenesis in liver and kidney. These considerations suggest that the pyruvate dehydrogenase complex is a likely candidate for metabolic regulation, and mechanisms for control over the activity of the complex have indeed been found.

Regulation by product inhibition

The activity of the mammalian pyruvate dehydrogenase complex is inhibited by the products of pyruvate oxidation, acetyl-CoA and NADH, and these inhibitions are reversed by CoA and NAD^+, respectively (19, 20, 21). Hansen and Henning (22) reported similar results with the *E. coli* complex, although the bacterial complex appeared to be much more sensitive to NADH than to acetyl-CoA. The site of NADH inhibition appears to be the flavoprotein component of the pyruvate dehydrogenase complexes. Acetyl-CoA acts, at least in part, as a feedback inhibitor of the pyruvate dehydrogenase component of the *E. coli* complex (23). The acetyl-CoA inhibition is competitive with respect to pyruvate, and it is reversed by nucleoside monophosphates and, to a lesser extent, by nucleoside diphosphates and inorganic orthophosphate (24). However, acetyl-CoA inhibition of the mammalian complex appears to be uncompetitive with respect to pyruvate (F. Hucho, M.W. Burgett, and L.J. Reed, unpublished observations). The activity of the *E. coli* pyruvate dehydrogenase is also inhibited by GTP (24). This inhibition is noncompetitive with respect to pyruvate, and it is reversed by GDP. Shen *et al.* (25) and Shen and Atkinson (26) have reported that the activity of the *E. coli* pyruvate dehydrogenase complex is regulated by a number of parameters: the concentration of glycolytic intermediates (stimulation), particularly fructose 1,6-diphosphate, the adenylate energy charge, the acetyl-CoA concentration, and the oxidation level of the NAD^+-NADH pool. Atkinson and co-workers suggest that modulation of the activity of the *E. coli* pyruvate dehydrogenase complex contributes to the stabilization of these regulatory parameters *in vivo*.

Regulation by phosphorylation and dephosphorylation

Another regulatory mechanism, involving phosphorylation and dephosphorylation of the pyruvate dehydrogenase component of the mammalian pyruvate dehydrogenase complex (but not the *E. coli* complex), has been uncovered in this laboratory. About three years ago, we observed that the activity of the pyruvate dehydrogenase complex in crude bovine kidney mitochondrial extracts was destroyed by incubation with α-ketoglutarate dehydrogenase complex. When the mitochondrial extract was filtered through Sephadex G-25 prior to incubation with α-ketoglutarate, the latter substance had no effect on the activity of the pyruvate dehydrogenase complex. These observations suggested that a low molecular weight compound, produced by α-ketoglutarate oxidation, but not by pyruvate oxidation, was involved in the inactivation of the pyruvate dehydrogenase complex. GTP, produced by succinyl-CoA synthetase, or another nucleoside triphosphate produced from GTP by the action of a nucleoside diphosphate kinase, appeared to be likely candidates. We found that ATP and, to a lesser extent, ADP inactivated the pyruvate dehydrogenase complex in the crude mitochondrial extract. Purified preparations of the complex were inactivated by incubation with 1-10 μM ATP (15). AMP, ADP, CTP, GTP, and UTP showed essentially no inhibition at concentrations up to 1 mM; ITP showed slight inhibition. Inactivated preparations of the pyruvate dehydrogenase complex were not reactivated by dilution, by dialysis, or by gel filtration on Sephadex. The apparent irreversible inactivation of the complex by ATP suggested that a part of the ATP molecule was covalently attached to the complex. Using radioactive ATP labeled with ^{32}P in either the α-, α,β, or γ-phosphoryl moieties, we established that the terminal phosphoryl moiety of ATP is transferred to the pyruvate dehydrogenase complex (15).

The next step was to determine which of the component enzymes of the pyruvate dehydrogenase complex underwent phosphorylation (and concomitant inactivation). Phosphorylation of the complex from bovine kidney did not affect its dihyrolipoyl transacetylase or dihydrolipoyl dehydrogenase activities. This observation pointed to the pyruvate dehydrogenase component of the complex as the site of phosphorylation. To confirm this possibility, procedures were developed to separate pyruvate dehydrogenase from the other component enzymes of the complex. This was accomplished by gel filtration on Sepharose 4B at pH 9, followed by fractionation with ammonium sulfate. When the phosphorylated (inactivated) pyruvate dehydrogenase complex was resolved, all of the protein-bound radioactivity was found in the pyruvate dehydrogenase (15).

Further investigation revealed that preparations of the phosphorylated, inactivated bovine kidney pyruvate dehydrogenase complex could be reactivated by incubation with Mg^{++} (15). A Mg^{++} concentration of about 10 mM was optimal. Restoration of activity was accompanied by a release of inorganic orthophosphate. The data presented in Fig. 5C illustrate the time course of the reciprocal changes in enzymic activity and protein-bound phosphoryl groups observed with a highly purified preparation of the pyruvate dehydrogenase complex from bovine kidney mitochondria. These data established that the kidney pyruvate dehydrogenase complex is subject to regulation by phosphorylation and dephosphorylation, and that the site of this regulation is the pyruvate dehydrogenase component of the complex, which catalyzes the first step in pyruvate oxidation. Subsequent studies provided evidence that phosphorylation and concomitant inactivation of pyruvate dehydrogenase are catalyzed by a kinase (*i.e.*, a pyruvate dehydrogenase kinase), and dephosphorylation and concomitant reactivation are catalyzed by a phosphatase (*i.e.*, a pyruvate dehydrogenase phosphatase) (15, 5). Wieland and co-workers (21, 27, 28) and Jungas (29) extended these observations to the pyruvate dehydrogenase complex from porcine heart muscle and rat adipose tissue, respectively. We have since demonstrated that preparations of the pyruvate dehydrogenase complex isolated from mitochondria of bovine heart, liver, and brain, procine liver, and rat kidney and heart are also subject to regulation by phosphorylation and dephosphorylation. The time course of reciprocal changes in enzymic activity and protein-bound phosphoryl groups obtained with preparations of the pyruvate dehydrogenase complex from mitochondria of bovine heart and brain and porcine liver are shown in Fig. 5. The differences in rates of inactivation and reactivation of the various preparations are apparently due to differences in the amounts and possibly the activities of the kinase and the phosphatase. We have found no evidence that the *E. coli* pyruvate dehydrogenase complex or the mammalian or *E. coli* α-ketoglutarate dehydrogenase complexes are regulated by phosphorylation and dephosphorylation.

Phosphorylation sites in pyruvate dehydrogenase

We have obtained both the pyruvate dehydrogenase and the phosphorylated pyruvate dehydrogenase in crystalline form (T.C. Linn, A.K. Woodman, and L.J. Reed, unpublished observations). Both forms of the enzyme from kidney and heart have a molecular weight of about 155,000. Pyruvate dehydrogenase contains two different kinds of subunits with molecular weights of approximately 40,000 and 30,000 as estimated by gel electrophoresis in the presence of sodium dodecyl sulfate (C.R. Barrera, G. Namihira, and L.J. Reed, unpublished observations). Only the larger subunit undergoes phosphorylation and dephosphorylation. The various data suggest that pyruvate dehydrogenase contains a catalytic and a regulatory subunit, and that the regulatory subunit undergoes phosphorylation. A radioactive tetradecapeptide has been isolated from tryptic digests of ^{32}P-labeled kidney pyruvate dehydrogenase, and its amino acid sequence has been determined (E.T. Hutcheson, J.R. Brown, and L.J. Reed, unpublished data):

Try-His-Gly-His-Ser(P)-Met-Ser-Asn-Pro-Gly-Val-Ser(P)-Try-Arg

The phosphoryl moieties are attached to seryl residues. The first seryl residue in this sequence is rapidly phosphorylated, and this phosphorylation results in inactivation of the pyruvate dehydrogenase complex. The third seryl residue is slowly phosphorylated. The physiological significance, if any, of this latter phosphorylation site remains to be determined.

Regulatory properties of pyruvate dehydrogenase kinase and pyruvate dehydrogenase phosphatase

Procedures have been developed for separation and purification of the pyruvate dehydrogenase kinase and the pyruvate dehydrogenase phosphatase from bovine kidney and heart mitochondria [T.C. Linn, F. Hucho, J.W. Pelley, D.D. Randall, and L.J. Reed, unpublished observations; (5)]. The two phosphatases are functionally interchangeable, as are the two kinases. Rabbit antiserum to the bovine kidney kinase crossreacts with the bovine heart and brain pyruvate dehydrogenase kinases (T.E. Roche, L.J. Reed, unpublished observations). Purified preparations of the bovine kidney and heart pyruvate dehydrogenase complexes contain variable amounts of the kinase and phosphatase. Assay of fractions at various stages of the purification procedure indicates that the two regulatory enzymes tend to dissociate from the pyruvate dehydrogenase complex during its extraction and purification from mitochondria. This observation may explain, at least in part, the difficulty some investigators have encountered in detecting kinase and phosphatase activities in purified preparations of the pyruvate dehydrogenase complex. Preliminary analyses indicate that the kinase, phosphatase, and pyruvate dehydrogenase are present in approximately equimolar amounts in bovine kidney and heart mitochondrial extracts.

Preliminary studies (T.E. Roche and L.J. Reed, unpublished observations) on the protein specificity of pyruvate dehydrogenase kinase indicate that this kinase does not catalyze a phosphorylation of histone, in the presence or absence of cyclic $3',5'$-AMP. A sample of skeletal muscle cyclic AMP-dependent protein kinase, kindly furnished by Dr. Edwin Krebs, did not inactivate preparations of the bovine kidney or heart pyruvate dehydrogenase complexes in the presence of ATP and cyclic AMP. The protein kinase exhibited little, if any, ability to phosphorylate the bovine kidney or heart pyruvate dehydrogenase complexes or the isolated bovine kidney pyruvate dehydrogenase. The protein kinase also did not crossreact with rabbit antiserum to the bovine kidney pyruvate dehydrogenase kinase.

Some of the kinetic parameters of the pyruvate dehydrogenase complexes and the kinase and phosphatase from bovine kidney and heart mitochondria have been determined. The results are summarized in Table 1. The apparent

K_m values for CoA are 13 μM and 12 μM, respectively, for the bovine kidney and heart pyruvate dehydrogenase complexes. The apparent K_i values for acetyl-CoA are 52 μM and 48 μM, respectively. These values are higher than the apparent K_m values of 6.7 μM and 5 μM for CoA, and apparent K_i values of 12.5 μM and 29 μM for acetyl-CoA reported by Garland and Randle (19) and Wieland et al. (21) for the pig heart pyruvate dehydrogenase complex. The differences in these values may be due to the different methods of assay and the different buffers used.

The true substrate for the kinase is $MgATP^-$, and the apparent K_m is about 0.02 mM for the pyruvate dehydrogenase kinases from both bovine kidney and heart. This was demonstrated by measuring the catalytic activity of the kinase as a function of varying concentrations of ATP and $MgCl_2$. The rate of catalysis by both the kidney and heart kinases exhibits a common dependence on the calculated concentration of $MgATP^-$ and not on that of uncomplexed ATP or Mg^{++} over conditions which produce wide variations in the relative concentrations of the three species. ADP is competitive with ATP, and the apparent K_i value for ADP is about 0.1 mM for both kinases.

Magnesium ion is required for pyruvate dehydrogenase phosphatase activity. We have observed considerable variation in the apparent K_m for Mg^{++}, due apparently to the nature of the phosphatase preparations and the assay procedure. Previous determinations of the apparent K_m for Mg^{++} gave values as high as 20 mM for both the kidney and heart phosphatases (F. Hucho and L.J. Reed, unpublished observations). We have recently developed an improved procedure for purifying and stabilizing the phosphatase (D.D. Randall and L.J. Reed, unpublished observations). Using such phosphatase preparations and different assay conditions, we have found that the apparent K_m for Mg^{++} is about 3 mM. Mn^{++} will also activate the phosphatase (apparent K_m about 1 mM), whereas Ca^{++} and Zn^{++} are inhibitory.

Pyruvate protects the pyruvate dehydrogenase complex against inactivation by ATP, and this effect is more pronounced with the heart pyruvate dehydrogenase complex than with the kidney complex (5). The apparent K_i values for pyruvate are 0.08 and 0.9 mM, respectively. Pyruvate is noncompetitive with ATP (F. Hucho and L.J. Reed, unpublished observations).

Our findings suggest that the intramitochondrial ATP/ADP ratio exerts reciprocal effects on the kinase and the phosphatase and thereby regulates the activity of the pyruvate dehydrogenase complex. Since ADP competitively inhibits ATP with respect to the kinase, the ATP/ADP ratio may regulate directly the activity of the kinase. The concentrations of free Mg^{++} in the matrix space should reflect the ATP/ADP ratio and, in turn, should regulate the activity of the phosphatase. In other words, Mg^{++} ion, acting as a feedback signal from the adenine nucleotide pool (30), regulates the activity of the phosphatase. We visualize that a decrease in the ATP/ADP ratio inhibits the kinase and releases free Mg^{++} which, in turn, activates the phosphatase. Consistent with this rationale is the fact that ADP and AMP form much weaker complexes with Mg^{++} than does ATP. Therefore, as ATP is converted to ADP in mitochondria, the free Mg^{++} concentration, as distinct from the total magnesium concentration in liver mitochondria is 25-35 mM. About 20% liver mitochondria is 25-35 mM. About 20% of this magnesium is in the matrix space, and an additional 70-80% is bound to macromolecules in the inner membrane (E. Kun, personal communication). Obviously, we need to know more about changes in the concentration of Mg^{++} in the matrix space under physiological conditions.

Effect of cyclic 3',5'-AMP

In our investigations no effect of cyclic 3',5'-AMP was found on the activity of either the pyruvate dehydrogenase kinase or the pyruvate detydrogenase phosphatase, even though preparations were examined at various stages of purification, including crude mitochondrial extracts. Jungas (29) reported that cyclic AMP did not facilitate the activation of the pyruvate dehydrogenase complex that occurs when Mg^{++} is added to crude extracts of rat adipose tissue. These results are contradictory to those of Wieland and Siess (28), who reported a cyclic AMP stimulation of the activity of pyruvate dehydrogenase phosphatase from porcine heart muscle. These investigators attributed this effect to a hypothetical cyclic AMP-dependent phosphatase kinase. Although the basis of this discrepancy is not known, it should be noted that the activity of bovine kidney and heart pyruvate dehydrogenase phosphatase is sensitive to small changes in the concentration of uncomplexed Mg^{++}.

Interconversion of active and nonactive forms of pyruvate dehydrogenase *in vivo*

Although isolated preparations of the mammalian pyruvate dehydrogenase complex have been shown to undergo phosphorylation and dephosphorylation, and the molecular basis of these interconversions has been established, there is as yet only limited evidence that these interconversions occur *in vivo*. Wieland and Siess (28) reported that purified preparations of the pyruvate dehydrogenase complex from pig heart muscle contain active and nonactive (presumably phosphorylated) forms of the complex. Thus, incubation of the purified complex with pyruvate dehydrogenase phosphatase and Mg^{++} resulted in a marked increase in pyruvate dehydrogenase activity. Wieland (31) has reported that the ratio of active to nonactive pyruvate dehydrogenase complex decreases markedly in rat kidney and heart during starvation of the animals. The activity of the pyruvate dehydrogenase complex was determined in homogenates of the organs by the pyruvate dismutation assay. The pyruvate dismutation activity of the untreated homogenate was attributed to the active form of the complex. Following incubation of the homogenate with 20 mM Mg^{++}, the pyruvate dismutation activity increased markedly. This increase in activity was attributed to dephosphorylation and concomitant reactivation of the phosphorylated (nonactive) complex. Jungas (29) has reported data suggesting the presence of active and nonactive forms of the pyruvate dehydrogenase complex in rat adipose tissue and modification of the ratio of these two forms by the hormones insulin and epinephrine. Soling and Bernhard (32) have reported that intravenous injection of fructose leads within a few minutes to a significant increase in the activity of the pyruvate dehydrogenase complex and a marked decrease in the concentration of ATP in rat liver without affecting the total activity of the complex (measured after incubation of the homogenate with 15 mM Mg^{++}). They interpret their data as indicating a conversion of nonactive to active pyruvate dehydrogenase complex under *in vivo* conditions. Although these reports are highly suggestive, a more direct demonstration of the occurrence of the phosphorylated form of the pyruvate dehydrogenase complex *in vivo* is desirable. There are technical problems to be overcome before this objective can be achieved.

Regulation of pyruvate dehydrogenase and pyruvate carboxylase

Liver and kidney mitochondria contain both the pyruvate dehydrogenase complex and pyruvate carboxylase. Thus pyruvate can undergo oxidation to acetyl-CoA and carboxylation to oxalacetate (*cf.* Fig. 4). It is apparent that metabolic control should be exerted over these two reactions and, indeed, the nature of this control has been under investigation in many laboratories. Metabolic conditions which stimulate pyruvate carboxylation, *e.g.*, oxidation of fatty acids, generally inhibit pyruvate oxidation (33, 34, 35). Attention has been focused on changes in the concentrations of acetyl-CoA and NADH, since these two compounds inhibit the pyruvate dehydrogenase complex, and acetyl-CoA is an absolute requirement for pyruvate carboxylase (36). The intramitochondrial concentration of acetyl-CoA and DPNH increases during fatty acid oxidation (34, 37), and it has been assumed that the increase in the concentration of one or both of these compounds causes inhibition of the pyruvate dehydrogenase complex. However, the possibility that the complex is inactivated by phosphorylation must also be taken into account. Obviously, there are technical problems to be overcome in determining the relative importance of regulation of the activity of the pyruvate dehydrogenase complex by product inhibition versus the phosphorylation-dephosphorylation mechanism.

A recent report by Walter and Stucki (38) points up the importance of the intramitochondrial ADP concentration in the regulation of pyruvate carboxylase activity in rat liver. It is known from the earlier work of Keech and Utter (39) that ADP is competitive with ATP, a cosubstrate for pyruvate carboxylase. It would appear that the intramitochondrial ATP/ADP ratio exerts reciprocal effects on pyruvate dehydrogenase and pyruvate carboxylase. Utter and Fung (40) have emphasized that a variety of factors affects the activity of pyruvate carboxylase. It is possible that a multi-layered system of controls regulates, in an inverse manner, the activities of the pyruvate dehydrogenase complex and pyruvate carboxylase. Thus acetyl-CoA activates pyruvate carboxylase and inhibits the pyruvate dehydrogenase complex; NADH inhibits the complex and relieves possible inhibition of pyruvate carboxylase by acetoacetyl-CoA by reduction of the latter compound to 3-hydroxybutyryl-CoA [an activator of pyruvate carboxylase (40)]; ATP (i.e., MgATP⁻) is a cosubstrate for pyruvate carboxylase and is competitively inhibited by ADP (39, 38), whereas ATP inactivates the pyruvate dehydrogenase complex (in the presence of the kinase), and this inactivation is prevented by ADP; changes in the intramitochondrial Mg^{++} concentration must also be considered. Another possible regulatory parameter is the intramitochondrial concentration of pyruvate. The apparent K_m of pyruvate carboxylase for pyruvate is about 0.4 mM (39), whereas the apparent K_m of the pyruvate dehydrogenase complex for pyruvate is about 0.04 mM (5). This marked difference in the apparent K_m values of the two enzymes for pyruvate would tend to favor pyruvate oxidation rather than pyruvate carboxylation, unless pyruvate oxidation were inhibited.

Presented by Lester J. Reed. The experimental work reported in this paper was supported in part by United States Public Health Service Grant GM-06590.

References

1. Koike, M., L.J. Reed and W.R. Carroll. α-Keto acid dehydrogenation complexes I. Purification and properties of pyruvate and α-ketoglutarate dehydrogenation complexes of *Escherichia coli.* J. Biol. Chem. 235:1924-1930(1960).

2. Jagannathan, V. and R.S. Schweet. Pyruvic oxidase of pigeon breast muscle I. Purification and properties of the enzyme. J. Biol. Chem. 196:551-562(1952).
3. Hayakawa, T., M. Hirashima, S. Ide, M. Hamada, K. Okabe and M. Koike. Mammalian α-keto acid dehydrognease complexes I. Isolation, purification, and properties of pyruvate dehydrogenase complex of pig heart muscle. J. Biol. Chem. 241:4694-4699(1966).
4. Ishikawa, E., R.M. Oliver and L.J. Reed. α-Keto acid dehydrogenase complexes V. Macromolecular organization of pyruvate and α-ketoglutarate dehydrogenase complexes isolated from beef kidney mitochondria. Proc. Natl. Acad. Sci., U.S. 56:534-541(1966).
5. Linn, T.C., F.H. Pettit, F. Hucho and L.J. Reed. α-Keto acid dehydrogenase complexes XI. Comparative studies of regulatory properties of the pyruvate dehydrogenase complexes from kidney, heart, and liver mitochondria. Proc. Natl. Acad. Sci., U.S. 64:227-234(1969).
6. Harding, R.W., D.F. Caroline and R.P. Wagner. The pyruvate dehydrogenase complex from the mitochondrial fraction of *Neurospora crassa*. Arch. Biochem. Biophys. 138:653-661(1970).
7. Koike, M., L.J. Reed and W.R. Carroll. α-Keto acid dehydrogenation complexes IV. Resolution and reconstitution of the *Escherichia coli* pyruvate dehydrogenation complex. J. Biol. Chem. 238:30-39(1963).
8. Hayakawa, T., T. Kanzaki, T. Kitamura, Y. Fukuyoshi, Y. Sakurai, K. Koike, T. Suematsu and M. Koike. Mammalian α-keto acid dehydrogenase complexes V. Resolution and reconstitution studies of the pig heart pyruvate dehydrogenase complex. J. Biol. Chem. 244:3660-3670(1969).
9. Glemzha, A.A., L.S. Zil'ber, and S.E. Severin. Isolation and characteristics of three components of muscle pyruvate dehydrogenase. Biokhimiya 31: 1033-1040(1966).
10. Gunsalus, I.C. Group transfer and acyl-generating function of lipoic acid derivatives. In: W.B. McElroy and H.B. Glass (Editors), The mechanism of enzyme action, Johns Hopkins Press, Baltimore, Maryland (1954), pp. 545-580.
11. Reed, L.J. Lipoic acid. In: P.D. Boyer, H. Lardy and K. Myrbäck (Editors), The Enzymes, 2nd Ed., Vol. 3, Academic Press, New York (1960), pp. 195-223.
12. Reed, L.J. and R.M. Oliver. The multienzyme α-keto acid dehydrogenase complexes. Brookhaven Symposia in Biology 21:397-411(1968).
13. Henney, H.R., Jr., C.R. Willms, T. Muramatsu, B.B. Mukherjee and L.J. Reed. α-Keto acid dehydrogenase complexes VII. Isolation and partial characterization of the polypeptide chains in the dihydrolipoyl transacetylase of *Escherichia coli*. J. Biol. Chem. 242:898-901(1967).
14. Vogel, O. and U. Henning. Pyruvate dehydrogenase component subunit structure of the *Escherichia coli* K12 pyruvate dehydrogenase complex. European J. Biochem. 18:103-115(1971).
15. Linn, T.C., F.H. Pettit and L.J. Reed. α-Keto acid dehydrogenase complexes X. Regulation of activity of pyruvate dehydrogenase complex from beef kidney mitochondria by phosphorylation and dephosphorylation. Proc. Natl. Acad. Sci., U.S. 62:234-241(1969).
16. Schnaitman, C. and J.W. Greenawalt. Enzymatic properties of the inner and outer membranes of rat liver mitochodnria. J. Cell Biol. 38:158-175(1968).
17. Brdiczka, D., D. Pette, G. Brunner and F. Miller. Kompartimentierte Verteilung von Enzymen in Rattenlebermitochondrien. European J. Biochem. 5:294-304(1968).
18. Smoly, J.M., B. Kuylenstierna and L. Ernster. Topological and functional organization of mitochondrion. Proc. Natl. Acad. Sci., U.S. 66:125-131(1970).

19. Garland, P.B. and P.J. Randle. Control of pyruvate dehydrogenase in the perfused rat heart by the intracellular concentration of acetyl-coenzyme A. Biochem. J. 91:6c-7c(1964).
20. Bremer, J. Pyruvate dehydrogenase, substrate specificity and product inhibition. European J. Biochem. 8:535-540(1969).
21. Wieland, O., B. von Jagow-Westermann and B. Stukowski. Kinetic and regulatory properties of heart muscle pyruvate dehydrogenase. Hoppe-Seyler's Z. Physiol. Chem. 350:329-334(1969).
22. Hansen, R.G. and U. Henning. Regulation of pyruvate dehydrogenase activity in *Escherichia coli* K12. Biochim. Biophys. Acta 122:355-358(1966).
23. Schwartz, E.R., L.O. Old and L.J. Reed. Regulatory properties of pyruvate dehydrogenase from *Escherichia coli*. Biophys. Res. Commun. 31:495-500(1968).
24. Schwartz, E.R. and L.J. Reed. Regulation of the activity of the pyruvate dehydrogenase complex of *Escherichia coli*. Biochemistry 9:1434-1439(1970).
25. Shen, L.C., L. Fall, G.M. Walton and D.E. Atkinson. Interaction between energy charge and metabolite modulation in the regulation of enzymes of amphibolic sequences. Phosphofructokinase and pyruvate dehydrogenase. Biochemistry 7:4041-4045(1968).
26. Shen, L.C. and D.E. Atkinson. Regulation of pyruvate dehydrogenase from *Escherichia coli*. Interactions of adenylate energy charge and other regulatory parameters. J. Biol. Chem. 245:5974-5978(1970).
27. Wieland, O. and B. von Jagow-Westermann. ATP-dependent inactivation of heart muscle pyruvate dehydrogenase and reactivation by Mg^{++}. FEBS Letters 3:271-274(1969).
28. Wieland, O. and E. Siess. Interconversion of phospho-forms and dephospho-forms of pig heart pyruvate dehydrogenase. Proc. Natl. Acad. Sci., U.S. 65:947-954(1970).
29. Jungas, R.L. Hormonal regulation of pyruvate dehydrogenase. Metabolism 20:43-53(1971).
30. Blair, J. McD. Magnesium, potassium, and the adenylate kinase equilibrium. Magnesium as a feedback signal from the adenine nucleotide pool. European J. Biochem. 13:384-390(1970).
31. Wieland, O. Paper presented at the 8th Int. Congr. Biochem., Lucerne (1970).
32. Söling, H.D. and G. Bernhard. Interconversion of inactive to active pyruvate dehydrogenase in rat liver after fructose application *in vivo*. FEBS Letters 13:201-203(1971).
33. Walter, P., V. Paetkau and H.A. Lardy. Paths of carbon in gluconeogenesis and lipogenesis III. The role and regulation of mitochondrial processes involved in supplying precursors of phosphoenolpyruvate. J. Biol. Chem. 241:2523-2532(1966).
34. Garland, P.B., D. Shepherd, D.G. Nicholls and J. Ontko. Energy-dependent control of the tricarboxylic acid cycle by fatty acid oxidation in rat liver mitochondria. Advances in Enzyme Regulation 6:3-30(1968).
35. von Jagow, G., B. Estermann and O. Wieland. Suppression of pyruvate oxidation in liver mitochondria in the presence of long-chain fatty acid. European J. Biochem. 3:512-518(1968).
36. Scrutton, M.C. and M.F. Utter. The regulation of glycolysis and gluconeogenesis in animal tissues. Ann. Rev. Biochem. 37:249-302(1968).
37. Williamson, J.R., R. Scholz and E.T. Browning. Control mechanisms of gluconeogenesis and ketogenesis II. Interactions between fatty acid oxidation and the citric acid cycle in perfused rat liver. J. Biol. Chem. 244:4617-4627(1969).

38. Walter, P. and J.W. Stucki. Regulation of pyruvate carboxylase in rat liver mitochondria by adenine nucleotides and short chain fatty acids. European J. Biochem. 12:508-519(1970).
39. Keech, D.B. and M.F. Utter. Pyruvate carboxylase II. Properties. J. Biol. Chem. 238:2609-2614(1963).
40. Utter, M.F. and H. Fung. In: H.D. Söling and B. Willms (Editors), Regulation of gluconeogenesis, Thieme Verlag, Stuttgart, in press.
41. Reed, L.J. and D.J. Cox. Multienzyme complexes. In: P.D. Boyer (Editor), The enzymes, 3rd Ed., Vol. 1, Academic Press, New York (1970), pp. 213-240.
42. Stahl, W.L., J.C. Smith, L.M. Napolitano and R.E. Basford. Brain mitochondria I. Isolation of bovine brain mitochondria. J. Cell Biol. 19:293-307(1963).

TABLE 1

KINETIC PARAMETERS[a]

Compound	Enzyme[b]	K_m	K_i
		mM	*mM*
CoA-SH	K-PDC	0.013	
	H-PDC	0.012	
Acetyl-CoA	K-PDC		0.052
	H-PDC		0.048
$MgATP^{2-}$	K-kinase	0.02	
	H-kinase	0.02	
ADP	K-kinase		0.09
	H-kinase		0.12
Mg^{2+}	K-phosphatase	∿3	
Pyruvate	K = PDC	0.044	
	H = PDC	0.035	

[a] Data compiled from the unpublished observations of F. Hucho, T.C. Linn, M.W. Burgett, D.D. Randall, and L.J. Reed.

[b] The abbreviations used are: K, bovine kidney; H, bovine heart; PDC, pyruvate dehydrogenase complex.

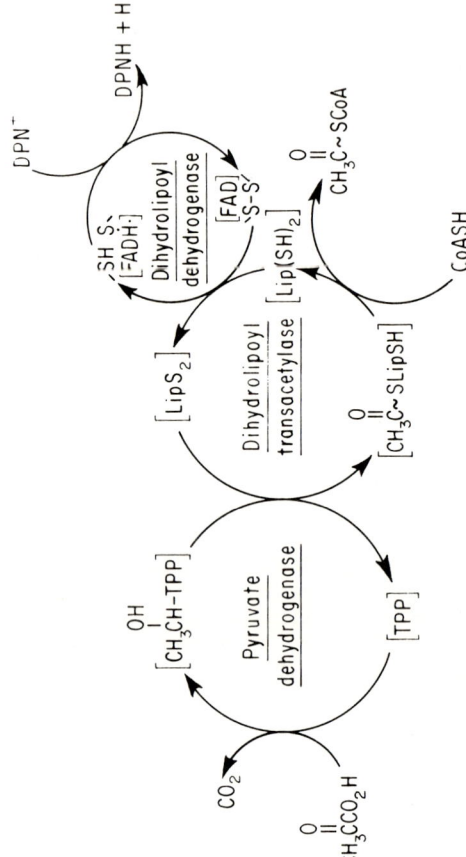

Fig. 1. *Reaction sequence in pyruvate oxidation*. The abbreviations used are: TPP, thiamine pyrophosphate; $LipS_2$ and $Lip(SH)_2$, lipoyl moiety and its reduced form; CoASH, coenzyme A; FAD, flavin adenine dinucleotide; NAD^+ and NADH, diphosphopyridine nucleotide and its reduced form.

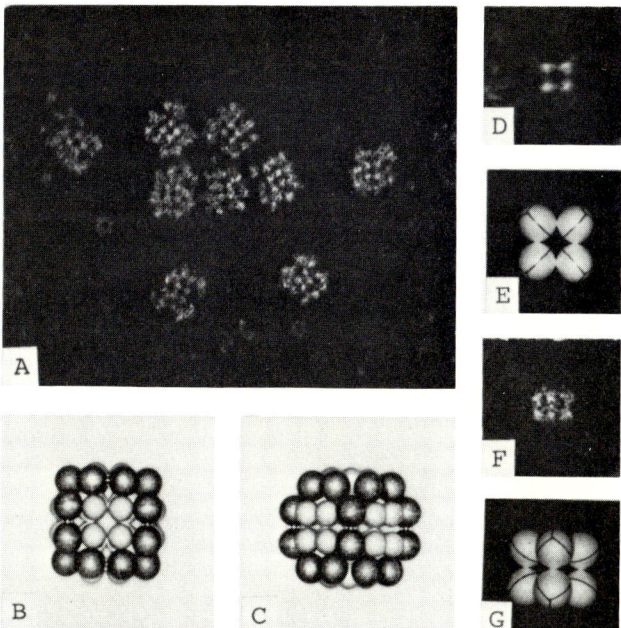

Fig. 2. *Electron micrographs and interpretative models of the* E. coli *pyruvate dehydrogenase complex and dihydrolipoyl transacetylase.* (A) Pyruvate dehydrogenase complex, negatively stained with phosphotungstate. X 200,000. (B, C) Model of the complex photographed down a 4-fold and a 2-fold axis, respectively, of the transacetylase cube. The 24 pyruvate dehydrogenase units (dark spheres) and 24 flavoprotein units (light spheres) are distributed in a regular manner along the edges of the transacetylase cube. (D, F) Individual images (X 350,000) showing two orientations of the transacetylase. (E, G) Corresponding views of a model of the transacetylase photographed down a 4-fold and a 2-fold axis, respectively. The model consists of eight spheres at the vertices of a cube.

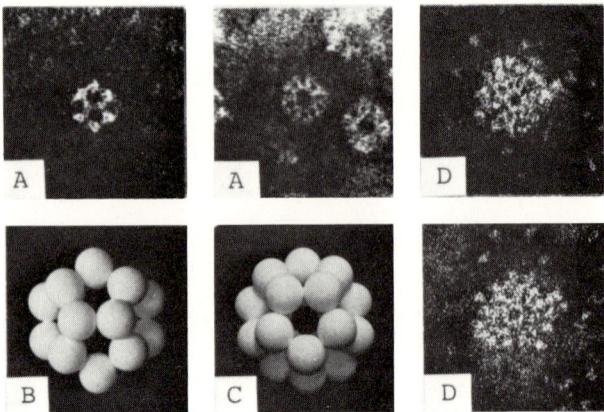

Fig. 3. *Electron micrograph images of the mammalian pyruvate dehydrogenase complex and its transacetylase component, and interpretative model of the transacetylase.* (A) Individual images (X 300,000) showing two orientations of the transacetylase. (B, C) Corresponding views of a model of the transacetylase photographed down a 2-fold and a 5-fold axis, respectively. The model consists of 20 spheres at the vertices of a pentagonal dodecahedron. (D) Electron micrographs (X 250,000) of the pyruvate dehydrogenase complex from bovine kidney mitochondria. The appearance and dimensions of the pyruvate dehydrogenase complex and its transacetylase component from bovine (R.M. Oliver, T.C. Linn, and L.J. Reed, unpublished observations) and porcine (8) heart mitochondria are very similar to those of the bovine kidney enzymes.

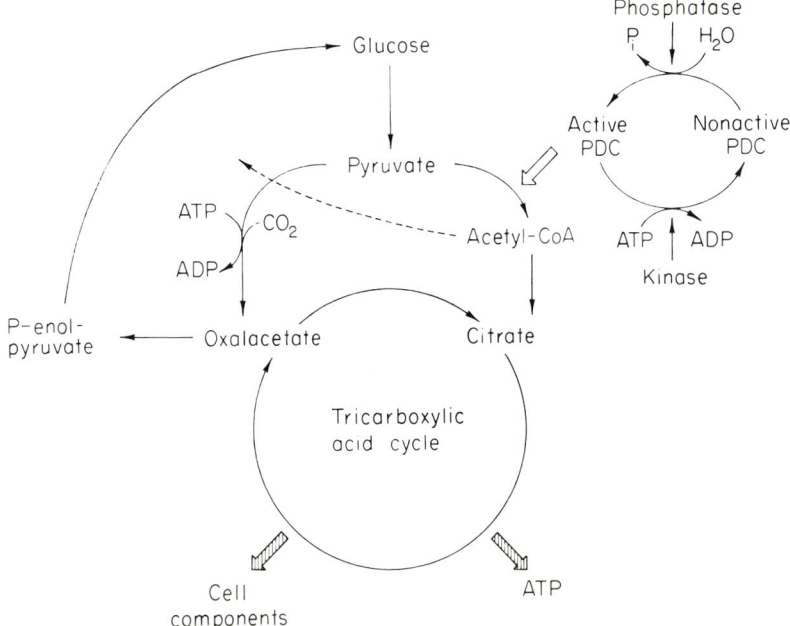

Fig. 4. *Regulation of pyruvate metabolism in animal tissues.* The open arrow indicates regulation of the activity of the pyruvate dehydrogenase complex (PDC) by phosphorylation and dephosphorylation. The site of this regulation is the pyruvate dehydrogenase component of the complex. The dashed line indicates activation of pyruvate carboxylase by acetyl-CoA (41).

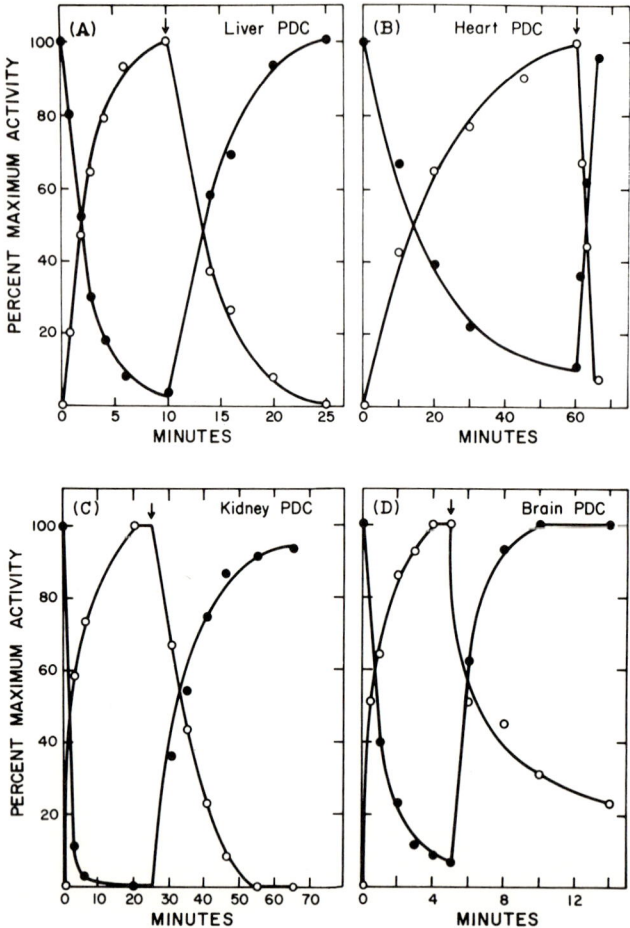

Fig. 5. *Time course of phosphorylation and dephosphorylation of purified mammalian pyruvate dehydrogenase complexes (PDC)*. The complexes were isolated from bovine heart and kidney mitochondria and from porcine liver mitochondria as described previously (5, 15). Bovine brain mitochondria were prepared by modification of the procedure of Stahl *et al.* [(42); M.W. Burgett and L.J. Reed, unpublished observations]. The reaction mixtures contained 20 mM phosphate buffer, pH 7.0-7.5, 0.5 or 1.0 mM $MgCl_2$, 2 mM dithiothreitol, 0.01-0.03 mM (A, B, C) or 0.5 mM (D) [γ-^{32}P] ATP, and enzyme complex in a total volume of 1.0 ml. The mixtures were incubated at 25° (A, C) or 30° (B, D), and aliquots were removed at the indicated times and assayed for NAD-reduction activity (●) and for protein-bound radioactivity (○). At the time interval indicated by the vertical arrow, sufficient $MgCl_2$ was added to give a final concentration of 10 mM (A, B, C) or 20 mM (D).

REDUCTIVE CARBOXYLATION OF α-OXOGLUTARATE BY MITOCHONDRIA FROM LIVERS OF NORMAL, DIABETIC AND FAT-FED RATS

Carl R. Mackerer

Introduction

The tricarboxylic acid cycle is most often pictured as the sequence of reactions which oxidizes acetyl-CoA to carbon dioxide and water. One does not usually consider the cycle as being reversible. However, in 1948, Ochoa (1-3) showed that NADP-linked isocitrate dehydrogenase was readily reversible and he suggested that this enzyme might be important for the fixation of carbon dioxide. Subsequently, D'Adamo and Haft (4, 5) and Madsen *et al.* (6-8) showed that the conversion of α-oxoglutarate to citrate by reversal of isocitrate dehydrogenase could be important in fatty acid synthesis. In fact, it is now believed that cytoplasmic isocitrate dehydrogenase normally functions in the reverse direction (9, 10). However, little is known about the reversibility of mitochondrial isocitrate dehydrogenase *in vivo*.

In the present report, the results of experiments dealing with α-oxoglutarate carboxylation via the reversal of isocitrate dehydrogenase in liver mitochondria are discussed. It was found that under aerobic conditions, with respiration partially inhibited by the presence of ATP, citrate accumulation was a reliable index of α-oxoglutarate carboxylation and that the rate of citrate accumulation was markedly increased by diabetes, fasting and fat-feeding.

Methods

Mitochondria

Mitochondria were isolated in 0.25 M sucrose by the method of Johnson and Lardy (11). The protein content of the final suspension was determined by the biuret procedure (12). All incubations were carried out in stoppered 25 ml Erlenmeyer flasks which were shaken in a water bath at 37°. The compositions of the reaction mixtures are presented in the legends and footnotes of the figures and tables. The methods of incubations, analyses of metabolites, and electrophoretic separations of radioactive organic acids have been previously described (13, 14).

Diabetes

Male Sprague-Dawley rats, weighing approximately 150 g, were fasted for 48 hr and made diabetic by administration of either alloxan (200 mg/kg, S.C.) or streptozotocin (75 mg/kg, I.V.). Purina Lab Chow was fed *ad libitum* until the animals were killed 30 days later. No attempt was made to increase rat survival by the administration of insulin. Mortality was quite high; 65% of the alloxan diabetic and 10% of the streptozotocin diabetic rats died during the 30 day period. Of the remaining diabetic animals, only those with blood glucose levels above 300 mg per 100 ml were used for the experiments in this report.

Fat-feeding

Male Sprague-Dawley rats, weighing approximately 150 g, were fed three prepared diets (Table 1) for 10 weeks. The rats were meal fed between 8 and 10 AM; on the day of sacrifice, rats were killed immediately after feeding. Mitochondria from these rats were isolated by the procedure described above except that the 0.25 M sucrose contained 5 mg/ml of bovine serum albumin (fatty acid poor).

Fasting

Male Sprague-Dawley rats, weighing approximately 150 g, were deprived of Purina Chow diet for 3 days before sacrifice. For the refeeding studies, a group of the 3 day fasted rats was refed with Purina Chow *ad libitum* for 3 days before sacrifice.

Results and Discussion

Reactions [1] - [4] show the mitochondrial dismutation of α-oxoglutarate oxidation and carboxylation.

[1] α-oxoglutarate + NAD^+ + ADP + P_i ⟶ succinate + ATP + NADH
[2] ATP + NADH + $NADP^+$ ⟶ ADP + P_i + NADPH NAD^+

[3] NADPH + CO_2 + α-oxoglutarate ⟶ $NADP^+$ + citrate
[4] SUM: 2 α-oxoglutarate ⟶ succinate + citrate

In reaction [1], α-oxoglutarate oxidation reduces NAD^+ and provides energy via substrate level phosphorylation. In reaction [2], this energy is utilized for the reduction of $NADP^+$ via energy-linked transhydrogenase. In reaction [3], NADPH, CO_2 and α-oxoglutarate combine to yield isocitrate, which is largely isomerized to citrate. Reaction [3] involves both $NADP^+$-linked isocitrate dehydrogenase and aconitase. This dismutation has been intensively studied in liver mitochondria from normal rats (13, 15-22).

When rat liver mitochondria were incubated under aerobic conditions in a medium containing α-oxoglutarate, ATP, Mg^{++}, P_i, HCO_3^- and triethanolamine (TEA) buffer (pH 7.2), the rates of α-oxoglutarate utilization and product accumulation were linear (Fig. 1). Malate and fumarate were the primary oxidative products obtained — succinate accumulated only to a very low level. Citrate accumulation amounted to about 20% of the α-oxoglutarate used. However, the mechanism of citrate synthesis was not indicated by the results of this experiment. Citrate could have been formed by forward metabolism in the tricarboxylic acid cycle as well as by reductive carboxylation.

The origin of the citrate was studied by examining the distribution of label in organic acids obtained from an experiment in which the medium was supplemented either with $^{14}CO_2$ or [1,4-^{14}C] succinate. Fig. 2A shows the distribution of label from $^{14}CO_2$ in an experiment identical to that of Fig. 1 and Fig. 2B, the distribution in the presence of malonate. It was apparent that the incorporation of $^{14}CO_2$ was primarily via α-oxoglutarate carboxylation because malonate did not prevent the accumulation of labeled citrate. The small peaks, representing succinate, malate and fumarate and which were eliminated by the addition of malonate, probably represented incorporation of label by exchange via pyruvate carboxylase. Fig. 2C shows the distribution of label when trace amounts (0.01 μmole) of [1,4-^{14}C] succinate was the source of radioactivity. Only a small amount of labeled citrate was formed, indicating that the synthesis of citrate via citrate synthase occurred only to a slight degree.

We attempted to use the total amount of $^{14}CO_2$ incorporated into organic acids as an index of the rate of α-oxoglutarate carboxylation; however, this was not reliable because the $^{14}CO_2$ was diluted by metabolic CO_2 produced through α-oxoglutarate decarboxylation (Table 2). In addition, experimental conditions were encountered in which the incorporation of $^{14}CO_2$ by exchange was increased while label in citrate was actually decreased. An example of this is shown in Fig. 2D and E. Fig. 2D shows the distribution of radioactivity from $^{14}CO_2$ in a control experiment, and 2E in the presence of ammonium chloride. The incorporation of $^{14}CO_2$ into citrate declined almost to zero in the presence of ammonium chloride, but total incorporation into other metabolites (i.e., fumarate, malate, succinate and aspartate) was slightly increased. Thus, it was possible to create a situation whereby $^{14}CO_2$ incorporation was entirely independent of α-oxoglutarate carboxylation; therefore, it can be concluded that the rate of mitochondrial α-oxoglutarate carboxylation cannot be accurately estimated from determination of $^{14}CO_2$ incorporation.

The results of the isotope experiments suggested that citrate was formed via α-oxoglutarate carboxylation, that citrate was not lost by cleavage to oxalacetate and that citrate was not formed via citrate synthase. Hence, it was likely that under aerobic conditions with respiration inhibited by 4 mM ATP, total citrate accumulation would be representative of α-oxoglutarate carboxylation. Isocitrate accumulation could either be ignored, since it was small, or estimated from the aconitase equilibrium constant.

In order to determine whether total citrate accumulation was representative of α-oxoglutarate carboxylation, mitochondria were incubated with fluorocitrate present to inhibit aconitase (Table 3). In the presence of fluorocitrate, citrate synthesis via α-oxoglutarate carboxylation is prevented and citrate formed via citrate synthase accumulates. Since fluorocitrate almost completely blocked the accumulation of citrate, it can be concluded (a) that citrate was not formed via citrate synthase and (b) that citrate accumulation in the absence of fluorocitrate occurred almost entirely via α-oxoglutarate carboxylation. Citrate accumulation was, therefore, an acceptable index of α-oxoglutarate carboxylation.

The accumulation of succinate in the presence of fluorocitrate (Table 3) was probably caused by inhibition of succinate dehydrogenase. A very high concentration of fluorocitrate (1.0 mM) was used to insure the complete inhibition of aconitase and 1.0 mM is higher than the K_i for succinate dehydrogenase [0.6 mM (23)].

The small contribution of citrate synthase to citrate synthesis under aerobic conditions in the presence of ATP was probably caused by low levels of oxalacetate and acetyl-CoA. Oxalacetate levels are known to be very low in mitochondria when the NADH/NAD$^+$ ratio is high (24). Respiration in the presence of ATP proceeded at a slow rate and the reduction level probably approximated that of "state 4" (25), which is quite reduced. In the experiment of Fig. 1, no source of acetyl-CoA was provided and it may be assumed that acetyl-CoA, as well as oxalacetate, was limiting for citrate synthesis by condensation. In order to anticipate the conditions under which citrate accumulation would not be representative of α-oxoglutarate carboxylation, substances which could facilitate the production of both oxalacetate and acetyl-CoA were added to mitochondria respiring in the presence of ATP. When fatty acids were added to produce acetyl-CoA via β-oxidation (Tables 3 and 4), there was no increase in the rate of citrate accumulation. From this experiment, it could not be determined whether or not reducing equivalents from the fatty acid oxidation lowered the oxalacetate levels by raising the $NADH/NAD^+$ ratio. However, since fatty acids did not stimulate citrate synthesis via citrate synthase, it can be concluded that low oxalacetate levels were limiting and that citrate accumulation was representative of α-oxoglutarate carboxylation.

Mitochondrial oxalacetate levels were increased indirectly by adding the uncoupler DNP to lower the $NADH/NAD^+$ ratio, thus facilitating the conversion of malate to oxalacetate (24). DNP caused citrate accumulation to be markedly decreased while α-oxoglutarate oxidation was markedly increased (Table 5). When fluorocitrate was added to inhibit aconitase, an accumulation of citrate was found which must have been derived from condensation via citrate synthase. Apparently, only an increase in oxalacetate was required to show citrate synthase activity; the necessary acetyl-CoA might have been derived either from oxidation of endogenous fatty acids or from the decarboxylation of oxalacetate to pyruvate and subsequent oxidation of the pyruvate to acetyl-CoA. Lowering the $NADH/NAD^+$ ratio enough to facilitate the conversion of malate to oxalacetate favors the loss of isocitrate via oxidation through NAD-linked isocitrate dehydrogenase irrespective of the presence of α-oxoglutarate carboxylation through NADP-linked isocitrate dehydrogenase. Thus, in the presence of a lowered $NADH/NAD^+$ ratio, loss of isocitrate (and citrate) through NAD-isocitrate dehydrogenase and citrate synthesis by condensation makes citrate accumulation an unreliable indicator of α-oxoglutarate carboxylation.

In order to be certain that α-oxoglutarate carboxylation proceeded at maximal rate under aerobic conditions in the presence of ATP, other substrates were added to see if they would have a stimulatory effect by functioning as supplementary hydrogen donors. Table 6 shows that when α-oxoglutarate decarboxylation was blocked by arsenite, succinate, octanoylcarnitine and β-hydroxybutyrate could act as hydrogen donors as indicated by the accumulation of citrate. However, none of the alternative donors was as effective as α-oxoglutarate oxidation alone (*i.e.,* α-oxoglutarate oxidation in the presence of malonate). When alternative hydrogen donors were added in the absence of arsenite (*i.e.,* with α-oxoglutarate oxidation permitted to occur), citrate accumulation was not enhanced (Tables 3, 4, 7, and 8); therefore, it can be concluded that α-oxoglutarate oxidation was capable of driving the carboxylation at maximal rate.

In 1965, R.H. Bowman (26) presented evidence which indicated that diabetic perfused hearts could convert α-oxoglutarate to citrate via carboxylation. This effect was duplicated in normal hearts by adding octanoate to the perfusion media. At about the same time, Wagle (27, 28) showed that the activity of mitochondrial NAD-linked isocitrate dehydrogenase, assayed in the direction of carboxylation, was increased by diabetes.

We have used the mitochondrial system described above to study α-oxoglutarate carboxylation in liver mitochondria from rats which were oxidizing fatty acids for energy. Three conditions were employed: diabetes, high-fat feeding, and fasting. Liver mitochondria were prepared from both streptozotocin and alloxan diabetic rats. As shown in Table 9 both diabetic states greatly increased the rates of citrate accumulation.

For the feeding studies, male rats were maintained for 5 weeks on the diets shown in Table 1. These diets were prepared on a weight by weight basis. The 70% fat diet contained no carbohydrate, thereby insuring a high *in vivo* rate of fatty acid oxidation. In accord with this, elevated levels of ketone bodies were found in the blood of the rats fed the 30% and 70% fat diets (Table 10). The dietary protein calories were 21% for the 10% fat diet, 17% for the 30% fat diet, and 13% for the 70% fat diet. Table 11 shows that increasing the level of fat in the diet also increased the rate of mitochondrial citrate accumulation via α-oxoglutarate carboxylation.

Table 12 shows that 3 days of fasting increased the rate of citrate accumulation by approximately 50%, and the rate was returned toward the control level by refeeding Purina Chow for 3 days.

The experiments of the present report show a positive correlation between the rate of citrate accumulation via mitochondrial α-oxoglutarate carboxylation and metabolic conditions which cause increased fatty acid oxidation. However, further experimentation must be performed before a possible *in vivo* role for mitochondrial α-oxoglutarate carboxylation can be suggested.

The experimental work presented in this paper was supported by National Institute of Health Grant AM 13782. The author would like to thank Professor Myron A. Mehlman for his assistance and support throughout this study.

References

1. Ochoa, S., Biosynthesis of tricarboxylic acids by carbon dioxide fixation. I. The preparation and properties of oxalosuccinic acid. J. Biol. Chem., 174:115-122(1948).
2. Ochoa, S. and E. Weisz-Tabori. Biosynthesis of tricarboxylic acids by carbon dioxide fixation. II. Oxalosuccinic carboxylase. J. Biol. Chem., 174:123-132(1948).
3. Ochoa, S. Biosynthesis of tricarboxylic acids by carbon dioxide fixation. III. Enzymatic mechanisms. J. Biol. Chem., 174:133-157(1948).

4. D'Adamo, A.F. and D.E. Haft. An alternate pathway of glutamate catabolism in the perfused liver. Fed. Proc., 21:6(1962).
5. D'Adamo, A.F. and D.E. Haft. An alternate pathway of α-ketoglutarate catabolism in the isolated, perfused rat liver. I. Studies with DL-glutamate-2 and 5-^{14}C. J. Biol. Chem., 240:613-617(1965).
6. Abraham, S., J. Madsen, and I.L. Chaikoff. The influence of glucose on amino acid carbon incorporation into proteins, fatty acids, and carbon dioxide by lactating rat mammary gland slices. J. Biol. Chem., 239:855-864(1964).
7. Madsen, J., S. Abraham, and I.L. Chaikoff. The conversion of glutamate carbon to fatty acid carbon via citrate. I. The influence of glucose in lactating rat mammary gland slices. J. Biol. Chem., 239:1305-1309(1964).
8. Madsen, J., S. Abraham, and I.L. Chaikoff. Conversion of glutamate carbon to fatty acid carbon via citrate in rat epididymal fat pads. J. Lipid Res., 5:548-553(1964).
9. Cleland, W.W. Enzyme kinetics. In: P.D. Boyer (Editor), Ann. Rev. Biochem., Vol. 36. Annual Reviews, Inc., Palo Alto, Calif. (1967), pp. 77-112.
10. Krebs, H.A. and R.L. Veech. Pyridine nucleotide interrelations. In: S. Papa, J.M. Tager, E. Quagliariello, and E.C. Slater (Editors), The energy level and metabolic control in mitochondria. Adriatica Editrice, Bari, Italy (1969), pp. 329-382.
11. Johnson, D. and H.A. Lardy. Isolation of liver or kidney mitochondria. In: R.W. Estabrook and M.E. Pullman (Editors), Methods in enzymology, Vol. 10, Academic Press, New York (1967), pp. 94-96.
12. Gornall, A.G., C.J. Bardawill, and M.M. David. Determination of serum proteins by means of the biuret reaction. J. Biol. Chem. 177:751-766(1949).
13. Walter, P., V. Paetkau, and H.A. Lardy. Paths of carbon in gluconeogenesis and lipogenesis. III. The role and regulation of mitochondrial processes involved in supplying precursors of phosphoenolpyruvate. J. Biol. Chem., 241:2523-2532(1966).
14. Somberg, E. and M.A. Mehlman. Regulation of gluconeogenesis and lipogenesis. The regulation of mitochodnrial pyruvate metabolism in guinea-pig liver synthesizing precursors for gluconeogenesis. Biochem. J., 112:435-447(1969).
15. Tager, J.M. Nicotinamide nucleotide-linked oxido-reductions in rat-liver mitochondria. In: J.M. Tager, S. Papa, E. Quagliariello, and E.C. Slater (Editors), Regulation of metabolic processes in mitochondria. Elsevier, New York (1966), pp. 202-216.
16. Klingenberg, M. Nicotinamide nucleotide-linked oxido-reductions (discussion). In: J.M. Tager, S. Papa, E. Quagliariello, and E.C. Slater (Editors), Regulation of metabolic processes in mitochondria. Elsevier, New York (1966), pp. 216-217.
17. Klingenberg, M. Anion effects on the reductive amination and carboxylation of α-oxoglutarate. In: E. Quagliariello, S. Papa, E.C. Slater and J.M. Tager (Editors), Mitochondrial structure and compartmentation. Adriatica Editrice, Bari (1967), pp. 216-219.
18. Papa, S. Control of the utilization of mitochondrial reducing equivalents. In: S. Papa, J.M. Tager, E. Quagliariello, and E.C. Slater (Editors), The energy level and metabolic control in mitochondria. Adriatica Editrice, Bari (1969), pp. 401-409.
19. Klingenberg, M. Discussion comments. In: S. Papa, J.M. Tager, E. Quagliariello, and E. Slater (Editors), The energy level and metabolic control in mitochondria. Adriatica Editrice, Bari (1969), pp. 431-434.
20. Tager, J.M., S. Papa, E.J. DeHaan, R. D'Aloya, and E. Quagliariello. Control of nicotinamide nucleotide-linked oxidoreductions in rat-liver mitochondria. Biochim. Biophys. Acta, 172:7-19(1969).

21. Papa, S., J.M. Tager, A. Francavilla, and E. Quagliariello. NAD(P)-linked oxido-reductions and the nicotinamide nucleotide specificity of glutamate dehydrogenase in rat-liver mitochondria. Biochim. Biophys. Acta, 172:20-29(1969).
22. Mehlman, M.A. Synthesis of precursors for gluconeogenesis and lipogenesis. I. α-Ketoglutarate metabolism and carboxylation by rat kidney mitochondria. N.J. Acad. Sci. Bull., 13:9-15(1968).
23. Fanshier, D.W., L.K. Gottwald, and E. Kim. Study on specific enzyme inhibitors. VI. Characterization and mechanism of action of the enzyme. Inhibitory isomer of monofluorocitrate. J. Biol. Chem., 239:425-437(1969).
24. Krebs, H.A. The regulation of the release of ketone bodies by the liver. In: G. Weber (Editor), advances in enzyme regulation, Vol. 4. Pergamon Press, Oxford (1966), pp. 339-353.
25. Chance, B. and G.R. Williams. The respiratory chain and oxidative phosphorylation. Advan. Enzymol., 17:65-134(1956).
26. Bowman, R.H. Fatty acid induced alterations in citric acid cycle intermediates. In: B. Chance, R.W. Estabrook, and J.R. Williamson (Editors), Control of energy metabolism, Colloq., Phila., Academic Press, New York (1965), pp. 357-359.
27. Wagle, S.R. Studies on the mechanism of gluconeogenesis in diabetes. Biochim. Biophys. Acta, 97:142-144(1965).
28. Wagle, S.R. Studies on the mechanism of glucose synthesis in diabetic and normal rat liver. Diabetes, 15:19-23(1966).

TABLE 1

FOOD VALUE OF DIETS USED IN THE FAT-FEEDING STUDIES

Components	Percent of components at various fat levels		
	10%	30%	70%
	weight percentage		
Protein*	22	22	22
Carbohydrate	60	39	–
Fat	10	30	70
Vitamins, minerals	5	5	5
Non-nutritive	3	4	3

*The protein values expressed as calories protein/100 dietary calories for the 10%, 30%, and 70% fat diets were 21%, 17%, and 13% respectively.

TABLE 2

THE INCORPORATION OF $^{14}CO_2$ INTO ORGANIC ACIDS BY MITOCHONDRIA INCUBATED IN THE CONTROLLED STATE

The reaction mixtures contained 4 mM ATP, 10 mM $MgCl_2$, 6.7 mM Pi, 13.3 mM TEA, 6.7 mM α-oxoglutarate, 13.3 mM HCO_3^- and 24.7 mg of mitochondrial protein. Initial pH was 7.2. Incubation was for 20 minutes.

Total $^{14}CO_2$[a] incorporated	$^{14}CO_2$ in citrate	$^{14}CO_2$ in isocitrate	Citrate accumulated
total μmoles/3 ml reaction mixture			
2.48	1.71	0.033	3.33

[a]The specific activity of the added $H^{14}CO_3^-$ (2.3 x 10^5 dpm/μmole) was used to calculate the μmoles of $^{14}CO_2$ incorporated.

TABLE 3

THE EFFECTS OF FLUOROCITRATE AND LAURIC ACID ON MITOCHONDRIAL α-OXOGLUTARATE METABOLISM UNDER AEROBIC CONDITIONS

The reaction mixtures contained 4 mM ATP, 10 mM $MgCl_2$, 6.7 mM Pi, 13.3 mM TEA, 6.7 mM α-oxoglutarate, 2.0 mM EDTA, 0.1 mM albumin, 13.3 mM HCO_3^-, and 14.4 mg of mitochondrial protein. Initial pH was 7.2. Incubation was for 40 minutes.

Additions[a]	Metabolite changes			
	α-Oxoglutarate consumed	Malate accumulated	Citrate accumulated	Succinate accumulated
	total μmoles/3 ml reaction mixture			
None	8.6	4.8	2.1	0.2
Laurate (0.53)	6.0	2.3	1.9	0.9
Fluorocitrate (1.0)	7.9	4.8	0.10	1.6
Fluorocitrate (1.0) + Laurate (0.53)	5.0	2.5	0.11	1.6

[a]Values are expressed as final concentration in millimoles/liter.

TABLE 4

THE EFFECTS OF FATTY ACID CHAIN LENGTH ON MITOCHONDRIAL α-OXOGLUTARATE METABOLISM UNDER AEROBIC CONDITIONS

The reaction mixtures contained 4 mM ATP, 10 mM $MgCl_2$, 6.7 mM Pi, 13.3 mM TEA, 6.7 mM α-oxoglutarate, 2.0 mM EDTA, 0.1 mM albumin, 13.3 mM HCO_3^-, and 11.2 mg of mitochondrial protein. Initial pH was 7.2. Incubation was for 20 minutes.

Additions	Metabolite changes			
	α-Oxoglutarate consumed	Malate accumulated	Citrate accumulated	Acetoacetate accumulated
	total μmoles/3 ml reaction mixture			
None	6.6	4.2	0.96	0.08
Hexanoate (0.5 mM)	5.8	3.6	0.98	0.48
Octanoate (0.5 mM)	4.8	3.0	0.98	0.52
Decanoate (0.5 mM)	4.6	2.4	0.94	0.56
Laurate (0.5 mM)	4.2	2.0	0.94	0.64

TABLE 5

THE EFFECTS OF DINITROPHENOL AND FLUOROCITRATE ON MITOCHONDRIAL α-OXOGLUTARATE METABOLISM UNDER AEROBIC CONDITIONS

The reaction mixtures contained 4 mM ATP, 10 mM $MgCl_2$, 6.7 mM Pi, 13.3 mM TEA, 6.7 mM α-oxoglutarate, 13.3 mM HCO_3^-, and 12.0 mg of mitochondrial protein. Initial pH was 7.2. Incubation was for 20 minutes.

Additions	Metabolite changes			
	α-Oxoglutarate consumed	Citrate accumulated	Malate accumulated	Succinate accumulated
	total μmoles/3 ml reaction mixture			
None	7.3	1.5	4.0	0.2
DNP (0.1 mM)	14.6	0.22	7.3	4.7
DNP (0.1 mM) + fluorcitrate (0.1 mM)	14.1	0.9	8.5	1.5

TABLE 6

THE EFFECTS OF ALTERNATIVE HYDROGEN DONORS ON MITOCHONDRIA α-OXOGLUTARATE METABOLISM UNDER ANAEROBIC CONDITIONS IN THE PRESENCE OF ARSENITE

The reaction mixtures contained 4 mM ATP, 10 mM $MgCl_2$, 6.7 mM Pi, 13.3 mM TEA, 2 mM EDTA, 1 μg antimycin A, 6.7 mM α-oxoglutarate, 20 mM HCO_3^-, and 12.5 mg of mitochondrial protein. Incubation was for 60 min at pH 7.4. The flasks were gassed for 1 min with 95% N_2 and 5% CO_2.

Additions[a]	Metabolite changes		
	α-Oxoglutarate consumed	Citrate accumulated	β-Hydroxybutyrate accumulated
	total μmoles/3 ml reaction mixture		
Malonate (20.0)	11.93	5.20	—
Arsenite (1.0)	0.33	0.08	—
Succinate (6.7) + arsenite (1.0)	0.98	0.68	—
L-octanoylcarnitine (1.0) + arsenite (1.0)	1.77	1.51	0.39
β-Hydroxybutyrate (6.7) + arsenite (1.0)	3.00	2.65	—
Malate (6.7) + arsenite (1.0)	0.30	0	—

[a]Values in parentheses indicate the final concentration in millimoles/liter.

TABLE 7

THE EFFECTS OF ALTERNATIVE HYDROGEN DONORS ON MITOCHONDRIAL α-OXOGLUTARATE METABOLISM UNDER ANAEROBIC CONDITIONS

The reaction mixtures contained 1 μg antimycin A, 4 mM ATP, 6.7 mM α-oxoglutarate, 10 mM $MgCl_2$, 6.7 mM Pi, 13.3 mM TEA, 2 mM EDTA, 20 mM HCO_3^-, and 12.5 mg of mitochondrial protein. pH was 7.4. Incubation was for 60 min. The flasks were gassed for 1 minute with 95% N_2 + 5% CO_2.

Additions[a]	Metabolite changes	
	α-Oxoglutarate consumed	Citrate accumulated
	total μmoles/3 ml reaction mixture	
None	12.50	5.20
Succinate (6.7)	11.25	5.28
L-Octanoyl carnitine (1.0)	9.95	4.37
β-Hydroxybutyrate (6.7)	11.69	5.22

[a]Values in parentheses indicate the final concentration in millimoles/liter.

TABLE 8
THE EFFECTS OF ALTERNATIVE HYDROGEN DONORS ON MITOCHONDRIAL α-OXOGLUTARATE METABOLISM UNDER AEROBIC CONDITIONS

The reaction mixtures contained 4 mM ATP, 10 mM $MgCl_2$, 6.7 mM Pi, 13.3 mM TEA, 2 mM EDTA, 6.7 mM α-oxoglutarate and 13.3 mM HCO_3^-. The quantity of mitochondrial protein used and the duration of incubation are as indicated for each experiment. Initial pH was 7.2.

Experiment	Additions	α-Oxoglutarate consumed	Metabolic changes		
			Citrate accumulated	β-Hydroxybutyrate accumulated	Acetoacetate accumulated
A			μmoles/20 min/15 mg of protein		
	None	7.1	1.42	—	—
	Succinate (6.7)	5.0	1.48	—	—
B			μmoles/20 min/15 mg of protein		
	Albumin (0.14)	5.3	1.1	—	—
	L-Palmitoyl carnitine (0.53) + albumin (0.14)	2.8	1.1	—	—
C			μmoles/20 min/15 mg of protein		
	Albumin (0.25)	4.5	1.02	—	—
	L-Octanoyl carnitine (1.0) + albumin (0.14)	2.6	0.92	2.68	0.149

[a]Values in parentheses indicate the final concentration in millimoles/liter.

TABLE 9

EFFECTS OF STREPTOZOTOCIN AND ALLOXAN DIABETES ON α-OXOGLUTARATE METABOLISM IN RAT LIVER MITOCHONDRIA

The reaction mixtures contained 4 mM ATP, 6.7 mM Pi, 13.3 mM TEA, 2 mM EDTA, 10 mM $MgCl_2$, 6.7 mM α-oxoglutarate and 20 mM HCO_3^-. The incubation flasks were gassed with a mixture of 95% O_2 + 5% CO_2. Incubation was at pH 7.4.

Treatment	No. of rats	Metabolite changes	
		α-Oxoglutarate consumed	Citrate accumulated
		nmoles/20 min/mg protein ± S.E.M.	
Normal	10	631 ± 38	91.8 ± 3.8
Streptozotocin diabetic	8	831 ± 81[a]	132 ± 9[b]
Alloxan diabetic	4	894 ± 85[c]	141 ± 14[b]

[a] $p < 0.05$

[b] $p < .001$

[c] $p < 0.025$

TABLE 10

BLOOD KETONE BODIES IN RATS FED 10%, 30% and 70% FAT DIETS

Fat in diet	No. of rats	Acetoacetate	β-Hydroxybutyrate	Total Ketone Bodies
%		*μmoles/ml of whole blood ± S.E.M.*		
10	3	0.049 ± 0.011	0.190 ± 0.011	0.239 ± 0.016
30	3	0.061 ± 0.021	0.311 ± 0.020	0.372 ± 0.029
70	3	0.218 ± 0.020	1.164 ± 0.112	1.382 ± 0.114

TABLE 11

THE EFFECTS OF FAT-FEEDING ON α-OXOGLUTARATE METABOLISM IN RAT LIVER MITOCHONDRIA

The incubation mixtures contained 20 mM HCO_3^-, 6.7 mM Pi, 4 mM ATP, 13.3 mM TEA, 2 mM EDTA, 10 mM $MgCl_2$, 6.7 mM α-oxoglutarate, 20 mM malonate and 10 mg mitochondrial protein. The incubation flasks were gassed with a mixture of 95% N_2 + 5% CO_2. Incubation was for 25 min at pH 74.

Fat in diet	No. of rats	Metabolite changes	
		α-Oxoglutarate consumed	Citrate accumulated
%		*total μmoles/3 ml reaction mixture ± S.E.M.*	
10	6	3.2 ± 0.2	1.21 ± 0.09
30	6	4.4 ± 0.2[a]	1.54 ± 0.08[b]
70	6	4.9 ± 0.1[c]	1.83 ± 0.08[a]

[a] $p < 0.005$
[b] $p < 0.05$
[c] $p < 0.001$

TABLE 12

THE EFFECTS OF FASTING AND FASTING-REFEEDING ON α-OXOGLUTARATE METABOLISM IN RAT LIVER MITOCHONDRIA[a]

The reaction mixtures contained 4 mM ATP, 13.3 mM TEA, 10 mM $MgCl_2$, 13.3 mM HCO_3^-, 6.7 mM α-oxoglutarate, 0.13 mM albumin (fatty acid poor), and 16 mg of mitochondrial protein. Incubation was for 20 min. Initial pH was 7.2.

Treatment	Metabolite changes	
	α-Oxoglutarate consumed	Citrate accumulated
	total μmoles/3 ml reaction mixture	
Control	7.4	1.25
Fasted (3 days)	7.9	1.85
Fasted (3 days)-refed (3 days)	7.2	1.40

[a] This experiment was performed with 3 rats from each group. The livers were pooled, mitochondria isolated, and the incubations performed in triplicate. The reported values are the means for the triplicate incubations.

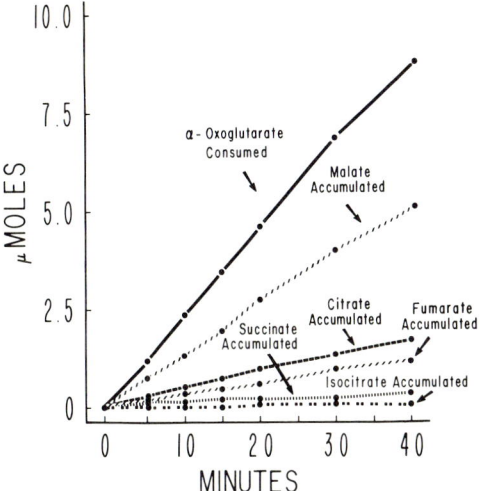

Fig. 1. *Formation of organic acids from α-oxoglutarate and bicarbonate by mitochondria incubated under aerobic conditions.* The reaction mixture (pH 7.2) contained 4 mM ATP, 10 mM $MgCl_2$, 6.7 mM P_i, 13.3 mM TEA, 6.7 mM α-oxoglutarate and 13.3 mM HCO_3^-. Mitochondria equal to 10 mg of protein were added at time 0 and the incubation times were varied as indicated.

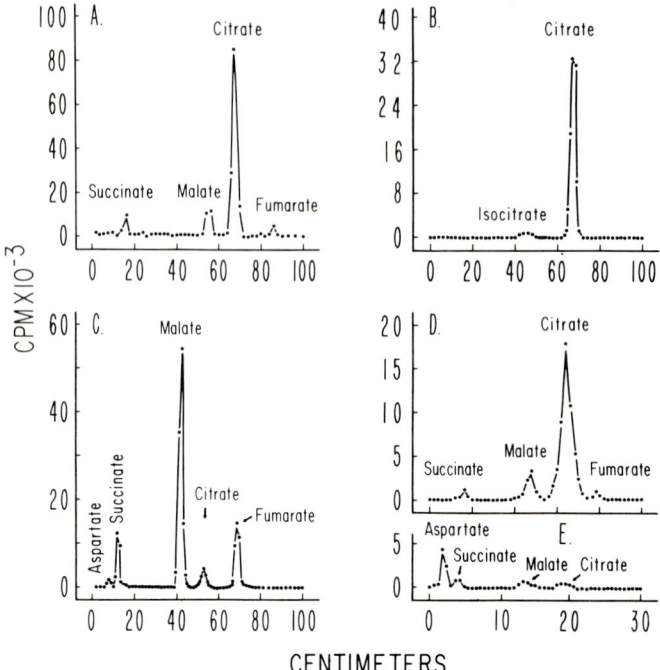

Fig. 2. *Determination by high voltage paper electrophoresis of the distribution of radioactivity in products of mitochondrial α-oxoglutarate metabolism under aerobic conditions.* A) incorporation of $^{14}CO_2$ into organic acid products of α-oxoglutarate metabolism. Except for the presence of $H^{14}CO_3^-$, the reaction mixture was the same as for Fig. 1. Mitochondria equalled 24.7 mg of protein. Incubation was for 20 min. Aliquots of sample were spotted on Whatman No. 3 MM paper strips (5 x 125 cm) 25 cm from one end. The strips were then soaked with 2 M acetate buffer which was adjusted to pH 3 with pyridine and placed in the electrophoresis tank. 4500 V were applied for 7 hours; the strips were dried, cut into pieces and radioactivity determined by liquid scintillation counting. B) incorporation of $^{14}CO_2$ after the addition of 20 mM malonate. C) distribution of radioactivity from (1,4-^{14}C) succinate into the organic pool derived from α-oxoglutarate metabolism in rat liver mitochondria. Except for the addition of tracer amounts of (1,4-^{14}C) succinate (0.003 mM) the reaction mixture was the same as for Fig. 1. Mitochondria were equal to 10 mg of protein. Incubation was for 20 min. Electrophoresis was at 4500 V for 6 hours. D) incorporation of $^{14}CO_2$ into organic acid products of α-oxoglutarate metabolism. Except for the presence of $H^{14}CO_3^-$, the reaction mixture was the same as for Fig. 3. Mitochondria equalled 17.2 mg of protein. Incubation was for 20 min. Electrophoresis was on paper strips (5 x 45 cm) for 1 hour at 4500 V. D) represents the experimental control of E. E) incorporation of $^{14}CO_2$ after addition of 13.3 mM NH_4Cl.

PROPERTIES OF MITOCHONDRIAL ATPase

Henry A. Lardy and David Lambeth

Several years ago we prepared a partially soluble ATPase from rat liver mitochondria (1). Because it retained the property of stimulation by 2,4-dinitrophenol, we felt it was part of the mitochondrial phosphorylation apparatus. We repeatedly attempted to purify the enzyme but made very little progress. Pullman, Penefsky, Datta and Racker (2) were able to purify the beef liver mitochondrial ATPase because they discovered the cold lability of the enzyme and devised procedures to avoid loss of activity.

Using the knowledge gained by the Racker group, it has been possible to purify rat liver mitochondrial ATPase to homogeneity and high specific activity. Briefly, the mitochondria are sonicated for 30 seconds at 0° and then centrifuged. Considerable protein with very little of the ATPase activity is released by this procedure. The suspended mitochondrial pellet is then sonicated at pH 8.5 and room temperature for a total of 25 minutes to release the ATPase. The soluble protein is fractionated with protamine sulfate and finally purified by chromatography on DEAE Sephadex. It is stored in the presence of ATP and 65% saturation of ammonium sulfate.

The procedure permits the isolation of 15 mg of the enzyme from 80 grams of rat liver in about 12 hours. The specific activity of the enzyme is 110 units (μmoles ATP hydrolyzed $\times$ mg protein^{-1} $\times$ min^{-1}). The ATPase is homogeneous by the criteria of sedimentation velocity and equilibrium. An unexpected finding is that the molecular weight is considerably greater than the 280,000 value reported for the beef heart enzyme (3, 4). Both the equilibrium ultracentrifugation procedure of Yphantis (5) and gel filtration on Bio-Gel P-300 with standard reference proteins, yielded molecular weights of approximately 360,000. This caused us to examine the size of the beef heart ATPase and we must conclude that it too has a molecular weight very near to that of the rat liver enzyme. On Bio-Gel P-300 the beef heart ATPase activity peaks at an estimated molecular weight of 345,000. Following centrifugation to equilibrium, plots of log fringe displacement *vs* r^2 are linear for the rat liver enzyme but curve upward when the beef heart enzyme is examined indicating that, for the latter, the molecular weight increases toward the bottom of the centrifuge cell. It seems therefore that the lower molecular weights reported by the Cornell group (3, 4) for the beef heart enzyme are the result of partial dissociation of that protein. Other evidence to support this conclusion will be presented below.

The subunit composition of the ATPases from these two sources is strikingly similar. We have compared our liver enzyme with the beef heart enzyme prepared by Senior and Brooks (6). There are 6 subunits with a molecular weight of approximately 53,000. These are not all identical for in 10% polyacrylamide gels containing sodium dodecyl sulfate they can be separated clearly into two distinct bands. In addition to the six large peptide subunits, both beef heart and liver mitochondrial ATPases contain 3 distinct smaller peptides with molecular weights of approximately 28,000, 12,500 and 8500 respectively.

Some years ago, in collaboration with Britton Chance, we found that aurovertin forms a fluorescing complex with mitochondria. In later studies, Dr. C.-H. Chiu Lin and I (7) found that the component of mitochondria responsible for forming the fluorescing complex was ATPase. Using beef heart ATPase the combining weight per mole of aurovertin varied from 245,000 gms to 312,000 gms with an average of 289,000 grms. Accepting a molecular weight of 284,000, we assumed these values indicated a stoichiometry of one molecule aurovertin per molecule of ATPase. With pure rat liver mitochondrial ATPase the stoichiometry is quite clearly 2:1. We have found that cold-dissociated ATPase does not form a fluorescing complex with aurovertin. Therefore, it seems likely that the high values for the combining weight of beef heart ATPase with aurovertin are the result of partial dissociation. We may infer from this finding that not only do the dissociated subunits of beef heart ATPase not form a fluorescing complex with aurovertin, but also that the subunits do not bind aurovertin with the same affinity exhibited by the intact enzyme.

References

1. Lardy, H., and Wellman, H., J. Biol. Chem., 201 (1953) 357.
2. Pullman, M., Penefsky, H., Datta, A., and Racker, E., J. Biol. Chem. 235 (1960) 3322.
3. Penefsky, H., and Warner, R. C., J. Biol, Chem., 240 (1965) 4694.
4. Forrest, G., and Edelstein, S. J., J. Biol. Chem., 245 (1970) 6468.
5. Yphantis, D., Biochemistry 3 (1964) 297.
6. Senior, A., and Brooks, J. C., Arch. Biochem. Biophys. 140 (1970) 257.
7. Lardy, H., and Lin, C.-H. C., in Inhibitors—Tools in Cell Research. Edited by Th. Bücher and H. Sies. Springer, N.Y., 1969, p. 279.

SUBJECT INDEX

A-particles
Acetate, 138, 146, 221, 242
Acetoacetate, 115, 147, 172, 175, 196
Acetoacetyl coenzyme A, 262
Acetyl coenzyme A, 57, 115, 117, 138-39, 141-43, 145, 147, 175, 186-89, 193-95, 198, 213, 220, 240-42, 254-55, 261, 271, 274-75
Acetyl coenzyme A carboxylase, 240
Acetyl coenzyme A synthase, 190
Aconitase, 79, 172, 273-75
Adenine, 260
Adenosine diphosphate, 215, 242, 256, 259-60, 262
Adenosine monophosphate, 54, 172, 174, 190, 214, 245, 256, 260
Adenosine triphosphatase, 6-9, 11, 80, 112, 242, 287-88
Adenosine triphosphate, 2-4, 8, 11, 12, 27, 32, 33, 39, 41, 43-48, 67, 79, 98, 111, 113, 118, 137-41, 143-44, 146-48, 172-74, 188, 190, 192, 195, 214-15, 242, 245, 254, 256, 258-59, 261-62, 271, 274-75
  exchange, 2
Adenosine triphosphate-adenosine diphosphate ratio, 64, 114, 190, 192, 242, 259-60, 262
Adenylate deaminase, 54, 55, 214

Adenylosuccinase, 214
Adenylosuccinate, 54
Adenylosuccinate synthetase, 54, 214
Adipose tissue
  enzyme activity, 56, 261
  lipolysis, 137-38, 140, 142-43, 147, 149
ADP (*see* Adenosine diphosphate)
Alanine, 97, 110, 148, 213-14, 221
Alanine aminotransferase, 212
Alcohol, 174
Aldehyde dehydrogenase, 174
Alloxan, 211, 272
Aminoxyacetic acid, 110
Aminotransferase, 115
Ammonia, 2, 5, 54, 174, 214-15, 220
Ammonium chloride, 274
AMP (*see* Adenosine monophosphate)
Antigenic, 111
Antimycin A, 28
Arsenite, 215, 275
Ascorbate, 10
Aspartate, 54, 57, 97, 110, 115-17, 190, 196, 212-14, 274
Aspartate-glutamate ratio, 55
Aspartate aminotransferase, 110, 115-16, 213, 221
Astrocytes, 93
ASU-particles, 6, 7
ATP (*see* Adenosine triphosphate)
Atractyloside, 64

289

SUBJECT INDEX

Aurovertin, 288
Bicarbonate, 110
Biguanides, 118
Biogenic amines, 96, 99, 100
Bovine serum albumin, 4
1,3-Butanediol, 171-75
  containing, 174
  feeding, 175
Butyrate, 242
Carbon dioxide, 217, 220, 239, 242
  radioactive, 64
Carboxylation, 241, 254, 273
Carnitine, 192
Chemiosmotic, 1, 64
  hypothesis, 1-3, 8-10, 12
  potential, 67
  theory, 67
Chloroplasts, 10
Cholate, 1, 2
Chromosome, 82
Chymotrypsin, 7
Citrate, 53, 65, 67, 115, 172, 185, 188-90, 194, 198, 211-13, 215, 221, 240, 271, 273, 274, 276
  synthesis of, 275
Citrate-isocitrate ratio, 172
Citrate lyase, 219
Citrate synthase, 64, 66, 79, 82, 84-85, 186-90, 192-95, 197, 274-75
Citric acid cycle, 53, 56, 63, 64, 67, 79, 80, 82, 85, 93-8, 100, 115, 147-49, 185-87, 189, 190, 196, 198, 211, 213-15, 221, 240, 254, 271, 273
Codfish muscle,
  enzyme from, 216
Coding, 93
Coenzymes, 93, 186
Coenzyme A, 4, 117, 174, 187-88, 190-91, 194, 255, 259
Coenzyme Q, 42

Compartmentation, 1, 93
Copper, 42
Cortisol, 240-41, 244, 246
Coupling
  factors, 2, 5-7, 11, 12
  process 8, 10, 12, 27, 30, 39
Creatine phosphate, 54
"Crossover", 239
3', 5'-Cyclic adenylic acid, 241, 242, 244-46, 258, 260
  dibutyryl derivative of, 244
Cyclic AMP (see 3', 5'-Cyclic adenylic acid)
Cytidine triphosphate, 256
Cytochrome (s), 27, 46, 81
Cytochrome a, 28-30, 41-44
Cytochrome a_3, 27-30, 34, 41, 44, 46, 47
Cytochrome a_3-CO compound, 28
Cytochrome b, 27, 28, 31, 32
Cytochrome b-ubiquinone-cytochrome c, 28
Cytochrome b_K, 31, 32
Cytochrome b_T, 31-33, 46
  oxidation, 34
Cytochrome c, 2, 28, 29, 32, 42, 43
Cytochrome c_1, 2, 28-33, 42, 43
Cytochrome oxidase, 2, 30, 42
Cytosol, 82, 109, 110, 116, 189, 217, 240
Deamination, 54, 55
Decarboxylation, 216, 218-20, 240, 241, 246
Dehydrogenation, 66
Dephosphorylation, 255, 257, 260-62
Diabetes, 137, 144, 148, 271, 272, 276
Dicarboxylic acids, 239
N,N'-dicyclohexylcarbodiimide, 5, 11

SUBJECT INDEX

Digitonin, 80
Dihydrolipoyl dehydrogenase, 253, 254, 256
Dihydrolipoyl transactylase, 253, 254, 256
Dihydroxyacetone, 175
Dihydroxyacetone phosphate, 172, 173
2,4-Dinitrophenol, 111-13, 145, 215, 275, 287
1,3-Diphosphoglycerate, 3
Dopamine, 96, 99
DPN (*see* nicotinamide adenine dinucleotide)
Durohydroquinone, 28, 32, 33
Electron
   carrier, 46
   donor, 39
   flow, 1
   microscopy, 82
   transfer, 28, 215
   transport, 33, 46
   transport-reversed, 47, 93
Embden-Meyerhof pathway, 138, 139
Energy
   balance, 143
   biological, 3
   conservation, 45, 93
   control, 63, 64
   coupling, 32-34, 39
   free, 47, 55
   states, 67
   production, 145
   transduction, 44-46
   transfer, inhibitors, 11, 12
Epididymal fat, rat, 137, 138
Epinephrine, 145, 146, 239-41, 244, 261
Escherichia coli, 253, 255, 257

Ethanol, 146, 171, 174, 175
Ethylenediaminetetraacetic acid, 4, 5
Faraday constant, 39
Fasting, 213, 271, 272, 276
Fat-feeding, 271, 272
Fatty acid(s), 53, 117, 138, 140, 142, 261, 276
   free, 144, 147-49, 189
   oxidation, 80
   synthesis, 85, 139, 141, 145, 211
   unsaturated, 12
Feedback inhibitor, 54
Ferricyanide, 10
Ferrocene, 10
Ferrocytochrome c, 7
Flavoproteins, 41, 42, 253, 254
Fluoroacetate, 53
Fluorocitrate, 53, 274, 275
Fructose 1,6-diphosphate, 255, 261
Fructose 1,6-diphosphatase, 56
Fumarase, 79, 98-9
Fumarase dehydrogenase, 55
Fumarate, 54, 55, 98, 111, 113, 214, 215, 240, 273, 274
GDP (*see* guanosine diphosphate)
Genetic disorder, 93
Glucagon, 118, 239-41, 243-46
Gluconeogenesis, 109, 110, 113, 115-19, 147-49, 211, 214, 239, 254
   precursors, 116
Glucose, 97, 111, 117, 137, 140, 141, 145, 147, 149, 189, 213, 215, 239
   radioactive, 139
Glucose-6-phosphate, 85, 172, 174
Glucose-6-phosphate dehydrogenase, 85
Glutamic acid dehydrogenase, 54, 55, 80, 97, 114, 197, 220

SUBJECT INDEX

location of, 55
Glutamic oxalacetic transaminase,
  186, 189, 196, 197
Glutamate, 54-57, 97, 100, 114,
  116, 174, 190, 197, 211-14, 221
Glyceraldehyde-3-phosphate, 3
Glyceraldehyde-3-phosphate dehydrogenase, 3, 4, 8, 97, 139, 145, 174
Glycerokinase, 144
Glycerol, 146
α-Glycerophosphate, 144, 172-74
Glycogen, 140, 245
Glycolysis, 3, 56, 97, 100, 143, 255
GTP (see guanosine triphosphate)
Guanine, 112, 186, 190
Guanosine diphosphate, 54, 55, 112,
  190, 214, 215, 255
Guanosine triphosphate, 54, 55, 111
  118, 190, 214, 215, 255, 256
Guanosine triphosphate-adenosine
  monophosphate phosphotransferase, 190
Guanosine triphosphate-guanosine
  diphophosphate ratio, 190
Heart, 53, 189, 211
  muscle, metabolism, 214, 220,
  221, 260
  perfusion of, 212, 276
Heme a
  components, 41
Hexokinase, 85, 113
  oligomycin inhibition of, 187
Histone, 258
β-Hydroxybutyrate, 114-17, 147,
  171, 172, 174, 175, 211, 275
β-Hydroxybutyrate-acetoacetate
  ratio, 196
β-Hydroxybutyryl coenzyme A,
  187
β-Hydroxybutyrate dehydrogenase,
  114, 196

3-Hydroxybutyryl coenzyme A,
  262
Hyperthyroidism, rats, 141
Hypoglycemia, 189
Inosine monophosphate, 54, 214
Insulin, 53, 137-41, 143, 144, 146,
  148, 149, 261, 272
Intramitochondrial
  phosphate potentials, 190
  processes, 55, 66, 94, 115
Iron, 42
Isocitrate, 53, 273, 274
Isocitrate dehydrogenase, 64, 80,
  173, 190, 193, 271
  cytoplasmic, 271
  mitochondrial, 271
  NAD dependent, 79, 83
  NADP dependent, 84, 271, 273,
  275
Isopotential, 43
α-Ketoacyl transferase coenzyme A
  187
Ketogenesis, 147-49
α-Ketoglutarate, 54, 56, 94, 96,
  97, 100, 111, 112, 115, 116,
  172, 175, 185, 190-96, 242, 255
α-Ketoglutarate dehydrogenase, 79,
  80, 83-5, 94, 186, 187, 190-93,
  198, 257
Ketone bodies, 53, 147, 173, 189,
  193, 276
Kidney
  cortex, gluconeogeneiss in, 109,
  110
  enzymes in, 55, 56
Kinase, 79, 254, 257-59, 262
Krebs cycle (see citric acid cycle)
Krebs cycle enzymes, 83, 84
Kynuramine, 96
  oxidation, 99

SUBJECT INDEX

Lactate, 56, 97, 110, 146, 173, 189, 217, 239
Lactate dehydrogenase, 173, 174
Lactate-pyruvate ratio, 144
Lipogenesis, 137-44, 147, 149
   de novo, 138, 141, 149
Lipoic dehydrogenase, 42
Lipolytic agents, 141
Liposomes, 1, 10
Liver
   enzymes in, 55, 56, 261, 262
   freeze-clamping, 114, 115, 174
   homogenation, 245
   metabolites, 171
   perfusion, 109, 117
Magnesium ion, 259
Malate, 53, 55, 65-67, 96-98, 110, 111, 113, 115-17, 141, 185, 187, 191-95, 196, 198, 211-12, 215-17, 219-21, 240, 273-75
   cycle, 141-43
   radioactive, 65
Malate dehydrogenase, 55, 64-66, 81, 84, 110, 115, 186, 196, 197, 216-19
Malate-fumarate ratio, 98
Malic enzyme, 139, 143, 216-18, 220-22
Malonate, 211, 273
Matrix, 81, 82, 84, 85
Membrane, 1, 12
   inner, 81, 85
   inner matrix, 80
   potential, 3, 10
Metabolic state, 148
Methylmalonic aciduria, 189
Methylmalonyl coenzyme A, 189
Michaelis constants, 193, 198
Microsome, 245
Midpoint potential, 32, 42, 46
Millipore filter, 95, 111

Mitchell hypothesis, 10
Mitochondria, 1, 3, 4, 8, 31, 40-43, 48, 93
   anaerobic, 28
   biosynthesis of, 85
   coupling of, 33
   DNA in, 82
   in brain, 94-100
   in heart, 40, 63, 66, 67, 81, 94-100, 185, 186, 189-91, 193, 195, 198, 211, 212, 215-20, 222, 257, 258
   in kidney, 216, 254-58, 261
   in liver, 63-65, 109, 110, 112-14, 116-18, 185, 212, 215, 216, 239-45, 257, 260, 261, 271-73, 275, 276, 287, 288
   in plants, 215
   Krebs cycle enzymes in, 79-85
   oxidation of pyruvate in, 194
   sonication of, 5
   uncoupling of, 41, 113
Mitochondrioplasm, 93
Monoamine oxidase, 99, 100
NAD (*see* Nicotinamide adenine dinucleotide)
NADP (*see* Nicotinamide adenine dinucleotide phosphate)
Nervous system, 95
Neurospora crassa, 253
Nicotinamide
   mitochondrial, 67
   nucleotides, 63
Nicotinamide adenine dinucleotide, 4-7, 27, 42, 48, 79, 80, 94, 97, 110, 113, 114, 116, 118, 140, 187, 189, 191, 212, 216-18, 220, 255, 261, 262
   dehydrogenase, 42
   oxidized-reduced ratio, 66, 85,

SUBJECT INDEX

115-17, 119, 173-75, 193-97, 219, 255, 274, 275
Nicotinamide adenine dinucleotide phosphate, 80, 142, 212, 216-18, 220, 222, 273
  dehydrogenase, 145
  oxidized-reduced ratio, 173, 175
Nigericin, 2
Nitrogen balance, 215
Nucleic acid, 93
Nucleoside diphosphokinase, 112-14, 190, 256
Nucleoside triphosphate, 256
Nucleotide, 67
  purine cycle, 54
  pyridine, 27, 185
Nucleotide diphosphokinase, 113
Octanoate, 117, 242
Octanoylcarnitine, 275
Oleic acid, 11, 147
Oligodendrocytes, 93
Oligomycin, 4, 5, 8, 10, 11, 66, 192-95
Oxaloacetate, 54, 56, 66, 79, 85, 97, 109-13, 115, 139, 148, 186, 188-90, 194-98, 213-14, 216, 218-20, 240, 261, 274-75
Oxaloacetate decarboxylase, 215, 216, 219
Oxidation, 3, 5, 46, 99
  control, 3, 12
Oxidation-reduction, 9, 12, 39-41, 43, 45, 46, 48, 66, 185
  potential, 114, 115
Oxidative phosphorylation, 1, 2, 8, 10, 12, 79, 118, 139, 140
α-Oxoglutarate, 64, 211-14, 220, 221, 271, 273, 276
  carboxylation, 271, 273-76
α-Oxoglutarate dehydrogenase, 64

Oxygen, 28, 29, 31, 39, 44, 97, 99
  consumption, 147
  tension, 48
"Oxygen control model", 29
Phosphate
  acceptor, 27, 64
  inhibition, 65
  inorganic, 44, 65, 172-74
  potential, 44, 46, 114, 185
Pentose cycle, 139, 144
Phenazines dyes, 43
Phenethylbiguanide, 118
Phosphatase, 257-59
  heart, 259
  kidney, 259
Phosphatase kinase, 260
Phosphoenolpyruvate, 54, 56, 109, 111-13, 115, 116, 118, 172, 174, 175
Phosphoenolpyruvate carboxykinase, 56, 109-11, 113-15, 118
  species found 110, 111
2-Phosphoglycerate, 175
3-Phosphoglycerate, 3, 172, 173, 175
3-Phosphoglycerate kinase, 174
Phosphokinase, 54
Phospholipids, 4, 10-12, 42
Phosphorylation, 255-57, 262
  coupling, 3
  sites, 28, 34, 44
  state, 171, 173, 174
  substrate level, 112
Photophosphorylation, 3
Polylysine, 7, 8
Potassium ion, 67
Potentiometric titration, 31, 32, 40
Propionate, 189, 220-22
Propionyl coenzyme A, 187, 189
Proteolipids, 11

SUBJECT INDEX

Protrons, 1, 9, 10
Purkinje cells, 93
Pyramidal cells, 93
Pyrophosphate, 5
Pyruvate, 56, 64-66, 94, 96, 97, 100
    110, 111, 113-16, 139, 141, 143
    172, 173, 175, 189, 191-94, 211-
    16, 218, 219, 221, 239-44, 255,
    259
    bovine, brain and heart, 258
    dismutation, 261
    oxidation in mitochondria, 194
    ratioactive, 65
Pyruvate carboxylase, 53, 54, 56,
    65, 110, 115, 117, 139, 239-41,
    243, 246, 261, 262, 273
    location, 56
Pyruvate decarboxylase, 80, 84
Pyruvate dehydrogenase, 57, 80,
    240, 241, 253, 257, 259-62
Pyruvate dehydrogenase kinase, 258
Pyruvate dehydrogenase phosphatase,
    257, 261
Pyruvate dehydrogenase phosphate,
    258
Pyruvate kinase, 54, 56, 57
Quinols, 28
Redox potential
    cytoplasmic, 144, 146, 171
    intramitochondrial, 116
Reducing equivalents, 67, 139, 140,
    171
Reesterification, 143
Respiration, 4, 97, 187, 215
    inhibition of, 5
    mitochondrial, 47
Respiration state 3, 27, 33, 63-67,
    96
Respiration state 4, 32, 33, 43-45,
    48, 63-66

Respiratory
    carriers, 29
    chain, 9, 27, 28, 33, 34, 39, 40,
    43, 67, 114, 189
Respiratory control ratio, 1-6, 10,
    12, 48, 63, 96, 140
Rotenone, 99
Rutamycin, 4-8, 11
Serotonin, 96, 99
Splanchnic bed, 147
Starvation, 215, 261
Steady state, 48
Streptozotocin, 272, 276
Submitochondrial particles, 2-8,
    10, 11, 42
Succinate, 6, 64, 65, 94, 95, 99,
    113, 185, 211, 220, 221, 242,
    245, 273-75
    metabolism, 189
    radioactive, 66
Succinate dehydrogenase, 2, 9, 42,
    79-82, 274
Succinate thiokinase, 80, 84, 190,
    191
Succinoxidase, 9
Succinyl coenzyme A, 186-95
Succinyl coenzyme A synthetase,
    256
Sulfur, 42
Synthetase, 148
Tetraphenylboron
    anion, 10
Thermodynamic(s), 27
    control, 39
Transacetylase, 254
Transaminase(s), 55, 80, 97, 214
    transamination, 190, 212
Transhydrogenase, 80
Translocation, 2
    of anions of calcium, 2

of potassium protons, 1-3, 7, 10, 12
Transphosphorylation, 111, 190
Tricarboxylic acid cycle (*see* citric acid cycle)
Trifluoromethoxycarbonylcyanide phenylhydrazone, 1, 4-7, 10, 64, 187
Triglycerides, 137
Triosephosphate dehydrogenase, 110
Triphosphate, 113
Trypsin, 6, 7
Tryptamine, 99

Tyramine, 96, 99
Ubiquinone, 31-33
 pool, 32
Uncoupling, 1, 2, 6, 9, 10, 12, 30, 31, 33, 64-67, 112, 141, 192, 215
 agents, 145
Urea, 213
Uridine triphosphate, 256
Valinomycin, 2, 67
Viologen dyes, 43
Vitamin B_{12}
 deficiency, 189

DATE DUE

NOV